Transnational governance through private authority

Transnational governance through private authority

The case of the Forest Stewardship Council certification in Russia

Maria S. Tysiachniouk

Environmental Policy Series – Volume 7

ISBN: 978-90-8686-218-4
e-ISBN: 978-90-8686-772-1
DOI: 10.3920/978-90-8686-772-1

First published, 2012

© Wageningen Academic Publishers
The Netherlands, 2012

This work is subject to copyright. All rights are reserved, whether the whole or part of the material is concerned. Nothing from this publication may be translated, reproduced, stored in a computerised system or published in any form or in any manner, including electronic, mechanical, reprographic or photographic, without prior written permission from the publisher:
Wageningen Academic Publishers
P.O. Box 220
6700 AE Wageningen
The Netherlands
www.WageningenAcademic.com
copyright@WageningenAcademic.com

The content of this publication and any liabilities arising from it remain the responsibility of the author.

The publisher is not responsible for possible damages, which could be a result of content derived from this publication.

Preface

This study is dedicated to the FSC's voluntary forest certification scheme as a transnational non-state governance institution, operating both globally and within the nation state. Russia is used as an example of the application of global rules and norms on concrete territories.

I began research on the transnational governance of natural resources in 2002, when, shortly after graduation from the Ms in Environmental Studies program at Bard College, NY, I received a MacArthur Foundation research and writing grant with my colleague Dr. Woijeich Sokolovsky from Johns Hopkins University on the topic 'Public-Private Partnerships and Making Environmental Policy in Russia'. In the framework of the MacArthur foundation project, I studied eight cases in both European Russia and in the Russian Far East, in Primorskii Krai and Kamchatka. From these case studies, three became part of this volume: Saving old growth forests in Karelia, the Pskov Model Forest and the Model Forest Priluzie. In 2002, I began to study institutions related to the Forest Stewardship Council (FSC) certification and started my longitudinal field research (2002-2012) on the above mentioned Model Forests, in which the FSC standards were implemented as part of the sustainable forest management model. Both Model Forest projects were funded by International Aid Foundations and implemented by the WWF, through engagement with other stakeholders. Interviews conducted on the federal level and in the regions were also used in this study as a historical background on the emergence of the FSC and relationships between NGOs and state agencies in the process of institutionalization of global rules in Russia. At that time, FSC certification had only begun in Russia and I was observing how transnational NGO networks, especially the WWF and Greenpeace, were promoting it in the Russian regions.

In November 2004, I became engaged in the PhD sandwich program of Wageningen University, and became affiliated with the Environmental Policy Group at Wageningen University. At that time, I started to combine working on different kind of research projects with my PhD study, guided by my advisers Prof. dr. ir. Arthur Mol and Prof. dr. ir. Gert Spaargaren. The base for my field research was the Centre for Independent Social Research (CISR) in St. Petersburg, Russia where I have chaired the Environmental Sociology Group since 1997.

I was interested in studying transnationalization in the governance of natural resources, and therefore became involved in multi-sited studies. I gained this opportunity by engagement in multiple projects received through the CISR and sponsored by different granting agencies. In addition, I was engaged in practitioners' and/or expert work as part of the FSC consulting group at the CISR.

My research would not be possible without the funding provided by several granting agencies. For conducting and updating my case studies in Russia, the most important projects were: 'Optimal Forest Resource Use in Northwestern Russia', sponsored by Stiftelsen Marcus och Amalia Wallenbergs Minnesfond (2005-2006); 'Governance of Renewable Natural Resources in Northwestern Russia' (2004-2007), sponsored by the Finnish Academy; 'The Role of Civil Society in Fostering Corporate Social Responsibility Within the Russian Forest Sector' (2006-2007), sponsored by the Moscow Public Scientific Foundation, as well as two projects: 'Making Democracy Work: Building Capacity of Ordinary Citizens in Russian Rural Settlements (2006-

2007)' and 'Promoting Democracy and Citizens' Rights in Resource-Dependent Communities (2010-2012)', sponsored by the European Union.

Several projects served for doing research in Europe. In Finland, interviews were conducted in the framework of the project 'Trust in Finnish-Russian Forest Industry Business Relations' (2008-2010), sponsored by the Finnish Academy, interviews in the UK, Netherlands, Austria, Belgium and South Africa were done as part of the project 'Transnationalization of Forest Governance' (2008-2011), also sponsored by the Finnish Academy.

Exchange projects with Finnish, Swedish and US universities and some of the EU mobility projects helped to reach representatives of international organizations, NGOs and top managers of TNCs located in different parts of the world. The interviews in Sweden were conducted in the framework of two exchange projects, sponsored by the Swedish Institute: 'An Interdisciplinary Network for the Study of Multi-Level and Multi-Stakeholder Forest Governance in Russia and Sweden' (2009-2010), coordinated by Umeo University and 'An Interdisciplinary Network for the Study of Company-Community Relations in Resource Extracting Sectors in Russia and Sweden' (2011-2013), coordinated by Lulea University. Interviews in the USA were done during two US-Russian exchange projects, coordinated by Yale University: 'Promoting Forest Certification in Russia and North America (2004-2005)' and 'Fostering Effective Partnerships among Russian and North American FSC Experts, Activists and Social Scientists' (2007-2008) sponsored by the Trust for Mutual Understanding. These projects also allowed me to work at the Yale University Program on Forest Certification and collaborate with the director of the program, Prof. Benjamin Cashore and his colleagues, who developed the theory of Non-State Market Driven Governance which was extremely important for the major concept of this study, the Governance Generating Network (GGN), which I developed in collaboration with my Finnish colleagues Prof. Jarmo Kortelainen and Dr. Juha Kotelainen. Involvement in different research projects was a direct advantage for this research.

The practitioners' work also gave me many advantages involving additional resources, access to the field and insider's opportunities to see certain aspects of the study. Practitioner's work provided excellent venues for participant observations. Since 2004, I participated in development of the Russian national FSC standard with its 7 versions, and since 2010, served on the Board of Directors of the FSC-Russia. I also worked as a FSC consultant for companies belonging to the holding Investlesprom, and for the Stora Enso Company as well as an expert auditor for SGS-Qualifor. It allowed obtaining an insiders view, establishing good relationships with gate keepers, key informants, and a large variety of stakeholders. Many of the colleagues with whom I was involved as a practitioner were essential for setting up logistics for the study, facilitating it, gaining access to high ranking officials, and reflecting on my writing. As part of the consulting work and the EU sponsored projects, I was involved in organizing local stakeholder consultations and public hearings related to the FSC process in different parts of Russia, allowing a closer look at the institutionalization of FSC rules on the ground.

Since 2007, I have been on the technical panel of the High Conservation Value Network, based in Oxford, UK. This engagement introduced me to many practitioners working on the transnational level and helped me to understand relationships between different kinds of networks and institutions involved in transnational governance.

Particularly important for this study was my engagement as a member of FSC-International. In 2008, I joined the global process of the FSC Principals and Criteria review, which is based on wide stakeholder input, involving multiple working group meetings in Bonn and consultations, including those at the FSC General Assemblies in Cape Town, SAR and Kota Kinabalu, Malaysia.

Working simultaneously on transnational, national and local levels helped me to observe in detail multiple challenges of the standard setting, adjustment and implementation of FSC rules. This effort resulted in the multi-sited ethnographic study of the FSC presented in this book. Participant observation of the FSC processes led me to develop the concept of governance generating networks (GGN), constituted by nodes of global governance design, forums of negotiations and sites of implementation. This concept became central in this book. Together with my colleagues from Finland, Prof. Jarmo Kortelainen and Dr. Juha Kotilainen, we further developed the concept from the perspective of human geography and governance literature. In this book I re-conceptualized it from the perspective of sociology of transnational processes. As a result, the GGN concept has become a new lens for analyzing decentralized non-state governance arrangements.

My library work, data analysis and writing was done in between of my field expeditions, predominantly at Wageningen University, the Netherlands; the Centre for Independent Social Research in St. Petersburg, Russia; the University of Eastern Finland in Joensuu, the Helsinki School of Economics in Finland, and Lulea University in Sweden. Some early work was done at Johns Hopkins University, Dickinson College PA, Buffalo University NY, and Yale University as part of different exchange projects, visiting scholar positions, teaching arrangements and fellowships. I am thankful to all the above mentioned universities and funding agencies that made my research and writing possible.

I am especially thankful to the Environmental Policy Group at Wageningen University and my advisors, Prof. dr. ir. Arthur Mol and Prof. dr. ir. Gert Spaargaren, who guided me through the whole research process, provided valuable insights and feedback, helped to theorize my findings yet allowed me to set up my field research and theoretical framework according to my own perspective. They were patient with me and understood the challenges that I was facing in combining my PhD research with professional responsibilities at the CISR, multiple other projects, and expert work. I would like to thank all my colleagues at the Environmental Policy Group, my peer PhD researchers and our beloved secretary, Corry Rothuizen, who created a lovely, enjoyable and relaxing atmosphere for my life and work at Wageningen. I appreciate the opportunity to have shared my happy life in Wageningen with all of you.

I would like to thank my research team at the Centre for Independent Social Research, for being a unique and supportive community, especially Dr. Antonina Kuliasova, Ivan Kuliasov, Svetlana Pchelkina, Dr. Svetlana Tulaeva, Dr. Laysan Mirzagitova, and especially my US intern Johnathan Reisman with whom I shared time in the research expeditions, discussed my findings and writing, and published together. I am thankful to my Finnish, Swedish, German and the US colleagues with whom we coordinated and shared research projects resulting in this book. I am particularly thankful to Dr. Soili Nysten-Haarala, Dr. Jarmo Kortelainen, Dr. Juha Kotilainen, Dr. Karina Keskitalo, Dr. Benjamin Cashore, Dr. Michael Conroy, Dr. Constance McDermott, Dr. Laura Henry, Dr. Errol Meidinger, Dr. Per Angelstam, Dr. Marine Elbakidze, Dr. Olga Malets, Dr. Tim Bartley and Dr. Margarett Shannon for their valuable contributions to my research, for being a source of inspiration to me, for feedback, insightful comments, fruitful ideas and editing help.

I appreciate help from my colleagues-practitioners, staff persons at FSC-International, Marion Karman, the director of the FSC-NIS Regional office, Andrey Ptichnikov and the FSC-Russia program coordinator Tatiana Yanitskaya, member of the FSC technical committee Michael Korpachevsky, chair of the FSC-Russia board of directors Yurii Pautov, Greenpeace activist Alexey Yaroshenko, who helped me with interviews, provided several graphs and maps, and reflected on my writing about the FSC operations in Russia. I am thankful to the director of the NGO SPOK Alexander Markovski, SPOK activist Olga Ilina, executive director of Silver Taiga Foundation Pshemislav Mayevski, top manager of the holding company Investlesprom, Fiodor Grabar, manager responsible for certification of companies belonging to Investlesprom Oleg Koniushatov, certification manager Ilia Teslia, certification manager Natalia Nikashkina, environmental director of Stora Enso company, Olga Rogozina, and researcher at St. Petersburg Forestry Research Institute Boris Romaniouk for facilitation my case study research, arranging interviews, giving interviews and reflecting on my draft writing.

I greatly appreciate the support of all my informants, in several countries and Russian regions who shared their time, their knowledge and their experiences with me.

Several people provided technical support for this book. I am greatly indebted to John Miller, who patiently and thoroughly edited my thesis, to Jiulia Landonio, who made maps for the case studies, Nina Krishtanovskaya and Katerina Kachavina, who helped with designing figures for this book and Katerina Guba, who helped with editing and arranging bibliography.

I appreciate support from my friend Raoul van Beeck, who was my entertainment 'director' at Wageningen, helping with accommodation, cooking, and graphic design, and my dearest friends Sander ter Meulen and Walter Castro, who made my life in the Netherlands endless fun, full of music, dance and delicious food. I am also thankful to my family, especially my mother Svetlana Plotnikova, my son Andrey Tysiachniouk and my granddaughter Renata Tysiachniouk for creating a supportive home environment for me back in Russia.

Table of contents

Abbreviations

AF&PA	American Forest and Paper Association
ASEAN	Association of Southeast Asian Nations
ASI	Accreditation Service International
BCC	Biodiversity Conservation Center
BEAR	Programme of the Barent's Euro - Arctic Region
CAR	Corrective Action Request
CB	Certification Body
CBD	Convention on Biodiversity
CEPI	Confederation of European Paper Industries
CIS	Commonwealth Independent States
CITES	Convention on International Trade in Endangered Species of Flora and Fauna
CISR	Centre for Independent Social Research
CoC	Chain of Custody
COFO	Committee on Forestry
COP	Conference of Parties
CO_2 emissions	Carbon dioxide emissions
CSA	Canadian Standards Association
DIY	Do it Yourself
EIDHR	European Instrument for Democracy and Human Rights
ENGOs	Environmental Non-Governmental Organizations
FAO	Food and Agriculture Organization of the United Nations
FLA	Fair Labor Association
FLEG	Forest Law Enforcement and Governance
FLEG-T	Forest Law Enforcement Governance and Trade
FLO	Fair Trade Labeling Organization
FM	Forest Management
FM-CoC	Forest Management Certification Chain of Custody
FMU	Forest Management Unit
FSC	Forest Stewardship Council
FSC GA	Forest Stewardship Council General Assembly
GEF	Global Environmental Facility
GFTN	Global Forest and Trade Network
GGN	Governance Generating Network
GINs	Global Integrated Networks
HCV	High Conservation Value
HCVF	High Conservation Value Forest
IFF	Intergovernmental Forum on Forests
IFOAM	International Federation of Organic Agriculture Movement
ILO	International Labor Organization
IOAS	International Organic Accreditation Service

IPF	International Panel on Forests
ISEAL	International Social and Environmental Accreditation Labeling Alliance
ISO	International Standardization Organization
ITFF	Interagency Task Force on Forests
ITTO	International Tropical Timber Organization
IUCN	International Union for Conservation of Nature
LEED	Leadership in Energy and Environmental Design
MFP	Model Forest Priluzie
MSC	Marine Stewardship Council
NEPCon	Nature, Ecology and People Consult
NIS	Newly Independent States
NGO	Non-Governmental Organisation
NSVFCR	National System of Voluntary Forest Management Certification in Russia
NSMDG	Non-state Market Driven Governance
NLC	Northern Logging Company
OECD	Organization for Economic Cooperation and Development
P@C	Principles and Criteria
PEFC	Program for Endorsement of Forest Certification
PMF	Pskov Model Forest
PPMs	Pulp and Paper Mills
REDD	Reduced Emissions from Deforestation and Forest Degradation in Developing Countries
RFTN	Russian Forest and Trade Network
RNCFC	Russian National Council of Forest Certification
RSPO	Roundtable on Sustainable Palm Oil
SAN	Sustainable Agriculture Network
SAR	South African Republic
SDC	Swiss Agency for Development and Cooperation
SEU	Socio-Ecological Union
SFI	Sustainable Forestry Initiative
SIDA	Swedish International Development Agency
TNC	Transnational Corporation
TRN	Taiga Rescue Network
UN	United Nations
UNCED	United Nations Conference on Environment and Development
UNCTAD	United Nations Committee on Trade and Development
UNDP	United Nations Development Program
UNEP	United Nations Environmental Program
UNFCCC	Unted Nations Framework Convention on Climate Change
UNFF	United Nations Forum on Forests
VOOP	Russia Society for Nature Protection[1]

[1] Translation of abbreviation from Russian.

VPA	Voluntary Partnership Agreements
WBC	World Business Council
WCS	World Conservation Society
WWF	World Wildlife Fund

Chapter 1. Introduction

1.1 Tunguda village, Russia

The ancient, 500 year-old Karelian village Tunguda, in Northwestern Russia, is considered by the state to be a declining settlement. Only 30 elderly retired inhabitants hunt, gather mushrooms and berries, fish and grow vegetables. However, in summer, the population grows by three times, as many relatives and artists come for vacations to this remote and beautiful place.

A young couple, Alina Chuburova and Alexey Tzikarev, are revitalizing the old traditions of the village and its museum. Both are trying to develop local tourism, attract artists and draw the attention of state authorities. Together with the Moscow artist Mikchail Lepyoshkin, who bought a house in Tunguda, they are designing the village's revival. The wilderness around the village is vital for both maintaining the traditional life style of the village residents and recreation of their relatives. The surrounding forests are essential for attracting tourists.

A couple of years ago, the villagers suddenly noticed that a company was designating plots for logging in the surrounding forests. Some inhabitants were crying, others were angry, but they all thought that nothing could be done against this logging operation as the territory was in lease, designated by the state authorities for logging. Usually in Russia, a zone of 1 kilometer around a village is prohibited for logging. However, in 1999, the state inventory agency decided that the Tunguda village was too small for warranting a zone excluded from logging. Alina Chuburova retells: 'We here have the mentality that it makes no sense clashing with authority. We'll grumble, we'll regret, and then we'll just give up'.[2] But Michail Lepyoshkin decided to investigate the possibilities of protecting the forest. He learned that the territory was leased by the Segezha Pulp and Paper Mill, and that logging was operated by the Northern Logging Company, which belonged to the holding company Investlesprom. He also found out that forest management on the leased land was FSC certified. He studied the FSC standards and learned that an FSC certified company is required to consult the local community and to exclude socially valuable forests from logging.

Lepyoshkin and Chuburova organized a citizen's campaign against logging. They wrote letters, signed by all the villagers, to the logging company, to the FSC office in Moscow, to the FSC auditors and to the top managers of the Investlesprom holding company. Rounds of negotiations were held between the company and the local community. After several rounds of consultations and negotiations, the company agreed to designate the majority of the plots as high conservation value and these were preserved as such. Mikhail Lepyoshkin wrote an acknowledgement to the company saying: 'Such a result is of great importance for the village and for its future, and I am sure that it will be appreciated by future generations of Tunguda people'.[3] Since 2010, the company and the local community have collaborated. The company has not only preserved the surrounding forests

[2] Interview with Alina Chuburova, Tunguga village, June 2010, the real name is used from the written permission of the informant.

[3] From the Mickael Lepyoshkin' open letter to the company, written in May 2011, used here from his permission.

but supplied villagers with timber to repair their old bridge and has become eager to provide other support for upgrading the village's infrastructure (Tysiachniouk, 2010d). Hence, global FSC standards helped local citizens in a remote Russian village to protect their forests. As such, this example of the Tunguda village exemplifies the subject of this study.

In introducing the background, the problem analysis and the research questions of this study, this chapter starts with explaining the importance of the intact forest landscape in Russia (Section 1.2). Subsequently, the major transformations in the Russian forest sector are highlighted, especially those during the transition from a nationally planned economy to a globalized market economy (Section 1.3). Next, it introduces private governance through international certification schemes, with special emphasis paid to the Forest Stewardship Council certification (Section 1.4). The following section provides a brief overview of existing perspectives to study certification schemes and identifies the need for developing a new perspective for analyzing certification as a private authority in governance (Section 1.5). After setting out the main objective and research questions for this study, the chapter closes with a brief outline of the volume (Section 1.6).

1.2 Russian forests on the world's conservation map

Preservation of old growth, ancient, intact forest landscapes is a priority of the world conservation movement (Wirth, Gleixner & Heiman, 2009). However, boreal forests, including those of the Russian taiga, are overshadowed by the focus of global media attention on the tropical forests (Mather, 2004; Lloyd, 1999). The tropical forests of the Amazon, Indonesia and the Congo Basin are at the forefront of global conservation initiatives. These forests are in danger of being converted to agricultural land for soy and palm oil, or cleared for cattle grazing grounds. Few people in the World are concerned about what is taking place in the Russian forests and forestry.

The Russian forests attract less attention as there is no conversion to agricultural land. On the contrary, forests are taking over the abandoned Soviet agricultural lands through reforestation. The role of Russian forests in preventing climate change is less obvious than in the tropics. The carbon sink function of the taiga is less studied as carbon is stored under ground (Carlson, Wells & Robertson, 2009; Carlson *et al.*, 2010), which makes that boreal forests perceived to be less significant in combating climate change. This makes Russia ineligible for financial support through the global program 'Reducing emissions from deforestation and forest degradation' (REDD).

Russia mirrors many of the world's problems with forest degradation, but in a much softer and less obvious way compared to the tropics. Dominated by pine and spruce, Russia's boreal forests have less biodiversity than the tropics. However, among forests of the Northern hemisphere, Russian forests are leading in biodiversity and ecosystem richness and these forests provide home to hundreds of plant and animal species, including the lynx, the black bear and the bald eagle. Although the endangered Amur tigers, the Far Eastern leopard, the red wolf and the Sikh deer are on the list of priorities for the world's conservation initiatives, they still attract much less attention than do orangutans, rhinos and forest elephants living in tropical countries, or the panda population in China. Although the indigenous people living in Russian woodlands face many problems with using their native land for subsistence, they attract less international attention than the indigenous populations of the Amazon, Australia, New Zealand and even Canada. The post-soviet indigenous population is well educated, mostly has forgotten its native culture and now lives in the same way as other Russians and, therefore, is perceived as less 'endangered', exotic and interesting.

Forests cover 67% of the Russian territory, accounting for 1,183 million hectares. From this area, 26 million hectares are specially protected areas, where logging is not allowed.[4]

What is remarkable about Russian forests is the large number of relatively large areas of intact forest landscapes, each consisting of 20,000-25,000 hectares, which make them the largest plots of old growth forests in Europe (Aksenov *et al.*, 2002; Bryant, Nielson & Tangley, 1997). The total area covered by old growth forests is 288 million hectares, which makes up 26% of all virgin intact landscapes remaining in the World. Many of them are under threat as they are located in areas designated for commercial logging, as only 5% of the intact forest landscapes are located in federal level specially protected areas (Yaroshenko, Potapov & Turubanova, 2001). Despite the development and implementation of state conservation programs of the Soviet Union, including the formation of strictly protected nature protection reserves (Zapovedniks), old growth forest were not articulated as being of intrinsic value, as they existed all over the Russian territory. On the contrary, old growth forests were perceived by the Russian forestry schools as a risk for insect infestation, diseases and fire hazards, and were considered in need of harvesting and logging. The remaining areas of intact forest landscape have not been logged yet because of the delay in road construction and because many of them were in military border zones.

In general, the European part of Russia is much more developed than that of the Russian Far East. The Russian Far East has enormous landscapes of virgin forests, which are home to endangered tigers and leopards, whose habitats are threatened by large and small logging companies (Tysiachniouk & Reisman 2005). The human population there is low in numbers, living mostly in traditional villages and indigenous peoples communities. The wood processing industry started developing rapidly there only since 2008, and created an additional threat to the surrounding high conservation value forests.

In the European part of Russia, many intact forest landscapes are located close to old pulp and paper mills, built in the beginning of the 20th century, which were modernized in the 2000s and are now searching for wood resources (Figure 1.1). Northwestern Russia represents a landscape of large scale industrial forestry for pulp and paper production, with mono-industrial towns built to serve the pulp and paper mills, a scattering of temporary forest settlements build in the 1960s especially for loggers, old traditional villages and remaining plots of virgin, old growth intact forest landscapes.

Therefore, both in the Far East and Northwestern part of Russia preservation of intact forest landscapes remains high on the list of the world's conservation initiatives, and represents a highly contested issue, both within Europe and globally. With the transnationalization of both the forest industry and the environmental movement, the old growth forest landscapes have become arenas for conflicts and negotiations between transnational and local environmentalists, state agencies, business and wider publics.

---

[4] Presentation of the head of Rosleshoz V.N. Masliakov, 'Major outcomes of forestry of Russian Federation in 2010 and goals for 2011' 21 March 2011, St. Petersburg, http://www.rosleshoz.gov.ru/media/appearance/57.

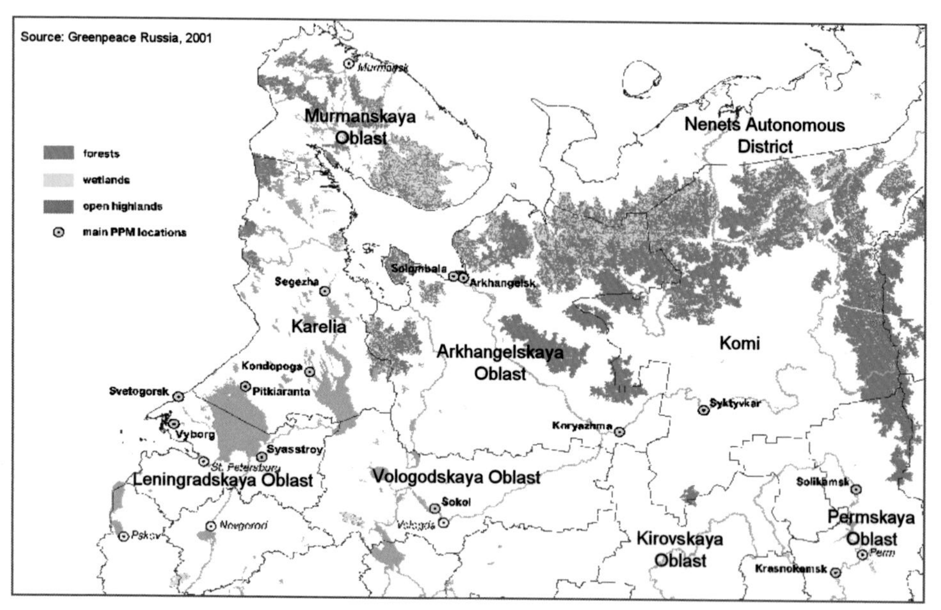

Figure 1.1. Location of the old growth forests and pulp and paper mills in Northwest Russia (Greenpeace Russia, 2001).

1.3 Transformation and internationalization regarding the Russian forests

1.3.1 Transformation and internationalization of the forest sector

During the era of a national planned economy in Russia, forestry was highly subsidized by the state and was perceived by Russians an area for expanding into new frontiers. It flourished, was successful and advanced with a stable international wood trade with Europe. Since 1990, forest production has declined greatly, with recovery in 2000's (Knize & Romaniuk, 2010). For example, in the period between 2000 and 2004, Russian timber exports tripled; in 2004 Russia exported 30% of its round wood, 60% of its sawn wood and 30% of its wood based panel (Malets, 2012). However, Russia still exports much less than its resource potential; its share in the world's export market is only 2.3%.[5]

Perestroika, initiated by Gorbachev in 1988-1989, was followed by large economic reforms in the 1990s, which resulted in privatization in all sectors of the economy, including pulp and paper mills and the state forest enterprises. Land continued to be the property of the state; therefore, newly formed private logging companies became leasers. Several industrial and quasi-state-industrial groups made an effort to build empires, and a new monopoly and many other holding companies were formed (Carlson, 2000; Lehmbruch, 1999). Russian legislation proved to be unfavorable for small and middle size enterprises and gradually logging companies and pulp

[5] Presentation of the head of Rosleshoz V.N. Masliakov, 'Major outcomes of forestry of Russian Federation in 2010 and goals for 2011' 21 March 2011, St. Petersburg, http://www.rosleshoz.gov.ru/media/appearance/57.

and paper mills became subsidiaries of holding companies. In the 2000s most of them became transnational, they formed joint ventures with TNCs or purchased companies abroad. Today, all pulp and paper mills are owned by TNCs either of Russian or foreign origin, and each TNC strives to obtain resources for its mills. The TNCs are active politically, lobbying for reforms in legislation leading to the intensification of forest management. Thus far, however TNCs cannot overcome the resistance of the state bodies.

Despite constant reform and reorganization, including the two revised Forest Codes of 1997 and 2006, the Russian forest sector paradoxically continues to be ineffective (Kotilainen, Kuliasova & Kuliasov, 2006) Russia accounts for 22% of the world's forests, yet uses only 26% of its own forest resource potential. Its revenue from cubic meters of wood is 5-10 times lower than in neighboring Sweden and Finland. Russia is fourth in the World in the amount of its wood stock, after the US, Brazil and Canada, yet exports very little compared to the other forested countries. All these inefficiencies are coming from poor forest management and the lingering effects of the Soviet past (Knize & Romaniuk, 2010).

1.3.2 Transformation in company-community relationships

The Soviet period was characterized by paternalistic relationships between logging enterprises and the forest settlements. State owned logging companies (lespromkhozes), in addition to being economic actors, served as substitutes for the government in providing services to local communities, including support of community infrastructure, housing, schools, kindergartens, libraries and cultural events (Pipponen, 1999). Privatization of forest enterprises in the 1990s started the process of a gradual disconnect between the companies and local communities. Many elements of community infrastructure became responsibility of the state, although privatized companies, responding to the expectations of local communities, continued their support to a lesser extent (Tulaeva & Tysiachniouk, 2008). The formation of holding companies and the TNCs' involvement brought the major shift in logging company-community relationships. When local logging enterprises became subsidiaries of TNCs, they lost the power of decision making as well as the capability of financial management, which lead to their inability to provide support for local communities unless the support was approved by corporate management. Further disconnects came from the modernization of the forest industry, which started to contract with mobile logging brigades working with harvesters and forwarders, instead of providing employment to local communities (Kotilainen, Tysiachniouk, Kuliasova, Kuliasov & Pchelkina, 2008). Ineffective state social policies, in combination with the disruption of logging company-community relationships, resulted in extremely disadvantaged forest-dependent communities (Sodor & Jarvela, 2007). Traditional villages survived the transformation more easily than the Soviet-build forest settlements, which were totally dependent on forest enterprises. However, both types of communities struggled from the privatization and modernization of forest industry.

1.3.3 Transformation of public participation in forest management

Two groups of civil society actors are essential for understanding public involvement in forest management in Russia today: first, representatives of village intelligentsia and other active citizens living in the forest settlements and secondly, environmental NGOs active in conservation issues.

Since the Soviet era, people active in rural areas mostly have been represented by the village intelligentsia, concentrated in schools, libraries and houses of culture. In Soviet times they organized cultural events in forest settlements, inspiring and energizing frontier loggers. These people have remained active despite harsh economic reforms, the hardships of transition and the present situation where social welfare in villages and forest settlements is poorly managed by the state and economic actors (Tysiachniouk & Meidinger, 2007; Tysiachniouk, 2010a). The study will show how transnational actors are engaging local activists in the process of institutionalization of the FSC's rules.

Contemporary Russian environmental activists have their roots in the 1960's. Soviet environmental activism, in the form of the Nature Protection Corps, emerged under Khrushchev during the post-Stalin 'thaw' in universities all over Russia, often under the umbrella of young communist organizations (Komsomol). They formed the first Russian NGO network, which in the 1990's joined multiple international NGO networks. Gorbachev's Perestroika in 1988-89, the collapse of the Soviet Union in 1991, and the resulting fall of the Berlin Wall fostered the fast development of environmental movements and NGOs along with overall world integration (Yanitsky, 1996). The activists trained in the Nature Protection Corp became leaders of major environmental NGOs, including national and regional organizations as well as transnational branches of Greenpeace and the WWF, in the post-Soviet period (Henry, 2010a,b; Tysiachniouk & Reisman, 2006; Tysiachniouk, Reisman & Kotilainen, 2006).

With the transnationalization of forest governance, the intact old growth forest landscapes remaining in Russia became the core issue for NGO cross-border networks. Since late 1980, their effort turned to designating specially protected areas involving the intact forest landscapes and/or boycotting companies buying wood from old growth forests. In the 2000's, another governing tool was used: NGOs fostered development of Forest Stewardship Council (FSC) certification as a new guarantor of old growth forest preservation and as a major instrument for promoting sustainable forest management.

1.4 Governance through certification schemes

Since the Earth Summit in 1992, sustainable development has been based on the three pillars: economic, social and environmental sustainability. The goal of sustainability has been incorporated into the development strategies of particular countries, in global, regional and local Agenda 21 and in many international conventions and innovative policy instruments. The intention to sign the Forest convention at the Earth Summit failed; however Forest Principles were developed and served as guidelines for both nation states and private actors in promoting sustainable forest management worldwide. Forest certification schemes also emerged on the basis of the above mentioned principles and became a new mode of private governance, installing sustainability through market incentives. The first was the FSC certification scheme; later, other competing schemes emerged worldwide, such as the PEFC and many others (Cashore, Auld & Newsom, 2004).

FSC certification became one of the innovative governance instruments based on the tripartite principle of sustainable development in forest sector, which advocates forest management that is socially beneficial, economically viable and environmentally appropriate. Therefore, transnational governance through the FSC means a transition from a forest management model in which maximizing the volume of production and profits is key to a model in which maximizing a different kind of values becomes equally important. The material quality of the wood remains the same; yet sustainable forest management adds value through preserving old growth forest landscapes and by protecting workers, local communities and indigenous peoples' rights. In short, certification represents a shift from sustainable yield to sustainable forest management with the involvement of multiple stakeholders in the decisions related to forestry.

Unfortunately, the FSC is not a panacea. Recent studies have demonstrated that the FSC can not prevent deforestation through the conversion to plantations, nor can it save forests from the rapidly growing expansion of agro-food businesses (Marx & Cuypers, 2010). Therefore, the world's deforestation continues at a growing pace. However, the most remarkable characteristic of the FSC is its ability to transform the market of wood products and to create new niches which convey the demand for implementation of corporate social and environmental responsibility on the ground. Therefore, the FSC can overcome unsustainable sourcing, production and consumption of wood products and significantly minimize the ecological footprint of commercial logging operations. The FSC is able to prevent forest degradation and improve forest management. It is assumed to be powerful in promoting responsible forest utilization, in fostering the preservation of biodiversity, and in protecting the rights of local communities and indigenous peoples.

There is little research available that examines these assumptions at the micro-level or that critically assesses the power of the FSC in implementing and enforcing its global rules, fostering concrete sustainable forest management practices on the ground. As a result, it remains an open question to what extent FSC implementation is able to institutionalize new practices on the ground. Noticeable little work has been done on countries in economic transition, such as Russia (Kuliasova, 2010; Keskitalo, Sandstrom, Tysiachniouk & Johansson, 2009; Malets, 2011; Tulaeva, 2010; Tysiachniouk, 2006a; Tysiachniouk, 2010b).

Russia in terms of its socio-economic characteristics can be said to represent a transitional society, while in terms of its forest ecology it is a boreal forest country, like Canada, Sweden or Finland. The country represents an ideal case for studying FSC related forms of transnational governance through private authority for three reasons. First, in Russia there is no deforestation, no conversion to plantations, as in other countries, yet severe forest degradation and poor forest management exist, as in a majority of transitional economies. Therefore, the FSC seems to be an appropriate instrument for improving forest management in Russia. Secondly, for understanding private authority in governance, it is relevant to investigate how in Russia the example of a non-state regulatory instrument, such as the FSC, is working in a statist transitional society. FSC implementation assumes multi-stakeholder engagement, including involvement of NGOs, companies, local administrations, different state agencies, forest management units, and local civil society groups from rural communities, in all processes related to the FSC's standards implementation. This criterion can be a challenge for FSC implementers in Russia, given the country's underdeveloped local civil society institutions and its 'managed' democracy on the federal and regional levels. Thirdly, Russia contains a fifth of all world's forest resources and is a leader in the world, second after Canada, in the amount of FSC certified territories. If the FSC's impacts are strong, this means that forest management on the world scale is improving.

1.5 Analyzing global private governance through certification: approaches

Certification and labeling is a booming research field. In this respect, scholars notice the transformation of modalities of governance toward the greater significance of private versus public actors in governance. Theorizing global governance, these scholars underline the need to accommodate new market and civil society actors in complicated governance arrangements (Arts, 2011; Arts, Leroy. & Tatenhove, 2006; Bartley, 2007b; Cashore *et al.*, 2004; Dingwerth & Pattberg, 2009a; Gulbrandsen, 2008a; Gulbrandsen, 2008b; Gulbrandsen, 2010; McDermott, Cashore & Kanowski, 2009; McDermott, Cashore & Kanowski, 2010; Reed, Utting & Mukherjee-Reed, 2012).

There are several ways to approach transnational governance through voluntary private regulatory schemes. First, by focusing on the emergence and institutionalization of new private regulatory tools driven by civil society actors (Auld, Bernstein & Cashore, 2008a; Bostrom & Gartsen, 2008; Cashore *et al.*, 2004) or by looking at certification schemes as private law created by civil society (Meidinger, 2007). The second approach is studying the standards-setting process in a particular context and comparing the standards produced (Tolleflson, Gale & Haley, 2008). The third approach is to study the institutionalization of new rules on the ground and the impact of implemented standards (Kulyasova, 2010; Malets, 2011; Tysiachniouk, 2006a, 2010c). Recently, many comparative studies have emerged examining FSC and other certification schemes (Auld *et al.*, 2008a; Hallstrom & Bostrom, 2010). Except for a few studies, (Oosterveer, 2007; Oosterveer & Spaargaren, 2011) scholars did not evaluate the network structure of governance arrangements through certification, analyzing them in terms of organizations interacting with each other.

To understand how global regulatory instruments impact the management of natural resources, it is methodologically relevant to study the roles, interplay and input of both transnational and local actors in governance arrangements. Scholars created the special term 'glocalization', which assumes the integration of global and local governance arrangements, resulting in new practices on the ground (Ritzer, 2004). This term reflects on crucial changes in the contemporary world, in which local practices became embedded in world capitalism and transnational politics (Hong & Song, 2010; Wellman & Hampton, 1999, Sassen, 2005). Global governance can be meaningfully applied to the development and implementation of voluntary certification schemes, as certification rules developed globally for regulating companies are introduced to individual local actors with the aim of changing their behaviors and practices.

Although there is plenty of literature on certification, there are no multi-sited studies which take an integrated or holistic approach to certification networks and which have captured simultaneously transnational and national standards-setting processes and processes related to the implementation and institutionalization of emerging standards on the ground, within circumscribed territories[6] or locales[7]. The phenomena of such governance arrangements are under theorized.

In order to understand transnational private authority in governance, and especially governance through the FSC, it is important to develop new concepts and a particular theoretical vocabulary, as the old ones can capture only part of the operation of certification networks. It is important that the new concepts capture both global and local processes and actors as well as their interplay. In this respect, the sociology of transnational processes, networks and flow seems to have great potential. These innovative approaches were applied to different kinds of global governance arrangements, such as new spaces of governance (Castells, 1997, 2009), social movements (Keck & Sikkink, 1998), global cities, migration, ethnicity, gender (Sassen, 2003, 2008), mobility (Urry, 2003), and environmental governance (Mol & Spaargaren, 2006).

Continuous debates are taking place in the governance literature around the changing role of the state in global governance and in particular to what extent the state continues to be important. However, there are still few theoretical insights on the role of non-state actors in governance (Bostrom & Gartsen, 2008; Haufler, 2009). Little attention is given to the development of new concepts that can explain private policy instruments operating on the global scale in the context of more fluid, dynamic and partly deterritorialized societies.

This book approaches global governance by private authority, particularly voluntary certification schemes, by building its foundations on the sociology of transnational processes, governance literature and neo-institutional sociology. It sets the goal of developing and testing new theoretical foundations that would explain voluntary certification schemes with their complicated agencies, involving the interplay of global and local actors.

Therefore, the aim of this study is to create a new theoretical and empirical lens that would help explain the transformative character of governance by private authority through certification schemes on the example of implementation of FSC certification in Russia. The key is in understanding and explaining the governance capacity and in-depth mechanism of the FSC operation, through capturing its global/local dimensions. The study focuses on how transnational stakeholders, representing diverse and conflicting interests, develop and negotiate global standards, how local actors provide input in global processes, how global standards are adopted and translated into national standards, and how the national standards are interpreted, implemented and institutionalized in Russia.

1.6 Objective and research questions

The central objective of this study is to understand the operation of the FSC voluntary forest certification scheme as a transnational non-state governance institution, both globally and within the nation state. Three research questions were formulated:

[6] Concrete geographical places.

[7] Space as setting for practices which in turn depend on the characteristics of the particular locale, term borrowed from A. Giddens 'Constitution of Society' (1984). Locale can be a transnational setting, dissociated from a particular place.

1. How to conceptualize the process of institutionalization of transnational private governance through certification schemes?
2. What is the role of global and local non-state governance actors, especially NGOs, in the process of installing new sustainable forest management practices in Russia?
3. How and to what extent do the FSC governance arrangements transform forest management practices in Russia?

1.7 Outline of the book

Chapter 2 introduces concepts used in the sociology of transnational processes and explains the emergence of the major analytical concept of this book: the governance generating network (GGN). Consisting of three major structural elements; nodes of design, forums of negotiation and sites of implementation, GGNs generate new global governance arrangements as well as institutionalize new practices on the ground in different parts of the world. The GGN concept is designed to analyze global governance from a local/global interface perspective. It draws, in a way, equally from sociology and political science, and adds new insights and arguments to the criticisms on nation-state based analyses of global governance. It is particularly well suited to analyze the role of 'wing-like' actors and organizations such as NGOs and TNCs in global governance.

In this chapter, the GGN concept is initially operationalized to explain the operation of major certification schemes, such as the FSC, the Marine Stewardship Council (MSC), and Fair Trade. Attempts are made to assess the role of private authority, especially NGOs and TNCs, in GGNs and to evaluate in what ways and to what extent they are able to regulate and transform markets towards a more sustainable pathway.

Chapter 3 is a semi-theoretical, semi-empirical chapter focusing on transnational processes and global regulatory networks. The chapter starts with a description of the history of global forest governance and the formation of some of the major GGNs related to forestry. The FSC is examined with the help of the GGN approach. The main components of the FSC-GGN, namely, the nodes of design, the forums of negotiation, and the sites of implementation are analyzed. The process of designing the standards, i.e. the major principles of the FSC's functioning network is explained. The chapter analyzes how different GGNs interact with each other in the context of global/local arenas.

Chapter 4 is on governance through the FSC in Russia. It first discusses the conditions of forests and forestry in Russia in a historical perspective. It characterizes the peculiarities of the state forest management authority in the context of socio-economic transition processes and institutional turbulence. Next, the chapter focuses of the emergence of NGO networks in forest governance. Further focus is on forest certification, including all the initiatives, existing in Russia with major emphasis on the FSC, its emergence, institutional design and infrastructure, and the operation of the national FSC office. The chapter develops the analysis of how the FSC global rules coincide with Russian national legislation and on how state and non-state actors interact in the FSC process. Next, the current state of FSC certification in Russia is presented.

The next four chapters are empirical case studies presenting different types of global/local networks and the different ways in which sustainable forest management practices were installed in Russia. All cases are in one way or another related to the FSC-GGN. Chapter 5 is dedicated to the NGO network that preserved the old growth forests in the Karelia Republic through the consumer boycott organized against the European companies that had planned to buy timber from these ancient forests. NGO-led market campaigns, despite being conflicts, offer a form of governance that enables business practices to become more environmentally friendly and that helps to promote and construct environmentally friendly niche markets. With European funding, NGOs involved Russian state agencies in the process of creating specially protected natural areas, in order to preserve the old-growth forests.

Chapters 6 and 7 represent cases in which NGOs, sponsored by development agencies, served as co-designers and implementers of the Model forests, first in the Pskov oblast and then in the Komi Republic. In both cases, the intention was to develop a showcase with the best sustainable forest management practices, educate local, regional and national stakeholders and to reproduce the lessons learned throughout the European part of Russia. FSC certification was a key for introducing sustainable forest management in both models. Although both projects had similar objectives and were successful in developing a model, their abilities to disseminate lessons learned were different. The Komi model project, where an NGO partnered closely with state agencies, developed best practices that project implementers could disseminate, at least in the Komi Republic. In the Pskov model, the NGO partnered with the TNC and poor governmental involvement prevented successful dissemination of the outcomes.

The case study presented in Chapter 8 is on the FSC-GGN as applied to a specific forest management certification case of the territory leased by the pulp and paper mill, belonging to a TNC of Russian origin and operated by its subsidiary the Northern Logging Company. In this case, the UK buyers of sawn wood were the major drivers for certification of the holding's leased territories. The role of NGOs in this case differed from the previous ones, although they continued to be essential agents of institutional change. They served as both consultants in the process of institutionalization of the FSC certification and surveillance agents, controlling the quality and depth of implementation of new practices. Barriers and trends of certification projects occurring without outside funding from foundations are also analyzed.

The concluding Chapter 9 highlights both theoretical and empirical findings and achievements of the study. It explores the major analytical concept of the book (the governance generating network), theorizes further and explains how it works in practice. It relates the GGN concept to other scholarly work, focusing on its explanatory capacity and contribution to social theory. It compares the case studies and highlights the role of NGO networks in fostering institutional change both globally and locally. The chapter proceeds with revisiting research questions and the conceptual framework for the study involving different types of networks, summarizes empirical findings and their diversion from the previously identified ideal types. Finally, it analyzes to what extent it is possible to change the forest management practices in Russia through FSC governance.

Chapter 2. Governance generating networks and the institutionalization of new practices for sustainable forest management

2.1 Introduction

In the 21st century, with the increasing speed of globalization, significant changes took place in all governance processes, including the management of natural resources. Both globalization and the development of information technologies generated a new era of informational governance (Mol, 2008). Scholars observe the growing importance of nongovernmental actors, both NGOs and TNCs, in the different kinds of governing arrangements, including agenda setting, global rule making and implementation (Bostrom & Gartsen, 2008; Haufler, 2009: 140; Campbell, 2004).

These new regulatory global arrangements result in more complex relationships between public and private actors, in which state-centric forms of regulation shift toward combinations of hard laws and private standards. Private regulation helps to narrow the governance gaps existing between states and global markets. Different kinds of transnational standards are developed by various combinations of actors, including NGOs, companies and state authorities (Vogel, 2008: 262, 267). This, to some extent, results in a transition of regulatory mechanisms, from hard law based regulation to soft law, self regulation and voluntary standards (Auld, Balboa, Bernstein & Cashore, 2009; Bartley, 2003, 2005; Bartley, 2010a; Cashore *et al.*, 2004; King & Toffel, 2009; Khanna & Brouhle, 2009). Scholars argue that governance is becoming more and more a hybrid, multi-stakeholder affair due to the tripartite pressures from increased globalization, decentralization and marketization (Delmas & Young, 2009: 29; Lemos & Agrawal, 2009: 73).

In order to better understand the (governance) dynamics of global modernity, Manuel Castells (1996, 2009) outlined the emergence of the network society, a new social morphology. Networks are seen as governing structures that organize and control key activities in contemporary society by connecting different actors and elements in distant locations. Networks, in the context of time and space compression and increased flows of information, money, goods and technologies fostered the transnationalization of environmental governance in general and natural resource management in particular.

Castells is joined by an increasing number of sociologists who regard it as a primary task of sociology to analyze the globalization and hybridization of governance and to reflect on the particular ways in which globalization is changing day-to-day practices on the micro level (Castells, 1996; Sassen, 2003; Spaargaren, Mol & Buttel, 2000).

In order to understand the global governance of natural resources, I am building my theoretical framework on the above mentioned sociology of transnational processes (Castells, 1996, 1997; Sassen, 2004; Urry, 2003) and further developing new categories and the vocabulary necessary for understanding some of the most crucial contemporary features of transnational governance. Following Castells, I distinguish between the spaces of places, in which day to day practices take place and transnational spaces, disconnected from concrete geographical settings, in which global rules and standards are being created.

Networks which play a crucial role in the development of global regulatory tools, products, or standards to be implemented in different parts of the world we suggest to be named Governance Generating Networks (GGNs). The three major components of such networks are (1) the nodes of global governance design, (2) forums of negotiations and (3) sites of implementation. The nodes of global design are transnational centers, which bring together and seek to unite stakeholders from around the globe working on new regulatory products, e.g. new tools, strategies and instruments for global governance. Sites of implementation are the concrete (also de-nationalized) territories or locales, where global regulation is institutionalized after being translated and made fit to the local circumstances. Forums of negotiation are part and parcel of the GGNs in the sense that they play a role in all phases of the governance process, from standards development, framing and translation to adoption and implementation (Kortelainen & Kotilainen, 2006; Tysiachniouk, 2006b).

The standards, frames and similar regulatory tools are designed and developed predominantly in the nodes of design. Next, they become translated, adopted, adjusted and transferred to the sites of implementation. GGNs and their forums of negotiation, as I will show in this chapter, can be initiated by different actors and organizations, yet their key characteristics are the aim to link actors of transnational spaces with actors in the spaces of places and to provide an infrastructure for their interplay.

Understanding the role of non-state, private forms of authority as represented especially by global NGO-networks involved in the governance of natural resources, is a key to my analysis. In this study I pay significant attention to NGO-driven or initiated GGNs, which I show to make significant contributions both to the development of global regulatory standards as well as to the institutionalization of new regulatory practices for natural resource management around the world. Of crucial importance for understanding the role of GGNs and their transformative capacities are the ways in which new regulatory designs reach the sites of implementation and result in more or less significant changes in the institutions and practices involved in the management of natural resources.

The chapter begins by focusing on recent developments within the sociology of transnational processes (Section 2.2). It describes the origins of the GGN concept, its major components, its explanatory power and its relationship with existing concepts in the literature on global governance (Section 2.3-2.5). It proceeds with further developing the GGN concept by analyzing it from the perspective of neo-institutional sociology and by looking empirically at NGO-driven GGNs involved in the construction of ecologically and socially 'sensitive' niche markets. Governance through certification is analyzed using the GGN concept. GGNs are shown to have different kinds of transformative potentials. They are involved in the reshaping of old path-dependencies while also creating new pathways or linkages between the transnational spaces of flows and the local space of places. NGOs operating in the context of these (market-based) GGNs are seen as major agents of change involved in the creation of new institutional fields and management practices. In short, GGNs are shown to be *public cum private* and *local cum global* networks that make global governance work in the context of the network society (Section 2.6). The chapter concludes with proposing three ideal types of GGNs involved in designing certification schemes for fostering sustainability around the globe (Section 2.7).

2.2 Transnational processes and the global governance of natural resources

This section describes the new sociological understanding of space and time as connected to the emergence and development of the global network society. It shows how the concept of GGN builds on the sociology of transnational processes, and looks at the literature discussing network governance. It explains how GGNs are linking transnational spaces with spaces of places and explores processes occurring in the nodes of design, forums of negotiation and sites of implementation.

Manuel Castells made a significant contribution to social theory by developing a new globalization paradigm based on the revolutionary inputs of information and communication technologies. Communication technologies enabled societies to become information societies, constituted by networks and flows (Castells, 1997). To make sense of the impact of these technologies in society, sociologists have to distinguish between the space of place and the space of flow respectively. Internet and other communication technologies create special 'spaces of flows', which allow actors to ignore and overcome geographical distance, and to interact simultaneously. In this respect, communication technologies provide 'material support for time sharing practices' (Castells, 1996: 412) and contribute to establishing the different time-space dynamics of the space of flows. Therefore, the 'space of flows is made up of personal micro-networks, which project their interests in functional macro-networks throughout the global set of interactions in the space of flows' (Castells, 1997: 416). Although people enter the cyberspace simultaneously from different localities of the globe, communication technologies create the transnational virtual public sphere, which enables multiple actors to enter global arenas and to participate in global processes (Sassen, 2004: 685-686). Through cyberspace, people enter the global flow of information, travel without the body's physical movement, globalize from within and participate in events beyond geographical boundaries (Urry, 2000: 74). Yet, people living in localities can both participate in the place-based practices and 'go out' to the space of flow, by watching TV, browsing internet sites, or participating in communities organizing or contributing to setting new global policies. Following Castells, I distinguish between geographical settings where place-based practices occur on the one hand, and 'transnational spaces' (Castells' concept of the space of flows) which people create in order to communicate across space and beyond particular geographical boundaries, on the other.

In Castells' work the concept of space of place continues to be important as the space of everyday life practices. In the context of the newly developing sociology of networks and flows, places are perceived as points of localization of many networks and flows (Mol & Spaargaren, 2006: 75; Urry, 2003). In this respect, places loose their traditional meaning and their new meaning comes to depend on global networks and flows that intersect at this place and on the process of the space of place becoming (not) linked to global networks and flows. The geographic landscape becomes fragmented by the fact that some places are included in global networks and flows while others are not. Networks tend to absorb those places that are valuable for the network as a whole and which can perform the tasks assigned to the network. By the same logic, they tend to exclude places that are regarded to be of little value for the network. Therefore, it can be argued that those people and places (made) external to the network are being excluded and discriminated against (Castells, 2009: 25-26).

In Castells' early work, places are viewed as victims of globalization, resulting at least in tensions between the spaces of flow and the spaces of places. The space of flows imposes power on actors in the space of places and changes their meaning. In this respect, it can be argued that the global overwhelms the local (Castells, 1997). In his later work, Castells discusses the axes of glocalization without necessarily blaming global actors. He even acknowledges the enabling role of global social movements. He claims that the local can exercise resistance; social movements can emerge locally, connect to alternative global networks and become empowered by this new network affiliation (Castells, 2009: 26).

In the work of Sassen, the tensions between the transnational space and the space of place are not characterized by negative connotations since they can result as well in empowering local actors. Growing digitization changes the meaning of the place and transfers them into 'micro-environments with global span' (Sassen, 2004: 663). Therefore, globalization and information technologies do not undermine the significance of places, but on the contrary, foster the arrival of new actors applying new forms of power for creating institutional changes which replace the old arrangements and institutions. For example, the internet can help establish trans-local connections and forms of information exchange which facilitate grassroot based globalization and the empowerment of local communities in disadvantaged places as well as anywhere. This can be regarded as a form of emancipation of the local. The ongoing process of dis- and re-embedding (Giddens, 1990) is the formal mechanism to describe what is happening, and in terms of outcomes this process can result both in the loss of power as well as in forms of empowerment when judged from the perspective of local actors in the space of place.

In this way, places have gained particular importance in the context of governance studies and a policy since there has come to being a multi-scalar form of politics which is at the same time transnational and differently embedded in the local. With the increased globalization of economies and the increased role of transnational, 'cosmopolitan' corporations, the world has become more denationalized and deterritorialized (Beck, 2005). To underline these tendencies in contemporary governance, the term glocalization has been introduced by scholars, which means 'the integration of the global and the local resulting in unique outcomes in different geographic areas' (Castells, 2009; Ritzer, 2004: 73). Local actors in contemporary politics, in order to implement social change, do not have to move 'through the set of nested hierarchies of scales' from the local to the national to the international (Sassen, 2003: 6). Thus, they can act globally without intermediaries. Localized struggles can engage global actors and create trans-local networks sharing place-based information. Therefore, global processes and conditions can be constituted and (re)produced sub-nationally, on micro-scales. Thus, local/global dynamics are conceptualized as a fluidity of scales and as variable hierarchies (Sassen, 1996; Sassen, 2003). Place making projects (Gille, 2006: 149) that win the territorial contestation have the capacity to attract specific types of global actors, networks and flows.

As a result, national and global elements and dynamics no longer have a binary character. Global features physically can be situated in a territory and fit the institutional setting of the local and the national while at the same time moving beyond the boundaries of what historically is understood as sovereign. In other words, the global can inhabit the national and form partial, sometimes highly specialized 'assemblages of bits of territory, authority and rights that begin to escape the grip of national institutional frames (Sassen 2008: 61-62)'. These assemblages shake and unsettle national institutions with their existing norms, producing spaces for new norms and rights. As a result, a process of the partial denationalization of territory, authority and rights takes place (Sassen, 2008: 61-62).

The new understanding of space triggers the new understanding of time, because both can not be separated and are simultaneously implied in the operation of the physical and social worlds (Urry, 2000: 107). Through communication technologies, information can become both instantaneously and simultaneously available across the globe, therefore the 'clock time', a category in which events are separated and information has to be delivered, looses its significance when actors interact in transnational spaces. To underline the innovations that communication technologies brought to society, Urry introduces the term 'instantaneous time' and argues that clock time is replaced in cyberspace by this new type of time, which has blurred the boundaries between day and night and between weekends and working days (Urry, 2000: 126-129). Different researchers are inventing different frames and terms for explaining the same phenomena. For example, they are talking about time-space compression, appearance of a new concept of 'timeless time', produced by cyberspace (Castells, 1997), or as it was described above, 'global instantaneous time', in which communication technologies play a crucial role in reducing distance between places and people (Urry, 2003: 2). Next to the concept of instantaneous time, Urry uses the concept of 'glacial time' to refer to the time-space dynamics of the world's global ecosystem.

The new spatio-temporal orders, brought about by the global economy, are partly based on hyper mobility and time-space compression. They are as well produced, however, by the time-space infrastructures which are characterized by the fixity of place-bound resources and by actors operating in real or clock time. So the use of cyberspace, with its instantaneous time, does not prevent local actors from conducting their day to day practices in clock time, e.g. to have a fixed work schedule, which is adjusted to the existing social infrastructure, such as child care centres, schools, centres for medical care and other social services. Therefore, the spatialities and temporalities of globalization assume the existence of 'assemblages that mix the national and the global', combining dynamics of mobility and fixity (Sassen, 2006: 383). The day to day practices of people continue to occur in real or clock time yet they tend to increasingly co-exist with the new categories of time and space as described above. They all are relevant for contemporary sociological analysis. What the sociology of globalization tells us is that global networks and their nodes and actors nowadays tend to operate in both clock time as well as timeless time when making decisions related to the environment in glacial time (Urry, 2003).

2.3 Characteristics of global networks

Globalization scholars emphasize that network logics penetrate institutions of production and consumption, social movements, and culture (Urry, 2003: 9-10). For understanding the role of governance in the network society, it seems crucial to get to know the structure and agency of existing network types, to analyze when and how they can become efficient governance-entities and to specify in some detail the power relations that are at work.

The networks have nodes, where key management and governance takes place, and information flows connect the nodes. The elements of the networks and their communications are coordinated in the hubs, specific network 'exchangers' (Castells, 1997: 413). The nodes that constitute the networks can be of different sizes, varying from super-state structures, such as strategic alliances of the states, to the size of the state, to sub-state units, such as global cities (Sassen, 2003). Global cities, such as New York, London, Sao Paulo, or Frankfurt constitute the nodes of multiple networks and flows and are essential in global governance (Bouteligier, 2011). Network nodes can be smaller in size, such as business districts, or offices of particular organizations. Although nodes are often located on a sub-state level, their significance for global governance can be considerable. For the proper understanding of the functioning of these nodes, an analysis of their relationships with the supra-state and nodes at the global level needs to be made.

Networks are open structures with different kinds of morphology resembling chains in which nodes are spread in a linear way, stars with a central node and a periphery and spider webs. An important characteristic of networks is their ability to rearrange and reorganize themselves, by phasing out nodes that have become redundant and by adding new ones that emerge as relevant (Arquilla & Ronfeldt, 2001, cited in Urry, 2003: 51, Anheier & Katz, 2005).

The efficiency of networks as organizational forms is determined by three major features, flexibility, scalability and survivability. Flexibility is the ability of the networks to adopt to changing organizational environments, to reconfigure themselves by changing their structure, connections and ways of communication. Scalability is the ability to stretch or shrink their size connecting scales from the global to the local. Survivability is the ability of the networks to continue operating even if some nodes are destroyed and dispersed (Castells, 2009: 23).

Networks show different dynamics of power which are determined by the power of the actors and organizations involved in the networks (Castells, 2009: 42; Castells, 2011: 773), the number of nodes involved, by the density of connections between nodes and by the number and density of connections to other networks. Similar to population growth, network growth might occur exponentially, as every node added triggers other nodes and connections. Improved connections and organizational learning also add to network's power (Urry, 2003: 54).

The networks can exercise power by controlling other networks and actors in the case they are well organized around some meta-program, oriented towards achievement of certain values and interests. Additional power comes from the ability of the network to build strategic cooperation with other networks, which share common goals, values and have the ability to provide additional resources (Castells, 2009: 45-46). Networks can exercise power through the mechanism of inclusion/exclusion as well as through the imposition of the rules of inclusion. In the case of the latter, the network power is the power of the core standards, which in most cases can be negotiated. These standards impose limitations on network actors' involvement and simultaneously became the rules of the game and gate keeping entities. The standards regulate interaction between actors within the network and power is exercised by the imposition of the rules of inclusion. Within the network, different actors also exercise power over each other, where the strongest win the battle (Castells, 2009: 43; Castells, 2011: 773).

Networks differ by the level of complexity, flexibility, and predictability.Two ideal types can be distinguished in this respect: globally integrated networks (GIN) and fluid networks. Fluid networks can be compared with 'global waves' taking different shapes along the way and therefore difficult to analyze and characterize in terms of their form, content and type of agency. Fluid networks are first and for all characterized by a high level or degree of non-manageability and non-predictability with respect to their pathways of change. GINs on the other hand are more structured, standardized and have a recognizable brand identity (Urry, 2003: 56, 60).

GINs can be represented by entities of different size and scale, for example, the supra national state structures, such as the UN, World Bank, CNN, European Union, World Health organization, International Union for the Conservation of Nature, and the WTO all have GIN like structures. They are constituted by highly controlled, predictable and efficient nodes with actors involved in decision making who are able to ensure manageability of the whole network (Urry, 2003: 81). Networks formed by chain businesses, such as McDonalds, Microsoft, Coca-Cola, or Federal Express, as well as major NGOs, including environmentals, such as the WWF and Greenpeace also belong to the category of GINs.

The agency of GIN networks, according to Castells, is represented by (groups of) programmers and switchers, which both in different ways determine the network-making power. Programmers set up the goals for the network, their 'design', their programmatic codes. They also work to maintain its efficiency and get involved in brand making for ensuring better competitiveness. Switchers are those actors who are able to reorganize the network and engage it in strategic cooperation with other entities, attract resources and enhance power. In the global world, if some networks exercise power, other networks can either become linked or develop into oppositional networks that exercise counter-power (Castells, 2009: 43-47). Brands bring both power and vulnerability to transnational networks. In the case of oppositional networks, counter-power can be organized with the help of consumer boycotts using wide media coverage to name and shame well-known brands and the companies behind them. As a result, the image of the company can be significantly affected and financial losses can follow. This fact is widely used by NGOs that strategically use naming and shaming as a tool for keeping corporate networks accountable and forcing them to change practices under the threat of consumer boycotts (Micheletti, 2003).

Successful naming and shaming can result in sudden and significant losses of power. However, non linear changes might also occur in the opposite direction, resulting in major and sudden increase in (both network and networking) powers. For example, this is the case when socio-technical developments and small scale innovations, including forms of organizational 'know-how', suddenly result in a huge increasing of positive feedback over time (Mahoney, 2000 cited in Urry, 2003: 55; Waldrop, 1994: 49). In this way, small patterns, which receive positive feedback, can break old path dependencies, create new pathways and initiate the exponential growth of institutional change (Urry, 2003: 56). For example, companies with labels indicating the sustainability of their production practices (such as FSC, fair trade) could receive positive feedback in the market and in civil society, this feed-back helping to suddenly upgrade their activities from niche markets to the level of global, mainstream markets. This kind of capacity inherent in the network structure and function can justify the hope that sustainable niche markets can at some point in time foster the overall greening of global trade.

2.4 Theoretical foundations of the GGN concept

Manuel Castells, John Urry and Saskia Sassen, as was described above, significantly contributed to the sociology of transnational processes by suggesting the new social morphology of contemporary society based on networks and flows. They presented a broad picture of the network society and its characteristics. The question arises how this emerging 'network, networked, and networking' society (Castells, 2009, 2011: 773) is engaged in designing, negotiating and implementing new global policies and standards? How are practices within particular territories affected by global policies and standards and in what ways are authorities and rights subsequently redefined both at the the transnational and local level? The concepts put forward by the contemporary literature on the sociology of transnational processes (Castells, 2009; Sassen, 2008; Urry, 2003) are rather broad and defined in general terms. They provide insights to the multi-level structures and dynamics and in the multiplicity of agents involved in the constitution of a global society of networks and flows. The sociology of networks and flows turns out to be useful for examining the overall characteristics and dynamics of environmental flows (Mol & Spaargaren, 2006; Mol, 2012). However, it seems not yet well equipped for analyzing how transnational policies and governance arrangements are being designed, negotiated and institutionalized in different parts of the globe.

To specify the sociology of networks and flows and to make it fit the analyses of global governance, this volume suggests making use of the concept of Governance Generating Networks (GGNs). It represents a special formulation of networks and flows theory that can explain how new forms of policy making, in between the global and the local, are happening in the global network society. The concept can be especially useful for analyzing governance beyond (inter) national regimes, and for cases of transnational policy making that involve private authority in particular. It gives a new perspective to the forms of trans-nationalized, de-territorialized, dynamic, fluid and post-national governance that emerge with the globalization of modernity. It provides an outlook which is able to better capture the global/local interplay of actors enacting their complicated forms of agency and power at different levels of scale.

GGNs are the networks that develop global regulatory products in the form of global standards, rules, norms or recommendations. These networks are organized mostly as GINs. The major components of such networks are the nodes of global governance design, the forums of negotiation and the sites of implementation (Kortelainen, Tysiachniouk & Kotilainen, 2010).

The concept of the GGN was developed on the basis of two existing bodies of literature, the sociology of transnational processes (Bouteligier, 2011; Castells, 1996, 2009; Sassen, 2004, 2006, 2008, 2011; Urry, 2003) and the literature on policy and governance networks (Borzel, 1998; Sørensen & Torfing, 2005, 2007a,b). The framework of neo-institutional sociology is used as a supplementary tool for understanding institutional change (DiMaggio, 1991; Fligstein, 2002; Fligstein & McAdam, 2012; Powell, White, Koput & Owen-Smith, 2005). Along with several other scholars (Dingwerth & Pattberg, 2009a,b), I expand institutional theory to the transnational level to explain the institutionalization of new norms and practices in both transnational spaces and the spaces of places.

In the sociology of transnational processes, as described above, the concept of 'global assemblages' is put forward by Saskia Sassen. These assemblages are neither completely global nor national, and situate themselves in the analytical borderlines between the global and the national, involving particular combinations of territory, authority and rights (Sassen, 2006, 2008). Global assemblages are represented by networks with global span, yet also involve territories belonging to the nation states as well as place based practices and infrastructures. Sociologists of globalization metaphorically envision networks as a new all-encompassing social morphology and architecture and analyze them through the lenses of complexity theory (Castells, 1996, 2009; Urry 2003). They talk about network structure, including concepts such as nodes, logic, flexibility, adaptability, and inclusion/exclusion. However, they do not give much focus to actors and networks as relevant for issues of policy making and governance. When issues of governance and management are at stake, the horizontalization of the global network society (Castells, 1996, 2009) and its fluid character (Urry, 2003) are emphasized and used to put forward a new set of problems and tasks for those involved in designing and implementing new forms of governance.

On the contrary, in the political science literature on network *governance*, scholars see networks as self regulating entities, made up of a variety of autonomous, but interdependent, actors who strive to resolve certain complex social problems. These scholars emphasize multiple stakeholders' involvement in decision making and the importance of negotiation among them. Negotiation among actors representing different interest groups is seen as a complex process which allows for persuasion and movement toward consensus and includes both elements of bargaining and elements of deliberation in which actors are oriented toward reaching mutual understanding (Bächtiger, Niemeyer, Neblo & Steenbergen, 2010: 37), as well as learning. Actors get involved in negotiation in order to pursue their own interests through information exchange and cooperation with others (Herting, 2007: 50). Negotiation is essential for the effectiveness of the network as regulating agent within a particular policy field (Sørensen & Torfing, 2005: 205-206; Sørensen & Torfing, 2007a: 9-10). In addition to incorporating negotiations in the process of decision making, network structure, actors, guiding norms, values and discourses are equally important for realizing the main aim of the governance network, e.g. delivering new policies, standards, and strategies that are both effective and legitimate.

The generative process of achieving social change by GGNs is explained, in this volume, using neo-institutional theory. Achieving social change by a GGN represents a range of shifting relationships between actors involved in the design, transfer and implementation of new rules in both transnational and local spaces. Actors belonging to a GGN exercise both coercive and normative pressures in order to break old path dependencies with old ways of decision making (DiMaggio, 1991; Powell *et al.*, 2005) and generate new paths involving new governance arrangements. By doing so they create new organizational and institutional fields (Bartley, 2003; Fligstein, 2002; Fligstein & McAdam, 2012), which become part and parcel of the new mode of governance.

The GGN concept involves both nodes (centers in transnational spaces) and sites of implementation (territories in the space of place). Here it builds on the sociology of globalization (Castells, 1997, 2009; Sassen, 2006, 2008) in particular. But GGNs also involve interdependent actors in forums of negotiation. This concept is part and parcel of GGNs and derived from governance theory (Sørensen & Torfing, 2005, 2007b) in particular. It also involved actors involved in introducing new rules and practices, here is builds on neo-institutional sociology (Bartley, 2003; Fligstein, 2002) in particular.

The aim of nodes and forums is to design, develop and implement new transnational standards, criteria, schemes and strategies which result in institutional changes in the space of place. In the phase of implementation they affect the rights and responsibilities of certain place-based actors operating at the sites of implementation.

There are many networks in the world with nodes and forums that make an effort to develop global regulatory products with the goal of changing practices in particular places. However, only few become recognized in world politics and develop into effective GGNs in the sense that their regulatory products are being used in the sites of implementation around the world and cause lasting institutional changes in the spaces of places. Some networks die, others grow and develop to the extent that they finally become effective GGNs. These globally legitimate networks develop regulation in transnational spaces and bring institutional change to the spaces of places that represent sites of implementation for these networks.

GGNs can be initiated by NGOs, business, and governments, but most of the times are not initiated by one single actor. Most of these networks are based on multi-stakeholder processes and participation, with complex relationships existing between different actors who operate at different levels of scale. For example, international conventions are driven by governments, but NGOs are actively involved as well. The World Business Council for Sustainable Development was initiated by major transnational corporations, but consultations with other stakeholders have been taking place as well. The Forest Stewardship Council (FSC) was initiated and supported by NGOs, although they involved business actors and declared that equal rights of participation in decision making should exist for all three FSC-chambers: the economic, the environmental and the social chamber[8].

Like all networks, GGNs can have different relationships with each other. They can form strategic alliances, for example, the International Social and Environmental Accreditation Labeling Alliance (ISEAL) involves networks that are based on multi-stakeholder processes and that focus on developing standards for forestry, fisheries, fare-trade, etc. GGNs can also compete with each other and strive for power, legitimacy and territory, especially if several GGNs operate in the same sector of the economy. For example, the FSC network is in intense competition with the Program of Endorsement of Forest Certification (PEFC), which is trying to attract business, interested in certifying as well as gaining legitimacy within different constituencies of stakeholders.

In summary form, I can conclude that GGNs are global assemblages (Sassen, 2006, 2008) that are particularly constructed for the goal of governing processes of change in the global network society. They are, most often, of the form of Global Integrated Networks (GINs) described by John Urry. They obey the laws and dynamics of network power as exposed by Castells, with key roles for standards and with groups of programmers and switchers operating as important actors in the global power game. They are particularly interesting and relevant for their explicit role in governance. They operate in between the space of place and the space of flows, and have particular relevance for describing the role of actors and organizations in the market and in civil society. In the next section, the different components of GGNs will be described in more detail.

2.5 Key components of GGNs: nodes of design, forums of negotiation and sites of implementation

Nodes of design, forums of negotiation and sites of implementation play different roles in GGNs and are defined differently, although the (network) boundaries between them may sometimes be unclear. Under the node of global governance design, I explain the association of actors in transnational space that initiate, design and develop rules and regulations that are intended to

[8] See Chapter 3 for details.

become applied in different parts of the world (Tysiachniouk, 2006b). Sites of implementation are the places where the new global rules, norms and regulatory regimes are being implanted and transposed in order to become more or less successfully institutionalized in local practices. They represent the 'de-nationalized' territories where global rules and norms are transferred and 'translated' into concrete practices. Forums of negotiation are important components of global governance networks since they involve different stakeholders that have different and often even opposite views and powers with regard to the issue to be resolved. Therefore, forums of negotiation often become arenas for dialogue and contestation.

2.5.1 Nodes of design

Nodes of global governance design are core centers, in which new global governance tools and products like rules, norms, models, frames, strategies, tactics, recommendations, and standards are being initiated, planned, developed and standardized (Tysiachniouk, 2006b, 2010a,e). Nodes of global governance design belong to transnational spaces yet they can involve actors from the space of places. They are not affiliated with a particular place, despite the fact that they have geographically bounded offices, or coordination centers. The coordination center is a hub-like node of governance design; however, actors and networks that are part of this key node, as they are contributing to the development of the regulatory product, can be spread all over the globe and often maintain only a virtual connection with the head office (Tysiachniouk, 2006b: 118). Actors of the node operate both in clock time and 'timeless time' and can change the mode of operation when necessary. To develop the regulatory product, the node of global design accumulates information from different places around the world, coming from different social, environmental, political and economic contexts. Actors of the node have to accumulate this information and respond to it. They analyze and use different kind of documents, research reports, and national legislations that help to combine, filter and choose information necessary for a global regulatory product. Nodes are open systems and use all kind of political, material, and discursive resources as well as resources of other networks. They compress and generalize all kinds of inputs while constructing new institutions applicable to many places. The process of standardization includes abstraction from the particular local circumstances and the development of generalized conceptual approaches and algorithms of changing local practices and institutions (Sayer, 1992: 87). As a result, global policies and/or standards developed in the nodes of design are usually too generalized and abstract and have to be adopted and adjusted to particular national, regional and local contexts prior to being institutionalized in the sites of implementation. The more general and flexible the global regulatory product, the easier it is to adapt it to different kinds of contexts (Kortelainen *et al.*, 2010).

Nodes of global governance design differ in scale. Mega nodes can include multiple networks and develop multiple regulatory products. For example, the Earth Summit in 1992 can be labeled as a mega node as it gave birth to 'Rio processes' (Rio+5, Rio+10, Rio+20, etc.) with the aim of fostering sustainable development around the globe. The Earth Summit became a mega node that gave birth to several GGNs, such as Agenda 21 and the Convention for Biodiversity (CBD). The CBD, for example, despite the fact that states are driving the network, allows wide NGO participation. The CBD node develops standards, norms and guidelines for biodiversity conservation all around the globe. It involved multiple forums of negotiation at regular meetings and different kinds of working groups. Territories of those nation states, which ratified the CBD, represent sites of implementation. The situation is different with forest governance. The effort to sign the Forest Convention failed, however, the Earth Summit developed the Forest Principles that laid the ground for Forest Stewardship Council (FSC) certification that became an NGO-initiated, yet multi-stakeholder driven GGN that will be described in the following chapters.

2.5.2 Forums of negotiations

Forums play an important role in the preparation of global regulatory products, in their adaptation to different contexts and during the stage of institutionalization at the sites of implementation. Both the adaptation and implementation of the regulatory products usually go through several rounds of negotiation (Callon, Meadel & Rabeharisoa, 2002). Forums of negotiations at different levels are guided by dominant discourses, as for example after the Rio Earth Summit in 1992 by the sustainable development discourse that united actors with economic, social and economic interests (Arts & Buizer, 2009). The dominant discourse in the process of negotiations provides a ground for the shared meeting that is necessary for consensual elaboration of the standard which accommodates the diverse interests. Forums of negotiation can be of a different scale from global to local and can be organized in the form of conferences of different levels, assemblies, working groups, and stakeholder meetings. Forums of negotiation in the FSC system, for example, can be subdivided into three major types[9]: (a) forums organized by the node of governance design for developing, discussing and negotiating the regulatory product; (b) adaptation and translation forums for adjusting the regulatory product to be applicable in local contexts, and (c) implementation forums for institutionalization of new standards, for transferring the new roles into concrete local practices.

Forums can be dispersed and conducted through communication technologies or condensed into the form of face-to face meetings. When forums are held in a particular place they create a borderline between transnational spaces and the spaces of places and switch their operation from 'timeless' time to real or 'clock time'. Such forums bring transnational discourses to a wider circle of national and local actors, fostering their transnationalization and learning processes regarding global concepts and frames. Face-to face forums, in addition, with their direct purpose of discussing and negotiating a certain issue through dense conversations, cement networks and contribute to social capital and trust building (Urry, 2007: 236). Transnational forums usually organize their conferences and assemblies at strategically chosen geographical settings in order to link actors from transnational spaces and spaces of places. For example, the FSC general assembly in 2005 took place in Manaus, Brazil to involve stakeholders from the Amazon region, which is strategically important for the world in terms of high conservation value forests (HCVF). When major world discourses and disagreements shifted to FSC certification of plantations, the FSC general assembly shifted its location to Cape Town, South Africa, where a new generation of plantations is the hot issue. In 2011, the general assembly took place in Kota Kinabalu, Malaysia with the hope of resolving issues of the conversion of tropical forests to palm oil plantations.[10]

[9] See more details in Chapter 3.

[10] Participant observation at the general assembly in Cape Town, 2008 and in Kota Kinabalu, 2011.

Depending on the structure of the GGNs and their strategic nodes, forums of negotiation can also be coordinated by actors and organizations that can be regarded as major drivers of the process. For example, Forest Dialogue is business-driven and allows the forest working group of the World Business Council (WBC) for Sustainable Development to discuss interesting issues related to forest management with their major stakeholders around the world, to meet experts of different levels and to learn about the dynamics of issues and problems. The outcomes of the dialogues are used for developing global corporate strategies. Forums of negotiation can include and/or be organized by several nodes of governance design. Using the above mentioned example, Forest Dialogue, involves not only WBC, but the World Bank and the World Resource Institute. Forums can be organized not only by 'official' and existing nodes of design, but in the form of alternatives to existing nodes, as well. For example, the World Social Forum was formed as an alternative to the World Economic Forum with its neo-liberal model of development. Alternative forums against the neo-liberal model of development, often including protest actions, are organized parallel to the meetings of such GGNs as the WTO.

Forums organized by the node of governance design are facilitating the 'disembedding' and standardization of information flows originating from localities. These forums usually involve several rounds of gathering, with the participation of different kinds of transnational actors. For example, the issue of plantations arose on several general assemblies in the FSC-GGN (Kortelainen *et al.*, 2010). These forums ensure feedback loops between actors of transnational spaces and the spaces of places necessary for making new standards applicable for implementation.

Forums of adaptation are translating and operationalizing global policies and standards to national and regional environments. They are usually organized on the regional or national levels. Forums of adaptation may be perceived to a large extent as topdown phenomena by the national stakeholders, as many of them did not participate in the transnational standard setting process. At these forums, the national and/or regional rules and standards are developed on the basis of global generic policies and standards. Operational procedures are developed at such forums as the global rules have to be adjusted to national legislation as well as to political, economic, environmental and social contexts. Such forums occur when governments ratify international conventions or as national standards are developed for the certification schemes.

Implementation forums are organized at sites of implementation as new rules and regulations have to be understood, renegotiated and/or accepted by local players. As the rules often change existing balances of power and authority, impose new obligations on certain actors and expand the rights of other actors, these forums are necessary for the institutionalization of new roles. The more stakeholders and local experts are involved, the more local citizens are empowered, the less social exclusion takes place, and the more probable it is that the new institution will change local practices and the regulatory product of the node of global governance design will work at the particular space of place.

2.5.3 Sites of implementation

At the sites of implementation, new formal and informal norms and roles, coming from the nodes and forums are being institutionalized. The process includes interpretation and de-coding of global designs (Kortelainen *et al.*, 2010). Sites of implementation can host several global regulatory arrangements and designs and they may be connected to several governance networks, formed by different kind of business or NGO actors. For example, on the FSC site of implementation of a particular certification project, ILO, CBD, FLEG, and other global rules and regulations can all be implemented.

Governance of natural resources in many places represents a combination of local and transnational relationships and interactions that are crossing different scales, both global trans-local, national and regional (Kortelainen & Kotilainen, 2006: 179).

Local social, economic, ecological, political and cultural contexts can provide both opportunities and barriers for the institutionalization of new global rules. For example, in the countries with high levels of corruption and poorly enforced national legislation, implementation of global rules faces difficulties. Standards that require consultations with stakeholders and inter-sectoral dialogue are harder to institutionalize in post-authoritarian regimes, where democracy is only emerging and where there are many lingering effects of the old ways of decision making. Global standards are difficult to implement when they contradict national legislation (Tysiachniouk, 2008a, 2010a,c), in this case, either the global standard or the national legislation has to be modified.

The product of the node of global governance design, when it reaches the local place, represents a pre-design, containing the necessary information for real implementation. The pre-design starts circulating through governance networks, expert networks and forums of negotiations as well as between stakeholders until it institutionalization on the ground in concrete practices.

In the process of adaptation to the local situation, the global standards are transformed into concrete practices and local actors become co-designers of transnational agents in the institutionalization of new roles. The power of co-design depends on the input of local stakeholders interpreting the new roles as well as on the agency of transnational actors.

There is some danger, however, that local actors and stakeholders will perceive new global roles as neo-colonialism, as external global power that imposes new rules from outside. In this case, local resistance to global rules takes place (Castells, 1997). In the case that local actors affected by new rules are involved in the decision making, they are more likely to develop joint responsibility and ownership for the decisions that facilitate the institutionalization of new practices (Sørensen & Torfing, 2007b: 13).

2.6 Putting the GGN concept to work: analyzing the creation of new modes of governance through markets for sustainable products

The aim of this section is to operationalize the GGN concept for understanding transnational private governance through certification. The core concepts of the GGN framework will be applied for a particular field of empirical studies related to responsible markets and supply chains for different kinds of sustainable products. This will help to further elaborate the GGN concept and relate it to the other approaches used by scholars for explaining governance by private authority.

As argued by scholars, the global economy can imply the transfer of production to developing countries with low living standards and weak legislation. As a consequence, diminishing of social responsibility and unsustainable resource utilization become possible (Bartley, 2007a; Vogel, 2008). Certification and labeling aim to cope with this issue by fostering the development of sustainable niche markets, which grow to become an alternative to the conventional regime of neo-liberal free markets (Newson & Hewitt 2005).

The construction of socially and environmentally 'sensitive' markets, based on ethically-produced, sustainable products has been analyzed by many researchers in many different ways. Certification and labeling, involving forests, fisheries, agri-food businesses, tourism, housing and even mining have become a booming field of study by scholars of different social science disciplines, including sociology, economics, political science and law (Auld *et al.*, 2008; Bartley, 2007a,b; Bozzi, Cashore, Levin & McDermott, 2012; Blackman & Rivera, 2011; Cashore *et al.*, 2004; Dingwerth & Pattberg, 2009a,b; Gulbrandsen, 2010; Vagneron & Roquigny, 2011).

For understanding how new market-based institutions are formed, this study extends the theory at stake and combines the GGN concept with the framework of neo-institutional sociology (Bartley, 2003; DiMaggio, 1991; Fligstein, 2002; North, 1990; Powell *et al.*, 2005; Scott, 1995) using in particular the concept of organizational fields (DiMaggio & Powel, 1983). The following sub-sections will explain the role of GGNs and their actors in creating new institutional fields for sustainable products. The added value of the GGN concept compared to other theories on certification and labeling will be discussed.

2.6.1 Fostering sustainable niche markets: the role of GGNs in creating new institutional fields

The organizational field of global market economy includes a large variety of actors and organizations involved in networks (Powell *et al.*, 2005) linking spaces of place and transnational spaces. Producers and consumers are geographically spread all over the world. Both NGO and TNC networks link all spaces of production with spaces of consumption and transnational spaces where the global rule making is taking place (Oosterveer, 2007). Production management, sales, advertising and other components of the market are materialized in transnational spaces. In spaces of place where the product is produced, the organizational field includes not only producers, but also the system of state control, and other groups of interest. Through the chains of custody, the field extends from the places of production to transnational spaces and finally to the places of consumption (Oosterveer, 2006). Therefore, the commodity chain is part of the institutional field of both production and consumption (McNichol, 2006: 351-353).

Stability of the organizational field enables a stable market over a period-to-period basis, and without such stability the companies could not exist. The theory of organizational fields assumes that 'actors try to produce a 'local' stable world where the dominant actors produce meanings that allow them to reproduce their advantages' (Fligstein, 2002: 29). Radical transformation of the market-based organizational field can happen only in case of presence of such factors as economic crisis, state intervention into the market or appearance of agents of social change that intervene in the market rules (Fligstein, 2002).

The formation of new niche markets, however, is impossible without the formation of new institutional fields (King & Pearce, 2010). Path dependence is the key factor affecting the supply side of the TNC network, while normative influences affect the demand side. Consumer pressure may create substantial obstacles for a producer and a distributor by limiting the marketability of products that are objectionable to some consumers. Of course, the adoption of standards or new ways of production that are consistent with consumer preferences also require an investment, but that cost is often outweighed by benefits in the form of greater marketability of the product.

Actors belonging to GGNs are involved in a collective action in society, in order to reconstruct the existing social order, through formation of the strategic action fields (Fligstein & McAdam, 2012: 9), which facilitate creation of new institutional infrastructure necessary for implementation of sustainable practices. New roles brought by actors, such as international foundations, conventions, NGOs and TNCs shape the local institutional fields and create new arenas for stakeholder interaction. In this process the role of private authority is crucial. Actors located mostly outside of the nation state contribute to translating global rules into local practices, and play the role of agents of institutional change (Meyer, 2010). NGOs represent key agents in GGNs that are using market dynamics for fostering institutional change in the production and consumption processes. Operating within both market campaigns and certification, GGNs and NGOs play roles of institutional entrepreneurs, who are using both coercive and normative pressures for breaking old and creating new path dependencies. In the case a niche market is in the process of formation, and market forces do not yet provide incentives for companies, there is a need for additional outside resources. Foundations serve as special types of networks that shape niche market formation in its early stage and facilitate the establishment of the new organizational field, allowing market forces to step in later (Bartley & Child, 2008, 2011, see Chapter 3, 6-7). Through this process new institutional fields, representing sustainable niche markets are formed, operating in both transnational and local spaces.

Consumer boycotts and the information-dissemination work organized by NGO networks design new market niches with new institutional fields (Bartley & Smith, 2008), in which the control functions are partly transmitted to GGNs. However, the changes are not extremely radical. Certification systems and the creation of socially and environmentally sensitive niches do not imply any revolutionary transformation and do not break the market, which usually occurs during crisis or abrupt governmental interventions. They are incorporated into existing market's interplay and do not destroy the principles of the free market. They imply that the actors voluntarily join the process of creation of 'the islands' of sustainable development inside of a milieu which is not intended to lead to sustainability otherwise. Socially and environmentally sensitive markets appeal to value concepts; they aim for gradual, non-violent transformation of social values, which should result in a gradual widening of such 'islands', and, finally, lead to a market which does not contradict sustainable development.

To create such a niche, three components are necessary. First, new socially and environmentally responsible producers should come into the market. Second, consumers should appear who care about the origin of the product and make the products of these responsible companies preferable. Third, these ideas of socially and environmentally responsible production have to be supported by society at large, which means they have to be legitimized. This latter concept is rather crucial because only wide social support can assure the critical mass of responsible producers, hence; enable the creation of socially and environmentally sensitive markets. The legitimacy of private governance through certification comes from evaluations of stakeholders, stakeholder representation in standards setting, and standard's representation third party audits and approval of the scheme by trustworthy actors such as NGOs (Auld *et al.*, 2009; Cashore, 2002; Cashore *et al.*, 2004).

As was explained above, the appearance of the GGNs designing certification implies the transformation of institutional fields. Two types of agents of institutional change are necessary. One agent should destabilize the old organizational field, and the other construct the infrastructure for new regulating institutions. Initially, new organizational fields of 'sensitive' markets were being formed by the two agents mentioned above: on the one hand, with the help of NGO driven consumer boycotts, where pressure was placed on corporations to force them to make their practices more socially and environmentally responsible. On the other hand, NGOs were organizing the information campaign and explaining to wholesale customers and retailers their active role and ability to influence the practices of corporations. In parallel, to support the companies that were striving for change of their practices in the direction of social and environmental sustainability, various certification systems were developed. It should be noted that in the cases where the transformation of organizational fields is induced not by crisis or state intervention, but by the agents of change, it is important not only to create the design of the new niche, but also to obtain the critical mass of producers and consumers (Conroy, 2007). It is a big challenge for the NGO-driven GGNs. Obtaining a critical mass of socially and environmentally responsible producers and buyers is directly connected to the society's habitual values and routines. In various countries, the stage of development of sensibility to such values is different. GGN will be able to function in an efficient way only if in the places of production and consumption there is sensibility to such values.

As was mentioned above, consumer boycotts can strongly affect stock markets and company's sales and often led to negative results in the spaces of places, such as closing the enterprises of the boycotted company and unemployment. Certification systems, on the contrary, solve the problems of 'ecologization' and 'socialization' of businesses in a gentler way: they stimulate producers to change practices and to voluntarily raise their responsibility. Certification has become a type of informational source for consumers. In the global economy, with its complex chains of custody, there is no other way to know how the product has been produced. For companies, on the other hand, certification is a sort of 'quasi-cartel' ensuring competitiveness for its participants (Bartley, 2007a).

In both certification and campaigning networks, market forces and dynamics are very important, and both types of networks serve to illustrate the analytical added value of the GGN concept as elaborated in Section 2.2 and further operationalized below.

I analyzed the GGNs through the lens of the new understanding of space and time (Castells, 1997; Castells, 2009; Urry, 2000). Market campaigns can be used to illustrate those GINs with a rather fluid structure, while certification systems tend to represent the more predictable and well organized GINs. In GGNs, when I refer to interaction of transnational space and space of place, both types of time were important: the timeless time of new communication technologies, in which a fair number of forums of negotiation and operation of the node of governance design take place and clock time of day to day practices in the spaces of places, in the sites of implementation. In the case of certification, the process goes on constantly, inter-twining timeless and clock times. Things related to certification and the market (sales and purchases) run mostly in the timeless time of communication technologies of transnational spaces. Everything related to the changes in practices according to certification requirements runs in clock time in the space of place.

2.6.2 The GGN concept versus non-state market driven governance (NSMDG)

In this study, the GGN concept is the master concept for understanding governance through certification. Benjamin Cashore developed a theory of 'non-state market driven governance' (NSMDG) to explain the same phenomenon. The GGN builds on NSMDG theory, acknowledging that governance through certification systems is based upon market incentives in which both producers and consumers involved in the supply chain are assessing the benefits and costs of joining one of the voluntary schemes. All parts of the system use market forces for gaining their own benefits. The companies use certification for legitimizing their practices, for obtaining additional advantages over competitors and for enjoying stability in the marketplace. The consumers support certified products as a way to participate in solving environmental and social problems via their purchasing power (Cashore *et al.*, 2004).

The aim of both theories is to understand the emergence and operation of certification schemes, such as the FSC, MSC, TSC, the apparel industry, and the agri-food business (Auld *et al.*, 2009; Bernstein & Cashore, 2007; McNichol, 2006: 349). Both GGNs and NSMDG were developed empirically on the example of the FSC certification scheme. NSMDG tested other schemes, such as those associated with the apparel industry, food industry, manufacturing, construction, mining and tourism (Auld *et al.*, 2008a,b, 2009; Cashore, 2002, 2004).

The GGN concept can be applied more widely, to other governance arrangements not necessarily involved in certification. The necessary prerequisites for the GGN concepts are only a transnational node of standard design, forums of negotiations and sites of implementation in a concrete territory. Therefore, the GGN theory can be applied to global governance arrangements not necessarily based on market instruments.

Both theories showed the essential role of NGO networks in setting up and driving the FSC scheme. The first stage of NSMDG was about 'converting' the companies, through both consumer boycotts, when the NGOs required the companies to change their practices into more socially and environmentally responsible ones, and by means of the creation of demand among customers for socially and environmentally responsibly produced goods. The second direction of NGO activity was their adaptation to the limited abilities of companies to change. To some extent, that means adaptation to the interests of business. The goal of this activity was to make the process of change not overly complex for the companies, as it helps to involve a larger number of business actors in the process. Here, NGOs as agents of institutional change, made some forced concessions, and their position became less radical. Third, a direction of NGO activity was informational work. For a new institution to be publicly accepted, it is necessary to prepare public opinion for innovations, in other words, to largely inform them about forthcoming and ongoing changes (Cashore, 2002; Cashore *et al.*, 2004: 33). One of the main purposes of this informational work is to ensure that the local interest groups become aware that they are stakeholders and encourage them to take part in the processes of change.

In the GGN theory, 'converting' companies towards a sustainable path takes place through the interplay of market campaign GGNs and certification GGNs; both use market forces and jointly build the strategic action field that results in establishing new global rules and a new organizational infrastructure for sustainable niche markets. At the beginning, GGNs, in a form of market campaigns, get engaged in a purposeful confrontational relationship with companies involved in unsustainable forestry. Big brand players were chosen for the role of a 'victim' of the boycott and an object of the NGO-led campaign that simultaneously affected the chosen company and destabilized the existing institutional field, by breaking down old path dependency. Not only the affected company, but other similar companies who feared similar boycotts became partners and constituents of the FSC-GGN. By doing so, they contributed to the formation of a new path dependency and a new institutional field.

Simultaneously, with widely spread market campaigns, the FSC-GGN, was developing the standards, accommodating them for adoption by the industry and by place-based stakeholders at the sites of implementation. A large number of forums of negotiation were involved, in all processes, including multi-stakeholder standard setting, standards negotiation and adoption, and special attention was given to the forums related to the information campaign both in transnational spaces and spaces of places.

In this respect, both GGNs and NSMDGs use different theoretical approaches to make a similar argument. The added value of the GGN concept compared to NSMDG is that it can capture all stages of certification processes, from standard setting and negotiation up to standards implementation on the ground; therefore, it captures the full extent of the network, down to localities. It allows analyzing the whole governance arrangement, from designing and developing the standard to translating it from global to the local, to finally implementing and institutionalizing it. Different stakeholders at different levels bring their interests and claims into the process, therefore forums of negotiations are indispensable. In concrete sites of standard implementation, it is possible to explain how the global/local interface affects concrete management practices. With NSMDG, it is hard to capture the feedback loops between the global and the local and vise-versa. Compared to GGNs, NSMDG is unable to capture peculiarities of the standard negotiations and implementation as well as standard's renewal. NSMDG is less useful in analyzing rules in use and practices on the ground when the certification standard is implemented in the concrete territory.[11]

2.6.3 NGO-driven market campaigns as an example of a GGN

A NGO-driven market campaign is a coordinated effort to attract the public's attention to problems in a company's supply chain with the purpose of damaging the company's market share through damaging its brand (Conroy 2001, 2007; Friedman, 2006). Market campaigns fall into the category of social movements as they mobilize groups of citizens for a collective action, which in this case is a consumer boycott against a corporation (Bartley & Child, 2008, 2011; Conroy, 2001: 2; O'Rourke, 2003, 2005: 117-119).

In the course of a market campaign, NGOs operate both in the space of places and in transnational spaces, although their role is different in each of these spaces. During the market campaign, the tensions and interactions between the space of places and the transnational spaces inevitably appear. These tensions can be dealt with and softened when actors of transnational spaces make efforts to translate, adapt and implement the new rules in the spaces of places.

In this study, market campaigns are perceived not only as social movements, but as NGO-driven GGNs. In their nodes of design, actors investigate markets, strategically set up targets for

[11] See more on comparison of NSMDG and GGNs in Chapter 9.

the campaign, choosing the most vulnerable companies as show cases. The campaign itself involves multiple forums of negotiation involving NGOs, TNCs, retailers and other actors. Forums of negotiations usually take place prior, during and after the market campaign. Sites of implementation may involve both places of production and consumption.

The GGN concept offers new insights in governance by pressuring for social change initiated by social movement organizations and explains mechanisms that are not covered either by social movement theory or by political consumerism approaches. Market campaign GGNs, which succeed in multiple attacks on TNCs, become well established and recognized entities with established nodes of design, in which strategies and tactics are produced by actors with well-developed negotiating skills that can enhance the overall network capacity. Such GGNs affect corporate offices to make them change their policies and operational standards. Therefore, on the one hand a NGO-driven market campaign is a form of governance; on the other hand it represents a conflict that is deliberately initiated to contribute to desirable social changes (Tysiachniouk, 2009). Social changes are expected in both transnational spaces and the spaces of place. NGOs act in transnational spaces and design international norms and practices to be implemented in sites of implementation in various countries.

Instead of lobbying governments for stricter regulations, market campaign GGNs directly target producers and consumers. It campaigns against TNCs; mass media are used as one of the ways of making impact both on consumers and corporations (Friedman, 2006: 46; O'Rourke, 2005: 118-119). These NGO-led boycotts involve blaming and shaming through attacking a well-known brand; therefore, scholars name these campaigns as brand damaging movements, or 'brand bombings' (Conroy, 2007; Klein, 1999; Raynolds & Wilkinson, 2007: 40). When organizing boycotts on a brand, NGOs carefully choose a 'victim' to get the largest response. Usually the object of criticism is not the company with the worst environmental performance, but an industry having the most valued established image. For effective planning and organization of their actions, NGOs investigate the chain of supplies and check all the buyers, from institutional ones to retailer stores, for susceptibility to the loss of image. They investigate the production practices in particular localities for framing effective strategies. These investigations are necessary for finding strong arguments in dialogues with the corporations, and to choose the most effective form of dialogue that could lead to negotiations or threats of boycotts.

The NGOs and social movements are more legitimate actors in wider society than are corporations, and usually the general public trusts NGO statements and judgments. The fact that NGOs need much less resources to fight against a brand than a corporation needs for its construction explains the vulnerability of the latter in this struggle (Bartley & Child, 2011: 444-445; O'Rourke, 2005: 119). Corporations percieve the threat to their brand as high and are eager to change their production practices. When the institutional infrastructure for the market campaign is well established, NGOs can simply threaten companies with possible boycotts to achieve their participation in a dialogue and to finally change practices. The main feature of consumer boycotts is that by organizing against one company, NGOs achieve the multiplier effect from many other companies which are also afraid of boycotts, and do not want to loose their image (O'Rourke, 2005: 125). Therefore, even a threat of consumer boycott can force industry to start negotiations with the movement's organizers and make commitments to changing practices. In the last decade, NGOs widened the sphere of their attacks, and found boycotts against the investors of corporations to be effective. City Group, Bank of America and many other banks investing resources into corporations with low social and environmental responsibilities have been boycotted amongst others.

While producers are pressured during such campaigns, by threatening their image, consumers are educated in purchasing practices and mobilized to participate in consumer boycotts of the company that has been chosen as a victim (O'Rourke, 2005: 118-119). The use of market mechanisms assumes two different processes which foster the construction of a 'sensitive' niche market. On one hand, it is the pressure on a corporation by means of consumer boycotts against the products produced unsustainably by this corporation, and discredit to its brands. On the other hand, there is mobilization of consumers to purchase the products of those companies that have changed their practices in accordance with the NGOs' requirements. This latter is education for the raising of consumers' civil awareness and stimulation of voting by purchasing, which is a form of public engagement in political processes called political consumerism (Follesdal, 2006: 3; Micheletti, 2003: 12; Micheletti, Follesdal & Stolle, 2006; Spaargaren & Mol, 2008). Research showed that consumers react to negative information concerning a specific corporation. It means that negative information acts more effectively on consumers than does positive information about good products (Elliot & Freeman, 2003). According to the studies, after participating in boycotts, consumers become more attentive to the products they purchase (O'Rourke, 2005).

Enabled by communication technologies, market campaigns require relatively little time and resources to obtain an effect. Modern technologies allow actors informing target audiences about a company's unsustainable practices using minimal resources and time. Network organization of these campaigns with permanent coordination of actions and management of informational flows from spaces of place to transnational space is a very important component. For example, during the campaign against Nike, information was sent daily to such organizations as Global Exchange[12], Fair Labor Association, Oxfam[13], USA National Labor Committee, and many other groups, networks and organizations. This exchange of information contributed to mobilization of organizations both directly involved in the campaign and those only supporting it (O'Rourke, 2003). Therefore, the campaign itself is conducted in 'timeless time' of modern communication technologies, and the brand can be destroyed in a very short time. Contrarily, restoration of the company's good image requires a long period of 'clock time', sometimes it can last for years. This fact defines the effectiveness of consumer boycotts as a very powerful tool in the hands of NGO-driven GGNs.

2.6.4 Certification programs as GGN's

Certification serves as an alternative to NGO-driven market campaigns in fostering sustainable production and consumption (Bartley, 2007b). It provides the tool for separating harmful companies from those that use resources in a sustainable way and exercise higher social fairness. Therefore, through certification systems, environmentally and socially responsible companies receive market incentives, stability in the market place, and positive recognition by the stakeholders

[12] Global Exchange is one of the key organizers of summits.

[13] Oxfam is a charity organization that sells second hand goods and uses the income for the realization of anti-poverty programs.

(Auld *et al.*, 2009: 183; Bartley, 2007b; Guthman, 2004, 2007; Haufler, 2010; Locke, Amengual & Mangla, 2009; Vogel, 2005).

Certification systems are voluntary governing arrangements (Potoski & Aseem, 2009). Standards are developed by private actors primarily. States do not require compliance, do not participate in standards setting, and do not provide companies with incentives to comply (Auld *et al.*, 2008a: 424; Bernstein & Cashore, 2004; McNichol, 2006: 371). This does not mean that government does not play any role in these processes, because national legislation makes a potential impact on implementation of certification, and sometimes governments even support it financially, initiating public purchasing programs of certified products, such as for the green Olympics (Cashore *et al.*, 2004). There is also an exception in certification systems, the Kimberly Process Certification, which is a hybrid and combines standards developed by the diamond industry with state enforcement through border controls. The aim of this certification is mainly peace keeping in Africa, where revenues from diamonds were used to finance wars and conflicts prior to certification (Haufler, 2010).

GGNs are aimed at the re-orientation of the markets for solving global social and environmental problems, which can be conceptualized as governing externalities (Cashore at el., 2004). In their absence, the companies would not have enough motivation to put the necessary investments in place. This feature differ multi-stakeholder certifications from soft self regulation, which simply strive for standardizing business-practices to obtain better co-ordination inside an industry. These latter are aimed at the optimization of production and making more profits (Bernstein & Cashore, 2007). As regulatory agents, certification GGNs are prescriptive, requiring companies oriented towards profit maximization to undergo profound changes in order to be able to implement social and/or environmental standards in the sites of implementation on the ground. In this respect, although certification is voluntary, it is in fact 'hard private law' (Meidinger, 2003, 2007, 2008).

Certification GGNs gain additional legitimacy as they have mechanisms for ensuring and verifying compliance to standards and for penalizing non-compliance. The most common mechanism of monitoring and enforcing compliance is the external audit, in which a third party (auditor) checks and confirms compliance of the company to the rules and norms, or defines the measures necessary to achieve such compliance. As distinct of the certification schemes, the standards of self-regulation, such as corporate codes of conduct, or policies of corporate social and environmental responsibility, have very loose enforcement mechanisms, and often are not implemented although self declared (Auld *et al.*, 2009, p.189; Bernstein & Cashore, 2007: 350; Cashore *et al.*, 2004; Taylor, 2005, 2007).

As any other GGNs, global certification networks have a transnational node of design, involved in standards and other policies setting, forums of negotiations of different levels and sites of implementation, where the standards are transferred into rules in use in particular places. As participation in the market of certified products is voluntary, the sites of implementation are limited by those places, where production sites of certified companies are located, or where consumption takes place (Oosterveer, 2007; Oosterveer & Spaargaren, 2011).

The distinctive future of certification GGNs is the multi-stakeholder engagement in all parts of the network (Hallstrom & Bostrom, 2010). Certification institutionalizes the system of involvement of interested groups, both in standards development and in the process of their adaptation in sites of implementation, hence it creates multi-stakeholder governance. The multi-stakeholder standard setting process involves a diverse range of interest groups and many rounds of negotiations, which ensures feedback around the world (Backstrand, 2006), but especially at the node of design during the standard setting procedure. The process of adaptation and implementation is more inclusive, democratic and transparent than in client-oriented state policy networks. Institutional and organizational learning take place across a wide range of stakeholders during the process of adaptation (Auld *et al.*, 2008a: 424; Auld *et al.*, 2009: 188).

Certification systems created by GGNs usually make use of flexible standards that allow stakeholders to interpret and negotiate them. Adaptation, modifications and education are permanent processes involving many interested groups (Bernstein & Cashore, 2007: 349). The renewal of standards, in case the standards are no longer relevant to the sites of implementation and challenged by stakeholders, is part and parcel of the operation of certification systems. So, in other words, the certification process assumes the functioning of forums of negotiation at all stages of development and institutionalization of new rules, namely during standards preparation, adaptation, and implementation on the ground and renewal. The dynamic character of such systems differentiates them from the more conservative initiatives of eco labeling, such as, for example, the ISO which usually supposes more prescribed requirements of environmental quality.

GGNs, producing global standards, may be more or less developed. When more developed they tend to include all three components: nodes, forums and sites of implementation, as in the case of FSC, the Marine Stewardship Council (MSC), Rainforest Alliance certification, Fair Labor Association, etc. When less developed, like in the case of emerging certifications the standards are in the process of negotiation, not yet finalized and implemented. This is the case in for example the Mining Certification Initiative and the Sustainable Tourism Stewardship Council. The latter GGNs have already developed nodes and forums, however, their designed and negotiated standards have not yet reached sites of implementation.

As was mentioned, all GGNs are based on a multi-stakeholder process of standard development, however, they can be initiated by different kinds of actors. Many of them are organized by private actors, NGOs or business associations, or by both at the same time. For example, the Fair Labor Association (FLA) was initiated by industrial companies producing footwear and textiles, their consumers, and trade union organizations, with the support of president Clinton's administration (Bartley, 2003, 2007b). The goal of this initiators' network was to close sweatshops, to solve the problem of implementation of workers' rights, and to ensure appropriate working conditions in developing countries where factories had been transferred from the USA. Their regulatory targets were the companies producing footwear and textiles. Certification provided the companies legitimacy in the eyes of western consumers and improved labor conditions for workers at the sites of implementation, e.g. the places of production were accommodating worker's interests and rights (Auld *et al.*, 2009: 216; Bernstein & Cashore, 2007).

Networks differ in their sites of implementation, policy issue to be resolved and regulatory targets. Even if GGNs' transnational standards regulate similar issues, they may differ in their strengths. This depends on the composition of the actors involved in the operation of the node of design. For example, several GGNs involving more radical or more moderate standards are at the same time promoting fair trade that excludes child labor and ensures decent labor conditions in developing countries. The more radical wings in these networks are eager to change market institutions and orient them completely towards social justice, and the less radical wings adhere to the existing market institutions, however, they still want to regulate social and environmental standards in the sites of implementation (Raynolds & Murrey, 2007: 223).

Certification GGNs represent constantly changing and evolving entities. This can be illustrated by GGNs involved in fair trade products. Three separate GGNs are involved in the certification of coffee. All three networks have multi-stakeholder nodes of design and forums of negotiation. First, the Fair Trade Labeling Organization (FLO), with roots in social movements against unjust free trade, was initiated by social NGOs, predominantly from Europe, consumer groups and alternative trade organizations (Jaffee, 2007: 12; Linton, 2008; Nicholls & Opal 2005: 128). It initially was for the regulation of coffee markets and later expanded to tea, cacao, sugar and bananas. The certificate is given to small producers and suppliers of these goods from developing countries marketing to the countries of Western Europe and USA, and predicates they stop the use of child labor and provide their workers with appropriate working conditions. Second, the International Federation of Organic Agriculture Movements (IFOAM) focused its standards predominantly on environmental issues. It was initiated by organizations of organic farming, expert associations and NGOs. The appearance of this regulatory network was due to the necessity to create a market of safe and healthy agricultural production, produced with no harm to the environment, without using pesticides, herbicides and other chemicals. The network regulates farmers' practices and processors of agricultural production. In later stages, the FLO and the IFOAM learned from each other and each of the networks included in their certification both social and environmental standards. Third, the Rainforest Alliance Certification, focusing on the certification of coffee, cocoa, citrus, flowers, chocolates and foliage, was initiated by NGOs and the Sustainable Agriculture Network (SAN), focused its standards on reducing the impact of agriculture on biodiversity, soils and forests. Their sites of implementation are tropical farms predominantly in Central America. Although biodiversity conservation was in the forefront of the standard setting, social and economic issues are included and balanced with the environmental in the standards in use (Auld *et al.*, 2009: 202, 207).

In the building industry in the US, a GGN appeared under the name of Leadership in Energy and Environmental Design (LEED). The US Green Building Council took the lead in creating this network. The network addressed the problem of the environmental footprint of construction, promoting among building companies resource saving technologies, energy conservation and the use of environmentally friendly materials, including certified materials. Over time, this GGN extended its sites of implementation from the US to other countries (Bernstein & Cashore, 2007).

One of the newest networks is the Mining Certification Initiative. It was developed by the WWF and several progressive mining companies. The network regulates companies involved in metal and gem stone mining. The goal of the network is the promotion of the practice of socially and environmentally responsible mining, with no impairment of the rights of local populations (Bernstein & Cashore, 2007).

In this field, the FSC-GGN was one of the first. Its positive experience and successes in forming new institutions inspired many certification systems in other industries. Some of them were created 'in the image and likeness' of the FSC: for example, the MSC (Gulbrandsen, 2005, 2009) and the Tourism Stewardship Council (TSC), where NGOs were also the main actors and 'driving force' of the governance arrangements.

2.7 A typology of transnational governance networks

In this chapter the GGN concept was first defined and discussed for its theoretical components, and subsequently operationalized for applying it to different cases, amongst which the case of governance through certification. FSC certification processes are the main empirical focus in this study and the FSC related transnational governance networks are used to show the relevance and functioning of GGNs. The aim of this section is to construct a typology of networks involved in FSC related transnational governance. As such, the section can be read as a preparation of the description and analysis of the empirical findings in the chapters to follow. The typology outlined in this section also served as a tool for the selection of the case studies in Russia.

As was described, GGNs link transnational spaces with the spaces of places, yet in different or diverging ways. The connections between the local and the global can be established in different ways, as was explained in Sections 2.3.3 and 2.3.4. In some cases, GGNs are initiated in a bottom-up way, by actors operating in the local spaces of places. They are represented, for example, by NGO-led market campaigns developed to change local practices by directing grievances toward TNCs and other global neo-liberal institutions. In other situations, actors operating in transnational space create and use GGNs to foster – in a top-down way – institutional changes in the space of place for example by imposing sets of new roles, norms and standards to be implemented locally. Third and last, there are GGNs which do not fit the bottom-up – top-down dichotomy since they are primarily characterized by a two-way mode of exchange of information based on the constant and organized interaction between local and transnational actors. Especially in the case of this third ideal type GGN I argue that new modes of global governance and new ways of decision making are created, which actively respond to the conditions of our time and show a better fit with the dynamics of the global network society.

Therefore, analysis of the GGNs involved in designing certification systems shows that we can point out three ideal types of transnational networks (Figure 2.1). The first ideal type is entitled the Bottom-up Market Campaign, led by NGOs, which in this study is seen as a special type of GGNs, directed bottom-up from the spaces of places towards transnational spaces in order to bring

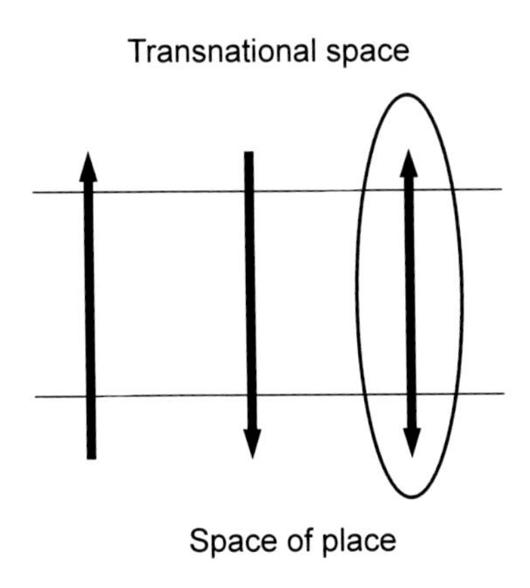

Figure 2.1. Ideal types of governance networks involved in governance through certification.

changes in practices in both spaces. As was shown, drawing from the literature on certification and labeling, NGO-led market campaigns turned out to be important for fostering governance through the FSC. The distinctive feature of the market campaigns, which makes them different from other protest movements, is the use of market mechanisms as a powerful tool of impact on corporations (Conroy, 2001; Cashore *et al.*, 2004). Market campaigns represent a special type of advocacy campaign which uses the forces of the markets and impose market based pressures on TNCs in order to foster changes of their practices into environmentally and socially friendly ones, in both the production process and trade relations (Conroy, 2001).

Such networks appear in response to concrete problems that emerge in the space of place, for example, cutting of old-growth forests, or the use of child labor or sweatshops. The network agents inform actors in transnational spaces about these problems and organize attacks against the brands of culprits, and, finally against the TNC's overall operations. In this type of network, very few people might be involved from the locality where the problem arises. The main operation exists in transnational spaces, where consumers become involved in the events and the market campaign is organized. After some period of time, when the producers become well acquainted with the scenario of action of such networks, there is no need to even organize a boycott. The threat alone of a boycott can be enough to start negotiations between the actors involved.

The second ideal type of transnational network is entitled a Top-down Guided Standard Implementation. It operates top-down from transnational spaces to the spaces of places. These networks come to localities with a certain idea of design, and create there the precedents of new practices or models. Networks of this type are created by project designers and implementers, often NGOs, co-designed and/or facilitated by charity foundations and international aid programs. These actors introduce social change in the space of place by imposing sets of new roles, norms and standards designed globally, yet implemented locally. This is particularly important when new governance designs are mostly initiated in industrialized countries and efforts are made to implement them in developing countries, or countries with economies in transition. Top-down networks are particularly important for building organizational and institutional infrastructure for governance through certification, as at the early stages market incentives are not yet in place for companies to become engaged. Therefore, certification GGNs, at early stages of their development, operate mostly 'top-down'.

As was explained in Chapter 2, at the early stages FSC implementation was widely sponsored by different foundations due to the fact that market instruments were not yet in place. For example, Model Forests built in the Baltic States and Russia were major prerequisites of the FSC and, therefore, chosen as an example of the top-down networks. In Model Forests, the NGO-implementer used resources from industrialized countries (granting agencies) for implementing sustainable forest management practices on the ground. Model Forests represent an arena, on which ideas and designs of global understanding of sustainable forest management can be tested and implemented on the ground in developing and/or transitional countries. In these networks, FSC certification is an essential part of building a model of sustainability on the ground. In such cases, FSC certification is facilitated by the availability of outside funding.

The third type is entitled the Hybrid Glocal Governance. It can be described as two fold GGNs, organized to operate in both directions: from transnational spaces to sites of implementations and back. Here, permanent interaction between the sites of implementation and the nodes of design is carried out, organizing at the same time a permanent interplay between actors at different levels. If a standard or product designed in the nodes of design does not work in an appropriate way in the space of place, the product is further developed and modified using the feedback loops connecting local and transnational actors involved in the networks of transnational governance. Therefore, this is dynamic, constantly changing constellation, presupposing the joint effors of actors both in transnational spaces as well as in sites of implementation.

The mature FSC certification system, with its feedback loops between global and local actors, can be argued to belong to this third type. Pure FSC certification, in which outside funding is not used, is supposed to be a sustainable long-lived entity, as it is built on incentives to producers to implement sustainable forest management practices in the logging sites.

Three ideal types of GGNs can represent the historical stages of development of certification schemes and represent a succession. However, in the real world, they usually co-exist, although the 'guide way' of their development tends towards the third type.

Intermezzo. Methodology

This chapter presents the overall research methodology, but each of the empirical Chapters 3-8 includes specific details on the number of field expeditions, locations of participant observations and the number of interviews. The chapter begins with the description of research strategy and design, including case study selection and description of the methods used (A). It explains the peculiarities of this study, in which the research was combined with practical engagement in the FSC system (B). It proceeds with the description of how the validity of the study was ensured (C) and concludes by explaining how the ethical issues related to the research were addressed (D).

A. General research strategy and design

This study combines elements of the grounded theory method (Charmaz, 2006; Glaser 1992; Glaser & Strauss 1967; Strauss, 1987) with a more traditional approach to the study, involving the development of the theoretical framework, conceptual model, and research questions following data collection (DeVaus, 2001). My involvement in research on the role of NGOs in transnational forest governance started in 2002, much earlier that the theoretical framework for this study was developed. Prior to engagement in the PhD project, I worked on multiple research projects using a variety of approaches, case studies and qualitative methodologies. Simultaneous with my research, I have worked since 2006 as an FSC expert-consultant on the FSC-Russia Board of Directors and as an FSC auditor. As an FSC member, I have participated in many working groups related to the FSC standard setting processes. On the local level in Russia I was involved in community capacity building activities in the framework of the FSC and worked as the FSC consultant for several companies, helping them to comply with the social aspects of forest certification. Observation at FSC General Assemblies, working groups andstakeholder meetings, as well as involvement in several certification projects in Russia, allowed me to form conceptual categories, more abstract neologisms, such as transnational nodes of design, forums of negotiation and sites of implementation that all together constitute governance generating networks (GGNs). Simultaneously, in the period 2006-2012, the GGN concept was developed theoretically and strategically tested in the field. The theoretical and conceptual frameworks for the study have been developed, research questions operationalized and case studies and data collection methods chosen in order to answer particular research questions.

This study used findings generated from many research projects conducted during the period of 2002-2012 involving: (1) the study of transnational governance networks, transnational actors and organizations, including those not necessarily directly involved in forest governance; (2) the study of the Forest Stewardship Council (FSC) on a transnational level and in Russia; (3) the study of transnational environmental NGO networks with links to Russia; (4) the study of the forest sector in Russia on the regional and federal levels; and (5), the study of commodity chains with links to Russia, involving several TNCs involved in forestry.

In the period of 2002-2012, nine case studies were conducted on the transnationalization of forest governance in Russia; however, only four of them were selected for the current project. Data from the other five cases and from studies listed above were used as secondary sources coming from the data-basis and as additional and/or background information. The field study of the FSC, coming from all projects both expert and participatory work, was used in full.

Longitudinal multi-sited research for this project (2002-2012), was conducted on the FSC-International and the FSC-Russia and on four case studies selected for this work involving participant observations at the events, expeditions to the field and regular interviews with key informants. Data were collected using qualitative methodology involving open-ended interviews, semi-structured interviews, direct observation, participant observation, and analysis of documents. Documents produced by governance actors, including NGOs and TNCs of different levels, such as leaflets, brochures, policies, reports and web sites were used.

A.1 Case study selection

The case study approach is considered the best strategy when certain phenomena are studied within the real world context. Case studies can be exploratory, descriptive and explanatory (Marbry 2008; Yin, 1994). My study design involved explanatory research questions, with the purpose of investigating casual links in complex real-life interventions (Baxter & Jack, 2008: 547; Yin, 1994); intervention by a transnational governance network. Therefore, the explanatory case study approach was selected in order to understand the phenomenon of governance by private authority through certification.

The theoretical approach outlined in Chapter 2 informed the case study design and the case study selection. The conceptual model, coming from the theoretical framework involved specific propositions (Baxter & Jack, 2008: 551) related to the interface between global and local, placed limits on the scope of the study, determining the case study boundaries. It helped to guide future data interpretations by identifying which actors and organizations would be included in the study and what kind of relationships could be between them.

As was explained, three ideal types of transnational networks were involved in the conceptual model for research, therefore multiple cases studies (Yin, 2003) were selected corresponding to the ideal types. All cases were in some way contrasting as representing different kinds of transnational networks operating differently, therefore with different results being expected for predictable reasons coming from theory. The fact that all cases were related to FSC certification facilitated comparisons of the outcomes of the interventions, e.g. comparison of impacts of different kinds of networks on concrete practices of forest management on the ground. From each of the explanatory case studies, analytic generalizations were drawn with certain conceptual claims. Each case informs the relationships between the network actors and processes and allows construction of certain claims which would be possible to generalize and apply to other similar situations and contexts (Yin, 2011: 13).

Generally, a case study represents an entity with boundaries defined by the researcher involving a spatially and temporally defined context (Yin, 2011: 9). In this work, each case represented a transnational network, linking transnational spaces with the spaces of places. The whole network was the unit of analysis. The structure of the network, the network's actors, and the events and activities in which the actors were involved determined the case study boundaries and research locations in this multi-level and multi-sited study. Distinguishing between the transnational spaces and the spaces of place allowed determining how contextual differences influenced both discourses and behaviour of the different kinds of actors. There were certain criteria developed for the case study selection. First, the socio-political, economic and environmental contexts in the spaces of place in Russia should be relatively similar in order to allow comparisons between cases. Second, the cases were chosen to fit one of the ideal type categories, e.g. to be a transnational network related to FSC certification, involving global/local interplay of actors within the network. The network needed to be directed bottom-up, top-down, or both ways linking transnational and local spaces. Third, the case would represent a long lasting project (around ten years) that would allow us to see the dynamics of the transnational network in operation and to analyse the impacts of particular mode of governance arrangements.

All case studies selected were in the North-western Russia boreal forests, relatively close to the European borders and markets. They were selected according to the above mentioned criteria and in correspondence with the ideal types of networks. The first case study was a market campaign for saving old growth forests in the Karelia Republic. The next two case studies were the Pskov Model Forest in the Pskov district and the Model Forest Priluzie in the Komi Republic. The forth case study was on the certification of forest management of the Segezha Pulp and Paper Mill lease, belonging to a TNC of Russian origin, Investlesprom, operated by a Northern Logging Company in the Karelia Republic (Figure A.1).

A.2 Methods used for the study

Participant observation

A number of methods were used in this project. As will be explained later, some methods were more appropriate for investigating certain activities than were others. Both direct observation and participant observation was used. Field notes were kept during both types of observation. Direct observation in field settings was used during field expeditions to the study areas in Russia, at different kinds of events at different scales and levels, including global (see Appendix A). The understanding of environmental, social and economic contexts and the relationships between people and organizations in different settings was important. Pilot visits to the study areas allowed for designing guides for future interviews.

Participant observation was essential in this study in global, national and local settings. As was mentioned, it was used not only as a field method, but for building the grounded theory. Participant observation is a methodology in which the researcher is simultaneously doing research and playing a certain role in the situation or event which is being studied (Douglas, 1976). The method is especially important when the research design involves a holistic understanding of certain phenomena (Kawulich, 2005: 3), as was the case in this study. This study used participant observation for studying international organizations and global negotiations as well as local communities. Local communities, compared to the global events, were explored in a more ethnographic, traditional way. In both global and local settings, informal interactions, including conversations and participation in the events were important. Participant observation was done during the author's involvement in practical work related to the FSC system. In particular, research for this project was combined with negotiating and voting as a member at the FSC General

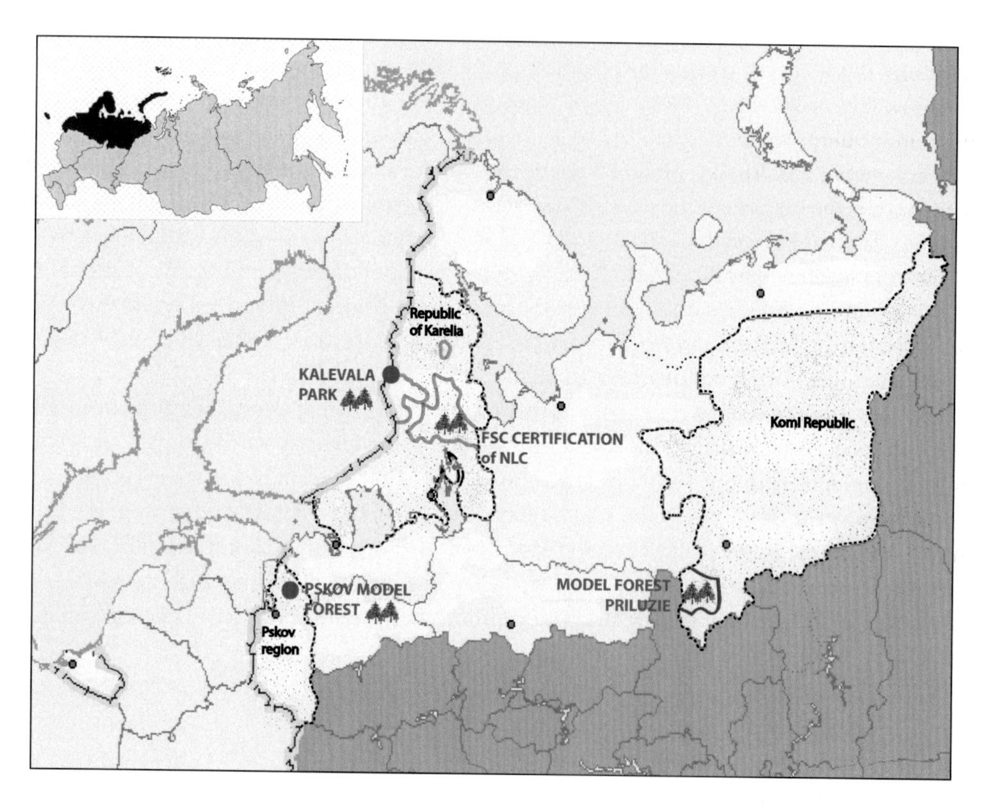

Figure A.1. Location of the case studies.

Assemblies, developing the FSC global and national standards, and being part of the national FSC initiative and part of the social chamber. Participant observation occurred during the author's participation in the FSC chamber meetings and working groups, at FSC audits in different regions of Russia, during FSC consultancy work, at the FSC-Russia board of director's meetings, and at local stakeholder meetings involving community engagement in the framework of the FSC.

The participant roles were different in the different settings and these roles affected the stance of the researcher. For example, participant observation in forest settlements, which was done for four years (2008-2011) in the framework of European Union-sponsored projects, required very close relationships with local communities because the goal was to help communities ensure implementation of their social and environmental rights in the framework of the FSC. Together with local activists, we investigated problems and issues and jointly developed solutions to those identified issues (Checkland & Howel, 1998). Such an approach to participant observation, a research method involving elements of action, was often criticized for containing tradeoffs between validity and practical significance, however in recent years it has become popular in academic communities and has developed in a way that adheres to research quality criteria (McNiff & Whitehead, 2009; Lindlof & Taylor, 2002). Measures were taken to reflect on the role of researcher in influencing the field of study (see Section C).

Interviews

Both semi-structured and open-ended interviews were done with international, national, regional and local informants. The interviews lasted from 40 minutes to 1-3 hours, with key informants, follow on-interviews were conducted step by step over several days. The length of the interview depended on the interview guide and on the informant's level of engagement.

Interviews with international actors involved managers of TNCs (#32), international NGOs (#46), representatives of granting foundations (#5), buyers of Russian wood products, particularly sawn wood (#17), publishers (#6), retailers (#4), representatives of certification bodies (#17), and FSC-International staff (#18) (see Appendix B, C).

In Russia, interviews were conducted with representatives of different kind of state agencies on the national and regional levels (#179), NGOs (#156), company managers (#149), FSC office representatives (#54), varying types of FSC experts (#73), and certification bodies (#23). On the local level, interviews were conducted in 82 villages and forest settlements, with local forest management unit representatives (#24), local administrators (#82), various community activists (#580), museum employees (#47), librarians (#78), teachers (#99), and House of Culture representatives (#72). Interviews were transcribed, coded and placed in the interview database for use when relevant (see Appendix B, C).

Semi-structured interviews were designed with the interview guides that allow for two-way conversations around each of the questions specified in the guide. The questions in the guide were not prescriptive or leading, only indicating topics and the sequence of what needed to be discussed. The discussion topics were set up in a way that contributed both to gaining thematic knowledge and to dynamic interaction with the informant. The interview procedure involved some questions that would serve for clarifying theoretical conceptions, but most discussion topics were formulated in a way that allowed for spontaneous description from the informant. The researcher's interpretations were constantly verified by asking additional questions. Many specifying questions were asked during each interview (Kvale, 1996: 29-130; Kvale, 2008). Different interview guides were developed for the different kinds of stakeholders, e.g. (a) buyers of FSC certified wood in Western Europe and the USA; (b) certification managers at different levels in the companies; (c) NGO representatives at different levels; and (d) government representatives at different levels. Similar interview guides were used for the same categories of informants throughout the cases. The interviewing process continued until the information being gathered become very repetitive and ceased providing new insights to the study. Semi-structured interviews allowed comparisons across cases between similar groups of informants.

The open-ended, non-structured interviews allowed understanding the phenomenon under study from the perspective of the informant and his/her social world. Open-ended interviews were usually conducted over several occasions: (1) as a follow-on with key informants with whom semi-structured interviews were previously done. This was necessary for learning about recent developments of the project and (2) in early stages of exploratory research when the interview guides were not yet developed. (3) Open-ended expert interviews (Meuser & Nagel, 2009) were important with the different kinds of insiders in the FSC system. Such interviews were held with different FSC staff members, board members, FSC consultants at different levels, FSC auditors, Foundation representatives, and many other experts involved in the network under investigation. Usually the conversation between the researcher and the informant was related to informant's involvement, responsibilities and experiences in issues related to their job. (4) Open-ended interviews were used with community activists, local administrators, teachers, librarians, museum workers and other stakeholders for finding contexts in a particular village, the history of interaction between the company and the local community, and for identifying the interests of local actors and their attitudes and experiences related to FSC certification. The process of data analysis involved explanation-building coming from the social world of the informants, linking data to propositions, and cross case synthesis (Baxter & Jack, 2008: 554)

B. Challenges of combining research with the practitioner's role

Implementation of research, involving multiple countries, regions and events of global, national and local levels became possible through combining research with the practitioner's role. Involvement in different research projects was a direct advantage as the grant applications took into accounts the interests and goals of this study. The practitioner'swork also provided many advantages, involving access to the field and insider's opportunities to see certain aspects of the study. It allowed for establishing good relationships with gate keepers, key informants, and a large variety of the stakeholders. Many of the informants with whom the author was involved as a practitioner were essential for setting up logistics for the study, facilitating it and gaining access to high ranking officials. However, combining research with practitioners' work was also a challenge, which required a combination of methods which could complement each other and provide additional verification of findings. Different forms of the practitioner's engagement affected the research in different ways; therefore the methods used were selected to offset the negative effects.

B.1 Combining the FSC member's duties with researcher obligations

The researcher's stances and level of involvement in practical and/or expert work determined the combination of research methodologies in order to reinforce advantages and offset disadvantages of combining participation in real life events and research activities, for example, being both an active FSC member and a researcher. For studying the FSC in the international setting, participant observation was combined with interviews and document analysis. Being part of the FSC network provided an opportunity to meet key transnational informants, such as company's top managers, lead FSC auditors, and heads of transnational NGOs. However, it was a challenge to combine FSC member's duties and research tasks. At the FSC General Assemblies, for example, as a member of the social chamber I was obliged to work within the chamber analysing, discussing and negotiating motions, therefore, because of that commitment, the events taking place in the economic and environmental chambers were missed. To offset one-sided investigations, during all possible opportunities such as lunches, coffee breaks and cultural activities I engaged in informal conversations with members of the environmental and economic chambers. Newsletters reporting on major developments at the General Assembly were used for creating guides for more formal interviews that were conducted later, through Skype, because of the time constraints at the General Assembly. Another challenge was related to the keeping of field notes. It was hard to simultaneously reflect on on-going negotiations, make voting decisions and keep research field notes. In this case, tape recording and analysis of many hours of records was done.

Combining FSC membership with research in studying FSC-Russia involved the same combination of methodologies, but was easier than research on the international level, mostly because there were less time constraints. Serving on the FSC-Russian Board of Directors required some limitations on the disclosure of information, but this information, organizational budgeting, for example, was beyond the scope of my study.

B.2 Working as an expert and doing research

Being an FSC expert-auditor and doing research

Working as an FSC expert had a different kind of dynamics, differing advantages and disadvantages for field research. I worked as an expert auditor with the certification body team in many FSC projects in different parts of Russia, and in 2006, in my case study setting on the lease of Segezha pulp and Paper mill, operated by the Northern Logging company (case study #4). The auditing work in different parts of Russia was very useful for developing guides for the interviews because it showed the pitfalls of FSC implementation elsewhere. It gave a background and a point of reference on how the FSC is implemented in Russia, what is in common and what is different across the various sites, depending on the contexts and the FSC's implementing actors. The auditing experience at my own research site was an interesting interplay of research and practitioner's' work. The audit allows access to all company documents and allows for interviewing staff and stakeholders involved in the FSC process. Company managers assist auditors with scheduling interviews within and outside the company. The auditor is allowed to ask all kind of questions related to the company's operations in the framework of the FSC. This allowed me to incorporate my research interviews into the set of conventional procedures that the auditor has to conduct for verifying compliance and to see certain aspects of the company operations that would otherwise never be possible to investigate through regular interviews. Therefore, the major advantage of auditor engagement for my research was deeper access to the field and facilitation from the company. The disadvantage of this researcher/practitioner activity was that both company staff members and stakeholders were taking into account that they are telling the story to an auditor, to a person with a controllers' assignment. The company's staff would try to hide incompliances with the FSC standards, stakeholders favourable to the company talk only about the positive sides of the company's operations, and the stakeholders that dislike the company try to tell the worst possible details of the company's failures in complying with the standards. To deal with this issue in my case study, I came back to the same site in the framework of the research project to conduct intensive interviews in 12 villages in the lease area with multiple stakeholders, involving both those who were previously interviewed during the audit process and those who were not.

Engaging in expert work with TNCs and doing research

After my field research at the site, I could not be involved in the auditing process any longer because of the conflict of interest, but the company offered me an FSC consulting job, which involved coaching the company in the FSC process, which was done 2007-2011 (data used for the case study #4). In my consulting role, new advantages appeared, including almost unlimited opportunities to revisit the area and a trust relationship was established between me and the company. Company staffs were eager to talk about all of their difficulties with the FSC process and to expose insider's drawbacks and challenges. As they knew I was a researcher, they were helping me immensely with my research, including setting up interviews with different kinds of company managers in Segezha, Petrozavodsk and the top managers in Moscow. The company facilitated my interviews with their buyers in the UK and the Netherlands, who were also very open after introductions from the top managers in Moscow. The company managers also provided feedback for my writing, while it was in Russian, and helped me improve my publications.

In a similar way, my engagement in training and consulting Stora Enso employees helped in setting up interviews with managers in Helsinki, Stockholm and Imatra. Stora Enso staff connected me with buyers of their products in the UK, Germany and the USA, which allowed me to do interviews there also (for the case #2). Stora Enso's manager for the environment was very much interested in reading and analysing all of my articles, including those written in English. This experience provided not only feedback to my writing, but a continuation of my research, as I had a ongoing dialogue on issues related to the study and interpretations of the outcomes.

I worked very closely with the top managers of Mondy Business Paper on the regional and international levels, which was important for case study #3. When a researcher and an informant are both experts and colleagues in certain processes, this allows many informal conversations, establishing trust and friendship, which fosters further access into the field of study. We were both involved in the High Conservation Value Network in Oxford, in negotiations of the FSC Principles and Criteria in Bonn, discussion of the FSC policies at the General Assembly in Cape Town and Kota Kinabalu, participated in the FSC field trips, met at international conferences and workshops in different countries, and in my study area in the Komi Republic. He introduced me to other top managers in the UK and Austria, whom I then interviewed. The company staff participated in my training workshop in St. Petersburg and gave feedback to my writing.

B.3 Engaging in collaborative projects with NGOs and local communities

The Centre for Independent Social Research (CISR), where I work, is primarily an academic research institution, yet the Environmental Sociology Group that I chair is getting engaged in some applied projects, predominantly sponsored by the European Union. The projects completed in 2008-2012 were implemented partly in my study areas on FSC certified territories, and contributed significantly to my research. In these projects, the NGOs that I studied, for example, SPOK (case study #1, case study #4) were our implementing partners and the WWF and the Silver Taiga (case study #2-3) were part of the collaborative network. The CISR and its partners were involved in promoting citizen rights in resource-dependent communities. Together with community activists, NGOs identifed cases of violations of citizen's or indigenous people's rights in the framework of the FSC and helped communities to negotiate contested issues with the companies. The peculiarity of this method is that the researcher's engagement is transforming the environment being studied as well as influencing the collaborating partners, who are also objects of the study. In order to investigate the influence of the researcher on the operations of an NGO, I extensively interviewed other NGO staffs, who were not involved in the action. Some of these interviews were conducted by my colleagues, for stronger objectivity. The transformation that our NGO network was bringing to the resource-dependent communities was addressed by comparative analysis with sites where no such intervention was.

C. Research reliability and validity

The research design and the case study strategy were developed in a way to allow comparisons between the cases and confirmation of the findings. Interviews with multiple stakeholders allowed for seeing the same process or event from different perspectives. Using a variety of qualitative methodologies and engaging in practical work ensured a deep familiarity with the actors and processes in networks that were being studied and contributed to the validity of the study. The data obtained in certain cases through observation were used as a check against the informants' narratives, outlining their own social world perspective. Longitudinal research allowed prolonged exposure of the researcher to the field; research was conducted with regular follow-on visits to the study areas, and detailed data were collected, managed and analyzed systematically, all of which contributed to the quality and trustworthiness of the overall project. The varieties of methods allowed triangulation (Meijer, Verloop & Beijaard, 2002), e.g. ensuring consistency of findings and convergence of evidence from similar and different sources. Draft papers with research interpretations were shared with the participants, in my case mostly with the FSC staff, NGOs and companies, so that they were engaged in active discussions and had the opportunity to provide additional perspectives to the study.

While doing active research, I was reflecting on my own role in transforming both NGO engagement and the dynamics in the study areas, including strengthening the FSC implementation in these specific areas. Prior to engaging in active research, I did direct observations (with detailed baseline notes being kept), and open ended and semi-structured interviews with the key informants and other stakeholders. Later, I compared the study areas where intervention was done with other FSC certified territories, where I did only direct observation and interviewing without directly engaging in the cause of events. Comparing these sites with those with my active engagement allowed reflection of my own role in transforming the field under study.

During field expeditions, observations and preliminary findings were discussed with colleagues, and selected interviews were read and discussed collectively to ensure peer examination of the data. Some parts of the project were published in peer reviewed journals, and the rewriting of drafts after peer review contributed to the quality.

D. Addressing ethical issues

For ethical reasons at all sites where participant observation was done, people were informed that I was doing research and not only participating in the activities. I had a deep commitment to protect the identities of the informants, and measures were taken to ensure that sources cannot be linked to the information they provided. Informed consent was obtained before the interviews. Debriefing was done about the study and its goals, informants were ensured that their names will not be used in any writing and it will be impossible to identify them from the interview citations.

When interview transcripts were typed and placed in the data base, instead of names special codes were used. Typically, names and locations were kept in the observation notes, but when written up, they were eliminated and only codes were used. The researcher was carefull to manage and never disclose any sensitive information related to the company's image, market competition, or insider's data that could affect the informants. When the author was interested in citing the interview, indicating the informant's affiliation and/or name, written permission was obtained from that particular person.

Chapter 3.
The organizational field of global forest governance: interplay of the governance generating networks

3.1 Introduction

This is a semi-empirical chapter focusing on transnational processes related to governance generating networks (GGNs) and their interactions. The chapter is focused on the concept of GGNs in the institutional field of the transnational forest sector. I am expanding the concept of the institutional field, developed mostly for nation states, (DiMaggio & Powell, 1983; North, 1990; Scott, 1995) to the actors of transnational spaces, engaged in interaction with each other (McDermott *et al.*, 2011b). As it was explained, GGNs link transnational spaces with the spaces of places. Nodes of global design are located predominantly in transnational spaces, with sites of implementation in the spaces of places.

In order to understand the institutional field of the global forest sector, the focus will be on the interaction between different GGNs. I will start the chapter by describing the history of global state and non-state driven international forest governance and the formation of the major transnational GGNs related to forestry (Section 3.2). Both fully developed GGNs and those only in the first stages of their formation will come into play. As the main interest in this book is to understand the role of private actors in global governance, special attention is given to the FSC, which represents the most well, developed NGO-driven GGN (Section 3.3). I will demonstrate how the FSC is constantly changing, developing, growing, and successfully bringing its rules to the sites of implementation. I explain how the FSC GGN operates, who the actors are, how they design and negotiate the standards, and adjust them to particular sites of implementation. I show the main components of this network; namely, the node of design, the forums of negotiations, and the sites of implementation.

Next I will switch to the interactions between GGNs. The certification-GGNs and their promoters: the drivers, brokers, and consultants that are necessary for the constructing and maintaining of niche markets for sustainably produced products are analyzed (Section 3.4). The following section will describe how the surveillance system of operation of GGNs is organized (Section 3.5)

Finally, I will show that relationships between GGNs linking transnational spaces with the spaces of places have both commonalities and differences, when compared with the interplay of organizations operating in organizational fields of spaces of places (Section 3.6).

3.2 State and private-driven GGNs: historical perspective

3.2.1 State-driven GGNs

In the 1970s and early 1980s, the issues related to global forest protection were poorly addressed, in part due to the perception that forest governance is a responsibility of nation states. Some attempts to regulate wood flows were made by the UN Food and Agriculture Organisation (FAO) and the International Tropical Timber Organisation (ITTO) with their agreement in 1983. The agreement was narrow in scope, addressing only trade related to tropical forests. Timber trade was connected with development issues, although there was some effort to make trade more sustainable, by promoting wood processing industries in developing countries, sustainable forest management and some conservation aspects (Field, 2001; Tarasofsky, 1999: 97-99; Tarasofsky & Downes, 1999).

In 1980s, despite many controversies, countries prepared a global convention which was to be signed during the mega forum at RioDeJaneiro in 1992, organized by the UNCED. However, differences in expectations between developing and developed countries, as well as within developed countries, made it hard to achieve the convention on forests. Developing countries blamed the industrialized countries for destroying forests through over-consumption, and were not satisfied by low commodity prices. They expressed concerns about sovereignty, inequality, debt and considered forests as their assets. Developed countries argued for the need to conserve biodiversity, care for endangered and rare species in the forest habitat and the need to preserve the forests in order to prevent global climate change. Among the developed countries, multiple disagreements existed between Russia and Canada. Transnational NGOs and their networks also opposed the convention. One of the arguments was that the Convention on Biodiversity (CBD), the UN Framework Convention on Climate Change (UNFCCC) and many other existing institutions could coordinate and achieve better results than would be achieved through a binding agreement (Rosendal, 2001). NGOs assumed that consensus could only be reached during negotiations with the significant lowering of standards of forest management and trade compared to the standards already existing in the Northern European countries. This would allow forested countries, such as Russia, to begin extensive international trade with low standards.[14] Therefore, the convention was not on the agenda of the Rio conference, due to multiple disagreements between parties, which were evident during the preparation process. The UNCED conference produced both non-binding documents that address issues of forest governance, and binding agreements indirectly affecting forest governance. Soft, non-binding agreements took the form of three major documents: the Statement of Forest Principles, Agenda 21 and the World Charter for Nature. Binding CBD and UNFCCC agreements, which were at the core of the UNCED conference, included elements of forest governance, and contributed to forest protection on the World scale. Later on, The Forest Principles were used for development of private voluntary certification systems, which became important regulatory tools in global governance. Therefore, since 1992, state-driven forest regulation was developed in parallel with private driven regulation.

Other international conventions contributed to forest protection, such as the Convention on International Trade in Endangered Species of Flora and Fauna (CITES), the Ramsar Convention, the World Heritage Convention, the ILO Indigenous Peoples Convention, and the Desertification Convention. Many regional agreements, such as the African Convention, the Western Hemisphere Convention, the Treaty on Amazonian co-operation, the Central American Forest Convention, the North American Agreement on Environmental Cooperation, the Berne Convention, the Apia Convention, and the ASEAN Convention also are part of global forest governance.

[14] Personal communication with NGO leader at the conference on Tropical Forest Management, 2004.

Under the UNCED, governing bodies related to forestry were developed quickly. The UN Commission for Sustainable Development was formed as a coordinating body for all major sustainable development initiatives, including sustainable forest governance. The Commission for Sustainable Development took the lead in establishing an open-ended Intergovernmental Panel on Forests (IPF) in 1995.

The IPF had the intention of developing a legally binding instrument. In Geneva, in 1995, under the FAO, the Interagency Task Force on Forests (ITFF) was formed to assist the IPF in linking initiatives related to forest governance undertaken by existing conventions and multiple international organizations (Rosendal, 2001). The IPF inherited all the controversies on forest issues from the pre-Rio process and did not achieve any substantial improvements in global forest protection during its four international meetings, although it developed a Plan for Action with the intention of reaching results on the ground. The IPF gave policy recommendations to the Commission on Sustainable Development in 1997. The UN General Assembly Special Session on Agenda 21 in 1997 analyzed these recommendation and the UN Economic and Social Council replaced the IPF with the Intergovernmental Forum on Forests (IFF).

The IFF had very similar goals to the IPF, and failed to implement the IPF proposal for action. Later on, the IFF developed its own proposal for action. Action plans of the IPF/IFF had the goal to implement Forest Principles and to clarify how conservation and utilization of forests should be balanced (Gulbrandsen, 2003). However, the forums of negotiation were not advancing, on one hand because of controversies between parties, and on the other hand, because the IPF was short of institutional and financial resources, and could not handle the institutional overlaps. Nevertheless, the IPF continued discussion on development of legally binding instruments. The final meeting took place in 2000, and the IFF asked the Commission for Sustainable Development to establish a new permanent intergovernmental institution: the UN Forum on Forests (UNFF). In addition, the IFF called on international organizations to partner with the UNFF and support global forest protection activities (Rosendal, 2001).

The UNFF introduced voluntary reporting for participating countries and organizations on the state of the forests.[15] The UNFF, in general, inherited the ineffectiveness of the IPF/IFF and their decisions were never implemented.[16] Therefore, the designs of this GGN did not reach the sites of implementation.

The FAO has played an important role in forest governance since its inception. The FAO also played an important role in the IPF/IFF/UNFF process. The FAO convened a great number of global and regional meetings and forums. One of its bodies is the Committee on Forestry (COFO), whose members are predominantly the heads of forest services of different countries. In addition, the FAO accommodates Regional Forestry Commissions, the Advisory Committee on Paper and Wood Products, the Advisory Committee on Forestry Education and several others (Tarasofsky, 1999). Despite its important role in forest governance, its designs do not always reach the sites of implementation. The FAO is criticized for failing to implement its Tropical Forestry Action Program. Developing countries criticized the FAO for supporting development of a new convention on forests, and NGOs criticized the FAO for being dominated by industry interests.

[15] http//www.un.org/esa/forests/reports/html.

[16] Interview with the Greenpeace Netherland's representative, February, 2005, Utrecht.

One of the reasons that most of the NGOs opposed the new forest convention is a lack of trust in the FAO, which would have become a head organization for the convention, had it been signed (Tarasofsky, 1999).

The UN Development Program (UNDP) links forestry issues with the more general goals of poverty alleviation and sustainable development. Since 1993, it has coordinated the Forest Capacity Program, which helps developing countries implement their national forest programs (Tarasofsky, 1999). The UN Environmental Program (UNEP) has limited responsibilities and modest capacity in forest governance. It usually deals with assistance to countries with little forest cover and contributes to the analysis of causes of deforestation (Tarasofsky, 1999).

In the state-driven GGNs, there is a place for multi-stakeholder involvement and private actors' participation. Official representation of TNCs during international negotiations is minimal. However, some representatives of TNCs participate as part of countries' delegations without disclosing their belonging to TNCs (Levi & Egan, 1998). Many TNCs are implementing sustainable forestry by supporting voluntary certification, taking progressive stands on the environment and heavily advertising their environmental policies in order to enhance legitimacy in the consumer community (Sears, Davalos & Ferras, 2001).

Contrary to TNCs, NGOs are actively participating in all negotiations related to the forest regime and over time their role in negotiations is increasing, although they do not have voting power. Large numbers of NGO network representatives attend meetings, give their recommendations, and organize side events and street forums.

3.2.2 NGO and TNC-driven GGNs

As I mentioned before, most of the NGOs are opposed the idea of a binding convention on forests. Disappointed by state authority in global governance, NGO networks themselves become global agents for institutional change (Bernstein & Cashore, 2004). As it was described in Chapter 2, the role of NGOs in global governance is much greater than is their involvement as stakeholders in transnational processes. NGOs not only support or oppose state-driven GGNs, but have become private authorities themselves and drive non-state GGNs. NGO-driven GGNs involve the diverse community of global designers and rule makers whose activities reinforce each other. Certification GGNs differ from NGO-driven GGNs as they all are based on multi-stakeholder processes for developing global voluntary standards. Certification GGNs in forest sector can be driven not only by NGOs, but by business and trade associations and may or may not be supported by nation states. Therefore, certification GGNs illustrate in the best way the emergence of private authority in global governance. They include global rule makers, ranging from the International Standardization Organization (ISO), to the Fair Trade Labeling Organization (FLO), to the Sustainable Agriculture Network (SAN) and many others (Bartley, 2003, 2007b, 2010b). GGNs developing forest certification schemes are also diverse.

Historically, the FSC-GGN was formed in 1993, driven by NGO efforts in order to promote sustainable forest management around the world. The foundations for the FSC standards were declared in the ten `Forest Principles' elaborated in the Earth Summit in Rio de Janeiro in 1992.

NGOs, first of all the WWF, the Rainforest Alliance and several supporting business coalitions managed to create a globally functioning FSC certification system, which reaches the sites of implementation, uses the forces of the market and leads to changes of the forest management practices in localities. The FSC was conceived as a 'marker' of companies' that pursues both environmental and social responsibility, which lets consumers, concerned by the situations in the world forests to choose products from well managed forests.

Following the FSC, other GGNs developing forest certification standards emerged. New certification systems were organized by those actors, which for varying reasons were not interested in joining the FSC system. In response to NGO and market pressures, several trade associations, such as the American Forest and Paper Association (AF&PA), the Canadian Pulp and Paper Association, and the Timber Trade Association of the UK developed certification schemes to legitimize their methods of operation. In the USA, the AF&PA created the Sustainable Forestry Initiative (SFI) certification system (Sears *et al.*, 2001). SFI compliance is monitored by the AF&PA, and around 15 companies have lost their membership for non-compliance.[17] Compared to the FSC, the SFI does not emphasize social indicators in forest practices. In contrast to the FSC system, which exclusively presumes performance based standards, the SFI certification includes system based standards. Other alternative certification systems to the FSC also have mostly system-based standards. The most developed of them are the Canadian Standards Association (CSA), initiated by the Canadian Sustainable Forestry Certification Coalition, consisting of 23 industrial associations, and the Pan-European Forest Certification (Cashore, 2002: 508-509), or PEFC, created in 1999 by associations of landholders, and in 2003 given the new name: the Program for Endorsement of Forest Certification.

The PEFC certification system was created to meet the needs of international accreditation for already existing national certification systems. Although the PEFC, to a large extent, involved governments as stakeholders, it represents a TNC-driven GGN. Contrary to the FSC, this system takes into account the interests of small private forest owners and landholders, and that is why it dominates in such countries as Finland, where most forest owners are small. In 2006, most certification schemes united into a coalition, with the PEFC at its head, by signing an agreement concerning mutual recognition and creating of a common market space. To date, it has endorsed 30 national standards, of which only four pertain to tropical forests, all others are for boreal and temperate forests[18]. This means that territories certified by their own national certification schemes could label their products as PEFC. Therefore, the PEFC united several certification GGNs under its umbrella and became a powerful force in private forest regulation and a major competitor of the FSC.

3.2.3 Emergence of new GGNs

The institutional field of the transnational forest sector is not constant or isolated from other institutional arrangements. It is represented by fragmented and overlapping state and private initiatives, involved in experimental governance arrangements (Arts, 2011; Sabel & Zietlin,

[17] Interview with SFI staff, Washington, DC, September 2005.

[18] PEFC News, May 2011, http://www.pefc.org/news-a-media/newsletters.

2012). It is subject to ongoing change; new GGNs are constantly emerging. New GGNs have been created at the convergence of several problems, during the course of interactions of the networks involved in solving those problems. For example, a new GGN is now being formed called Reduced Emissions from Deforestation and Forest Degradation in Developing countries (REDD). This GGN was formed for connecting the issues related to climate change with forest issues (Bozzi *et al.*, 2012; McDermott, Cashore & Levin, 2011a). Several GGNs united their efforts to create the new REDD network, among them FAO, the UN Development Program (UNDP), and the UN environmental program (UNEP). REDD+ was created to supplement the reduction of CO_2 emissions with Sustainable Forest Management. REDD+ cooperates with the FSC, the PEFC and other certification-standardization GGNs to ensure human rights, poverty reduction and biodiversity conservation in the sites of implementation (Merger, Dutschke & Verchot, 2011).

In contemporarily forming GGNs, designs are not yet reaching the sites of implementation, but they attract maximum human, financial and other resources.[19] As a result, many other GGNs reorganize and reorient their work to get access to these resources. For example, the FSC, joining with REDD focuses on HCVF designation for reducing CO2 emissions. They also plan development of a new the FSC standard for reduced emissions.

Creation of REDD provoked reorientation of resources toward tropical countries where the problem of forest conservation is very acute because of the large conversion into agricultural plantations. This reorientation of donors' attention, to some extent, leads to discrimination of the non-tropical countries where the problems of forest conservation also exist. In response to such discrimination new alliances of actors are being created, such as 'Beyond REDD'. They also link the problems of deforestation with climate change, but augment this issue with other content. It has to be noted that it is difficult to organize an effective program for a new GGN when it is formed by other well developed actors with their own priorities and goals, because they place their preferred subjects at the fore (Levin, McDermott & Cashore, 2008: 3).

3.2.4 Inter GGN forums

Forums play an important role in the interaction of the GGNs in transnational spaces. For example, numerous GGNs related to the forest sector gather at the Global Forest Congress, organized by the FAO. All kinds of representatives attend: from state-driven GGNs: UNFF, NGO-driven GGNs: the FSC, High Conservation Value (HCV) network, the WWF and others. From business-driven GGNs: the Working Group on the forests of the World Business Council (WBC); and other intersector networks: Forest Dialogue, etc. In these forums they make connections, interact with each other, and elaborate new ideas, which can later initiate new alliances and GGNs.

If the forum is organized by other (non-forest) GGNs, the actors engaged in forest issues can take part as 'adjacent' GGN representatives. For example, they participate in the COP conference organized by the climate change mega-process UNFCCC, which in turn was born out of another mega-process, the Earth Summit in Rio de Janeiro. The 'forest' related GGNs take part in these conferences as participants of main events and as organizers of side events, to draw other's attention to the forest issues.

---

[19] Interview with researcher from the Rainforest Center, Oxford, June 2009.

3.3 The FSC-GGN: node of design, forums of negotiation and sites of implementation

The FSC represents a voluntary certification system based on the principles of tripartite sustainable development, which presumes a balance of economic, environmental and social aspects in forest management. The FSC designs and implements global voluntary regulatory standards. The FSC develops several types of standards and delivers two major types of certificates: the certificate of forest management (FSC-FM) and chain of custody (FSC-CoC). FM certificate guarantees that logging and other forest operations are done in compliance with the Principles and Criteria of the FSC Standard, taking into account economic, ecological and social components of sustainable forest utilization. The FCS-CoC guarantees timber legality and shows that the path of wood along the chain of custody has been monitored from the moment of logging through all the stages leading it to the customer, including transportation, processing, and the manufacturing of goods using this wood. If a company purchases only certified wood for processing, it obtains the FSC-COC-PURE certificate. Another kind of certificate is the FSC-CoC-MIXED, which along with certified wood includes 'controlled wood'. This term means that the wood was legally produced and it did not originate from the High Conservation Value Forest (HCVF), which was duly verified. The FSC-GGN node of design, forums of negotiation and sites of implementation represent a complex and dynamic system (Figure 3.1).

3.3.1 FSC node of design

The FSC-GGN has a multi-level, multi-actor node with a democratic form of governance. It is coordinated from the International Center, located in Bonn, Germany, in which the FSC president and staff are located. Staff is recruited from all over the world. The FSC node includes national and regional offices (Figure 3.1 – FSC in grey rounds). National offices are opened in countries having large forest territories, such as Russia, Canada, USA, Mexico, and China. Other countries are coordinated by FSC regional offices. There are several in number: FSC Europe, FSC-CIS, FSC Latin America, FSC Central Africa, and FSC Asia.

The legislative body of the FSC is its General Assembly, and the International Center is its executive body. The FSC is governed by the Board of Directors, consisting of an equal number of representatives from each chamber. The Board of Directors defines the policy and strategy of the node operation, and makes day to day decisions in the time between the General Assemblies, which are organized every three years. The Board of Directors is democratically elected by all members.

The system of coordination and governance in the FSC strives for maximum democracy and wider participation of representatives of different countries, involving members and stakeholders at different levels. The FSC node of global design is a highly organized structure, in spite of the large number of its members, 891 members from 82 countries. All the members are divided into three chambers (Figure 3.1 – internal grey circle) which they join based on self-identification: economic (405 members), environmental (321 members) or social (165

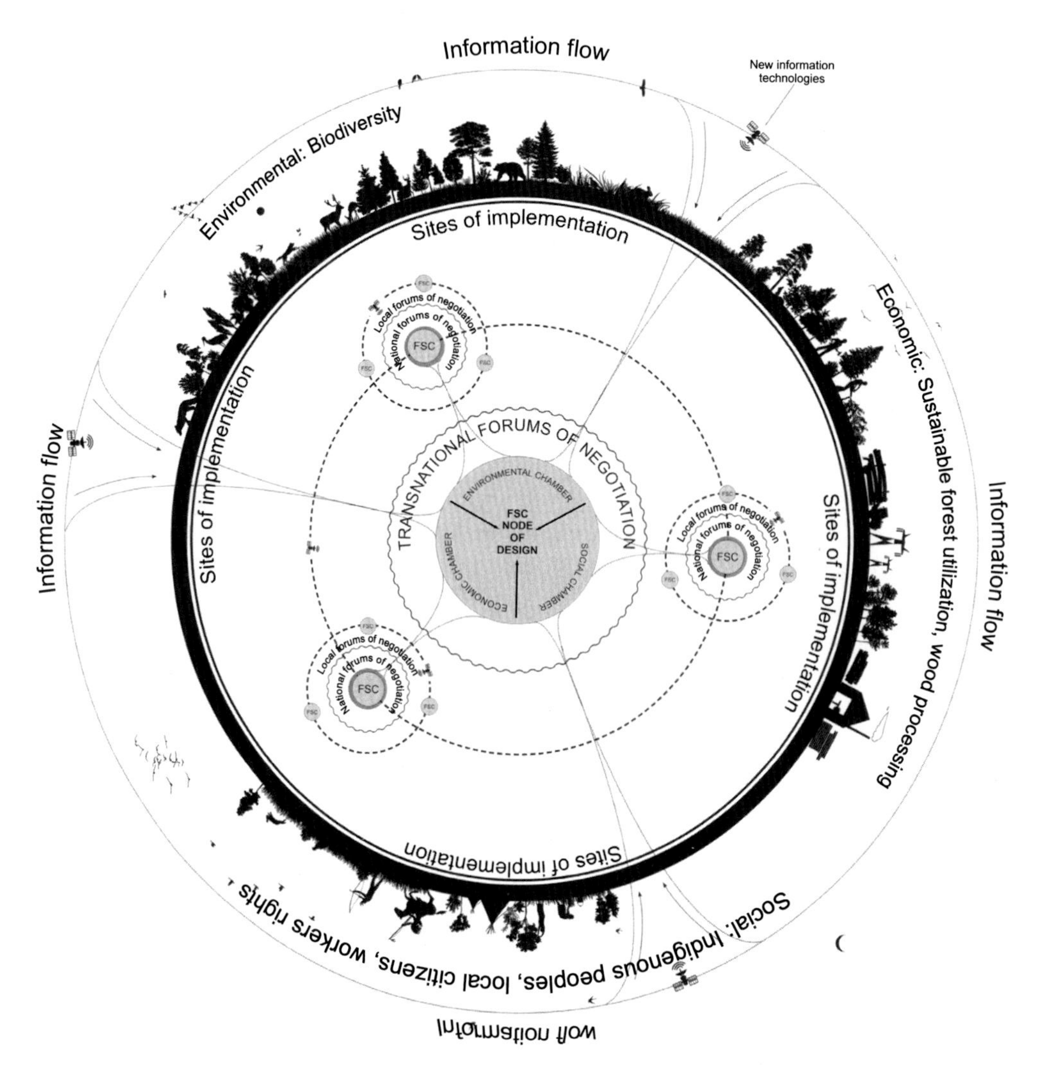

Figure 3.1. Major components of the FSC-GGN. The figure is explained in the text.

members)[20]. Governments are not represented in any of three chambers. The FSC tries to involve equal representation in all three chambers. However, economical and environmental chambers are more powerful and contain more members than the social. The Environmental chamber is represented by strong environmental NGOs, and the Economic chamber by large, well-recognized TNCs and by certification bodies which are mostly for-profit organizations. In the Social chamber, representatives are not consolidated because each of them is interested in some specific aspect of the block of social issues (Hallström & Boström, 2010). There are representatives of trade unions,

[20] Data on membership provided by FSC coordinator Sarah Banda (numbers on membership updated July 1, 2011), personal communication, 2 August, 2011.

who are interested mainly in workers' rights. There is a large representation of indigenous people, but they also are concerned with their specific issues; very few social NGOs are involved. There are many individual members in the Social chamber, who have a small voice if compared with organizations; hence, little ability to influence the FSC's final decisions. Because of the above, it proves impossible to provide equal representation in all three chambers[21].

The FSC strives for equal geographic representation from all regions. To contribute to this type of equality, the FSC has fixed different membership fees for developed and developing countries. To define which type a country is, the FSC uses the UN classification, which conditionally defines the developed countries as 'Global North', and the developing countries as 'Global South'. According to this classification, Russia is one of 'Global South' countries, in spite of its Northern geographical location. Members of social and environmental chambers receive subsidies to attend the General Assemblies. Due to the FSC's good will and this financial indulgence towards developing countries, the membership in the Global Southern environmental and social chambers is higher than in the Global North. However, language barriers and a lack of motivation to participate prevents equal involvement in decision making by members of Global North and Global South.

The FSC national offices, serve as a link between the node, which operates mostly in transnational spaces, and sites of implementation in a particular country. The FSC strives to create a National Initiative in each forested country, but it does not always manage to do that because of various reasons. The FSC National Initiatives have to be accredited by the Accreditation Service International. In 2011, there were more than 40 accredited National Initiatives[22] and about 10 candidates for accreditation.[23] Creation and, especially, accreditation of the National Initiatives is a very difficult and long procedure, which takes from fine to ten years.

The National Initiatives (FSC-offices) are organized in the same way as the FSC International and consist of social, environmental and economic chambers with equal representation. Their major purpose is to develop national standards, and to govern the FSC process within the nation state. Each country where the FSC office is created has its specific political regime with its structure of political opportunities, hence with its own context. In the countries with underdeveloped democracies it is more difficult to initiate a democratic process and to achieve equal representation of economic, social and environmental actors, which would engage in a constructive dialogue.

Besides the FSC offices, members, and staffs, all other stakeholders at different levels and scales take part first-hand in the activities of the FSC-GGN. Stakeholders of transnational spaces are those individuals and organizations, which in one or another way are engaged in this certification system: they develop it, criticize or oppose. In the nation states the FSC stakeholders are state bodies, at all levels from national to local, and all those who show an interest in forest utilization issues: NGO's, researchers, and local populations. Interested parties are involved in the development of standards, in monitoring and control over the system, and in commenting on all issues (Tollefson *et al.*, 2008). Stakeholders are involved in the process of certification of the companies; the certification bodies

[21] Participant observation at the meeting of Social chamber, April 2010, FSC General Assembly 2008, Cape Town, October 2008, FSC General Assembly 2011, Kota Kinabalu, June-July 2011.

[22] In 2011, the reorganization took place in the FSC structure, National Initiatives were reorganized into National FSC offices in order to facilitate sharing finances.

[23] Report of FSC Russia director, November 02, 2011, www.fsc.ru.

and Accreditation Service International (ASI) consult with them. The issues raised by stakeholders and members are reflected in motions, which are further discussed at the FSC General Assembly. The FSC makes an effort to cooperate with researchers, and hence, to involve the world's expert community in its system. Many researchers combine their studies with expert work in the FSC. Some of them are stakeholders of the FSC system.

Within the FSC-GGN it is difficult to combine democracy with effectiveness; it takes time to come up with the decisions and to implement them[24]. The slowness of decision making is heavily criticized by members of the Economic chamber. Business needs consistency and effectiveness[25] for management of their companies, and in the FSC, the urgent questions are solved as slowly as are all the others. Employees hired all over the world also leave their mark on the style of the work, as representatives of very different cultures have different understandings of work ethics and style[26]. However, all these features do not change the fact that the FSC-GGN is one of the most democratic organizations in the world.

3.3.2 FSC forums

Organized by the node

The largest forum is the FSC General Assembly, which is organized every three years, and where economic, environmental and social chambers from the 'Southern' and the 'Northern' countries participate in decision making. The General Assembly decides how the FSC-GGN will develop and how the FSC standards will change. Therefore, the General Assembly is the main strategic driver of the FSC (Figure 3.1 – transnational forums). Smaller forums are working groups and conferences, chamber meetings, inter-chamber discussions as well as permanent Internet forums for discussing various documents, electing candidates to the Board of Directors and discussing other issues.

All FSC members are encouraged to meet at the General Assembly to design strategies of development for the next three years and to make general decisions concerning the operation of the GGN. Prior to the General Assembly's gathering, stakeholders from all over the world send motions concerning standards or policies. Discussions, negotiations and voting on the motions are part of FSC's core forum that leads to the decision making. Prior to the General Assembly, members submit motions to the Motions Committee. In case they overlap, the committee advises submitters to merge their motions, and some motions are withdrawn. At the General Assembly, motions are discussed separately in each chamber and later negotiated between chambers. Some organizations organize additional forums around the motions. For example, during the General Assembly 2011, the FSC-Board member from the WWF initiated daily meetings with WWF representatives from all over the world to elaborate common positions on motions prior to

[24] Participatory observation, 2008-2010.

[25] Interview with representative of Mondi Business Paper, October 2008, Cape Town. SAR Interview with representative of UPM Kummene, February 2009, Helsinki, Finland.

[26] Interview with representative of the FSC Russia, August 2008, Moscow.

participating in chamber meeting and inter chamber negotiations.[27] Prior to the General Assembly business session, motions are reformulated and arranged according to their importance, and informal agreements on voting are made. Sub-committees can be organized for discussing and reducing the number of motions. In the General Assembly business sessions, the motions are presented by their authors, and short debates take place and votes follow on each motion. Usually, less than half of the originally submitted motions are passed[28]. For example, 69 motions were originally submitted to the General Assembly of 2011, 56 presented to the General Assembly, 38 were voted on, and 27 passed while11 failed[29].

The motions are classified as statutory motions and policy motions. The statutory motions concern the operations of the node itself. For example, in the General Assembly 2008 statutory motions such as a motion on the mandatory translation of all FSC documents into Spanish (in addition to the English version) were approved, along with others, such as improving the FSC management structure, improving dispute resolution procedures, harmonizing the FSC membership on transnational and national levels, increasing the efficiencies of the Board of Directors and the FSC International Center. In the General Assembly 2011[30] statutory motions were related to changes in the FSC's statutes and the FSC's by-laws.

The policy motions concern the design of the node, in particular its policies and standards. For example, revisions of the Principles and Criteria of the FSC Standard, the chain of custody or trade mark, controlled wood, certification of small forest owners, the general policies of the FSC, and issues concerning accreditation.

If a motion is accepted by the General Assembly, changes are usually implemented over the next three years. The General Assembly only makes decisions on changes, while introducing changes into the policies and standards is a long, multi-stage process that can last for years. The FSC members introduce proposals on the realization of the proposed modification. This proposal must be accepted by the Board of Directors. On the basis of the proposal, the FSC staff designs the draft document, which is then submitted for discussion by the stakeholders. After receiving feedback from the stakeholders (in the form of comments to the draft) the staff formulates the final draft document, which is submitted for discussion. For the final revision of the modification, a technical working group is created with participation of the FSC members elected via the internet. The technical working group designs a final draft of the modification, which has to be approved first by the working group and then by the Board of Directors. Only after passing all these stages is the modification introduced into the FSC standards.[31]

Between the periodic General Assemblies, debates continue, and some actors even organize whole processes for this purpose. For example, in May 2008, on the eve of the General Assembly, a process was initiated dedicated to the problem of incorporating small forest owners into the FSC

[27] Participant observation at the General Assembly 2011, July 2011, Kota Kinabalu, Malasia.

[28] Participant observation at the General Assembly of 2008, Cape Town, SAR; at the General Assembly 2011, July 2011, Kota Kibnabalu, Malasia.

[29] FSC report of GA motions passed, http://motions.fsc.org/wp-content/uploads/2011/07/FSC_Report_GA-Motion-Passed_English.pdf.

[30] Participant observation and the FSC General Assembly 2011, 25 June-1July, 2011, Kota Kinabaly, Malasia.

[31] Http://www.fsc.org.

certification system. Prior the General Assembly in 2011, several working groups were engaged in preparation for the Assembly; for example, a carbon working group was making an effort to elaborate recommendations regarding the FSC's involvement in combating climate change.

The process of developing and correcting the standards is dynamic and continuous. In the course of their implementation, feedback is given by local actors, and if the standard does not work well in many locations, the FSC designs a new international standard to replace the existing one. From the beginning of the FSC in 1993, the same Principles and Criteria (P@C) have been applied. Among the issues to be revised were reinforcement of the social part of the P@C, the plantation review, and many others. An effort was made to include carbon storage and sequestration into the new standards. At the General Assembly 2011, a motion on the FSC's stance on carbon by the P@C passed, thus it will be included in next three years.

The process of development of the new FSC P@C demonstrates how forums are important for creating new global designs. The first decision on changing P@C was made in the General Assembly 2005, which took place in Brazil. At the General Assembly 2008, in Cape Town, the working group on revision of P@Cs was organized and started working in January, 2009. Since then, four drafts have been developed and circulated by the P@C review group. In addition to hundreds of phone, Skype, and e-mail forums, multiple face-to-face meetings have been organized. Among them, five meetings of the P@C review group, two social chamber meetings, five indigenous peoples meetings, meetings of the National Initiatives, a two-day inter-chamber workshop in Bonn, and a workshop of FSC members with the HCV network to discuss Principle 9 have occurred. Prior the General Assembly 2011, an on-line survey on the P@C was organized. It demonstrated that there are disagreements on certain issues not only between chambers, but also within chambers. Two days were allocated at the General Assembly 2011 for a P@C review workshop to discuss issues raised by membership in the on-line survey of the fourth draft of the P@C. The idea behind the workshop was to reach at least a 'soft consensus' if a 'positive consensus', proved hard to reach.[32] The facilitated workshop involved plenary discussion on the 23 issues raised in the survey, one-on-one discussions, discussions of cross-chamber teams and reports back to the plenary. The P@C review group actively participated and clarified the rationales of formulations. After the General Assembly, the report on the P@C review workshop was presented to the membership for comments. [33] Membership voted positively on new P@C in January, 2012. In July 2012 the working group consisting from members of all three chambers started developing the generic indicators for new P@Cs.[34]

In each General Assembly there are issues that do not provoke much discussion and are easily accepted by the majority. Along with these issues, there are others which are regarded and approached in different ways by the members. Such issues can create peculiar multilevel and multi-actor negotiations, where actors gather different layers of information from various parts

[32] 'Positive consensus' is a term used in FSC discourse to explain that during negotiations, members of one chamber help the other chambers to reach agreement, 'soft consensus' means an agreement without sharp contradiction, in which final formulations are left to the working group (participant observation in the FSC processes 2007-2011).

[33] Participant observation at P@C review Workshop, 25-26 June, 2011, Kota Kinabalu, Malasia.

[34] Participant observation at the working group discussions via Skype, July 2012.

of the network and from various spaces of places. For example, in the years 2008-2011, many stakeholders made attempts to toughen the controlled wood standards. On the initiative of the Rainforest Alliance, with its NEPCon subsidiary in collaboration with the FSC, they created the Global Risk Register, which classified all countries depending on the risk of illegal wood penetrating into the supply chain. Only three European countries in the Global Risk Register (Great Britain, Ireland and Spain), were defined as low risk zones, in which timber production does not require any additional verification. All the other countries were classified as zones of uncertain risk, and the FSC offices from different regions of the world, in co-operation with the stakeholders in their countries, made attempts for the creation of national and regional Risk Registers. This means that a new global process has been initiated, which can result in the gradual exclusion of controlled wood from the market due to the toughening of requirements to the FSC-CoC-MIXED type of certification. Hence, this can stimulate development of the FSC-FM certification.

The FSC standard originally was developed for middle size companies, but with development of the certification system, a high demand was created to certify both small forest operations and mega-forest operations. This created a new issue to be negotiated at the FSC. Some NGOs, including Greenpeace, insisted that special standards would have to be designed for each of these categories of companies, because the existing standard largely takes into account the needs of middle sized operations. Greenpeace observed that in the case of certification of a large holding by only one certification body, both parties can have too close a connection with each other, which can result in softer audits and increasing compromises between the certification body and the company. Also, in certification of the mega-forest operations, it is difficult to organize consultations with interested parties in many localities, and as a result, the consultations are mainly conducted on the regional level, and the local stakeholders are neglected. Another side effect of such an approach is that it is difficult to measure the social impacts in the sites of implementation. Certification bodies have the opposite opinion. They believe that, on the contrary, it is easier to work with the mega-forest operations because they usually have developed both environmental and social policies and emphasize corporate social responsibility in their practices. Good management systems allow the TNCs to easily spread good practices and lessons learned throughout all the sites of implementation where the corporation has branches. They think that in mega-forest operations the certification rules are easier to institutionalize, because such large operators are vulnerable to social movement pressures and care about their image and brand. Because of the differences and approaches to mega-forest operations, the FSC membership has continued to negotiate on this issue for years. In the General Assembly 2011, a motion to assess the local level impacts of big operations was finally passed[35]. In some countries, like Russia, such mega-forest operations dominate the market, so this issue is of great importance.

Another FSC 'hot' issue concerns plantations. The case in point is mainly pine and fast growing eucalyptus plantations, so-called plantations of a new generation. The issue is the question whether plantations can be certified as forests, or if they are agriculture? The FSC certifies plantations, because due to the certification requirements, the companies managing the plantations assume large obligations of conserving ecosystems located around plantations. Due to this fact, some

[35] Participant observation at the General Assembly, 25 June-1 July, Kota Kinabalu, Malasia.

of the environmental NGOs support the idea of certification of plantations while other NGOs strongly object to it.

Another long contested issue concerns indigenous people and local communities. The FSC standard provides priority rights to indigenous peoples in the framework of certification. That means that no activity 'shall take place without the free, prior and informed consent of the indigenous peoples'.[36] This definition means that the indigenous people should realize what the long term consequences can be of industrial forest utilization by the forest companies on the territories of their traditional, natural use. Thus, the issues concerning indigenous people are not simple, and the process of decision making is sometimes very complex. In the General Assembly 2011 the decision was made to form a permanent indigenous people's committee within the FSC.

Many stakeholders argue that local communities should not be discriminated against in preference to indigenous peoples. Since August 2009, consultations with stakeholders on the issues of indigenous peoples have begun in four regions: Europe, Latin America, Africa and Asia. Innovation consists not only of the fact that the consultations were conducted in a form of face-to-face meetings, but also that the FSC made a special effort to involve researchers, experts and indigenous people's representatives in the discussions. The first consultation on the indigenous people of Europe was conducted in August, 2009. In the course of this consultation a new issue arose. It concerned extending the concept of indigenous people by including the concept of 'traditional people'. Under 'traditional people' the FSC members mean people pursuing traditional natural resource utilization over a long period of time, but not being indigenous people as defined by the UN. For example, they refer to hereditary peasants of Latin America, or other regions or countries, as traditional people. At the above mentioned FSC meeting, the opinion of the experts and the representatives of indigenous peoples differed. The representatives of indigenous peoples were against a unified approach to indigenous and traditional peoples. They argued that these two groups have different interests. Besides, they were afraid of losing the governmental support that indigenous people have in some countries, such as in Sweden.[37] In April 2010, at the meetings of the Social chamber in Bonn and in Latin America, standards protecting the rights of indigenous and traditional peoples were allocated under different principles; indigenous under Principle 3 and traditional under Principle 4.[38] The term 'traditional people' was finally not introduced to the new P@C, but the rights of local communities in FSC framework were strengthened significantly and became almost equal to those of indigenous peoples.

The issue concerning the quality of certificates continues to be a hot one at the FSC. The majority of the members at the General Assembly 2008 agreed that it was necessary to increase the quality of certificates. Discussions arose around the question of by what means can the quality be improved under the existing conditions of a low budget for the Accreditation Service International that monitors and controls the work of the certification bodies. The situation improved in 2009-2011, ASI surveillance improved, and grievances from the membership became less intense.

[36] UN definition of indigenous people's fundamental rights.

[37] Interview with the expert on indigenous people from Russia, participating in the consultations on indigenous people of Europe, that took place in Bonn, Germany, in August 2009, and in October 2009, St. Petersburg, Russia.

[38] Participant observation at the meeting of the Social Chamber in Bonn, April 2010.

Forums of adaptation

Adaptation of the FSC standards to specific country conditions is the responsibility of either the Certification Bodies (before accreditation of the National standard), or of the National Initiatives (since 2011, FSC offices).[39] P@C remains the same for the whole world, while indicators and verifiers are developed nationally in order to adjust the standards to the local context. When Certification Bodies are engaged in adaptation of the standards to a local context, there are usually very few forums of adaptation initiated. Forums are limited to consultations with stakeholders on national, regional and local levels. One of the FSC certification requirements is compliance with both national legislation and certification standards. In the course of these consultations, the participants correspond and harmonize the FSC standard with the national legislation.

Adaptation forums of the members of three chambers belonging to the national FSC offices differ greatly from that of the certification body. The main responsibility of the FSC national office is the democratic development of the FSC national standard, with involvement of all three chambers, which, when developed, should be accredited by the FSC (Figure 3.1 – national forums). If the matter concerns a large country having different types of forest ecosystems, regional standards are developed along with the national. The FSC national office designs the standards through a forum. Like the FSC General Assembly, the national FSC office forum gathers more or less regularly to discuss the standards. The dynamics of intercommunications between stakeholders resemble those on the global level.

Accreditation of the National Standard is an even more complicated process than the accreditation of the national FSC office described above. By the beginning of 2006, only 26 National Standards had been accredited all over the world, and seven standards from 21 countries were in the middle of the accreditation process. Certification Bodies had adjusted 16 global standards to local contexts by the same date.[40] When the National standard is accredited, Certification Bodies must follow. The success of the forums organized by the National Initiatives in the process of elaboration of the National standard depends upon the level of development of democratic institutions in the specific countries. In places having a tradition of inter-sectoral dialogue, the negotiation processes go more smoothly. In other places, it can be more complicated to organize and conduct the forums. Some countries do not even start developing the standards – Belorussia, for example, although the major part of their forests have already been FSC certified using the standards adjusted by Certification Bodies[41], has not yet begun. In many countries the FSC, in fact, stimulates the development of democracy and civil society. So, it can be stated that in those countries the FSC acts as an agent of institutional change, which stimulates the processes of democratization via certification forums. Success of the forums also depends on the degree of the stakeholders' involvement, which differs from country to country. The level of involvement can be determined by the system of ownership, by the presence of indigenous people or by other

[39] In 2011, the reorganization took place, the National Initiatives were renamed into FSC offices. The new scheme was introduce in order to facilitate sharing financial resources between FSC International and FSC in countries and regions.

[40] Report of director of the FSC Russia, April 2008, Moscow.

[41] Participatory observation in audit in Belorussia, October 2008, Minsk, Belorussia.

causes. Composition of stakeholders participating in the forums also can vary. For example, in the Ukraine, where only state forest enterprises (leskhozes) are being FSC certified, governmental structures participate in the work of the national FSC office, despite the fact that formally they have no vote.[42] In Canada, where the issue of indigenous people is very acute, a fourth chamber has been created in their national FSC office, which is engaged in solving problems related to the indigenous people.[43] Thus, national FSC offices are the key organizers of forums of adaptation of the FSC standards to local conditions.

Forums of implementation

On the local level, where the adapted standard is transformed into concrete practices, many forums of implementation are organized for this purpose (Figure 3.1 – local forums). In these forums, local stakeholders dominate, though the actors from transnational spaces also participate. In these local forums the issues of HCVF designation, creation of specially protected natural areas, realization of the needs of local populations, and many other issues are discussed. Most of these forums are organized by the companies seeking certificates in order to comply with the FSC standards, which require local stakeholder consultations.

3.3.3 Sites of implementation

By July 2012, a total of 155,641,072 ha of the forests in 80 countries had been FSC certified, and 1,124 FM certificates have been delivered.[44] Canada leads in the number of FSC certified territories, next are (in decreasing order): Russia, USA, Sweden, Poland, and Brazil.

Regional statistics for June 2012 give the following data: in Europe 42.75% of the forests are FSC certified (461 certificates delivered), in North America 40.47% (204 certificates delivered), in South America and Caribbean countries 7.33% (231 certificates delivered), in Africa 4.58% (48 certificates delivered), in Asia 3.22% (145 certificates delivered), and in Oceania 1.65% (35 certificates delivered)[45]. As we see, the FSC certified territories are located mainly in specific parts of the world, and this has several causes. There must be some preliminary conditions in a country for FSC certification. The government should not be too corrupted, forest businesses should operate mainly legally, NGOs should be legitimate actors, and there should be environmentally and socially sensitive markets or companies which regularly export wood to such markets abroad. So, the process of certification in a specific country strongly depends on the structure of political opportunities, and, often, on governmental support for the certification process in the form of a governmental purchasing program.

FSC certification initially was created for the conservation of tropical forests in the Amazon River basin and in other problematic regions of the world. But, as it turned out, the forests of

[42] Interview with representative of National Initiative of FSC Ukraine, November 2008, Cape Town, SAR.

[43] Interview with representative of National Initiative FSC Canada, November 2008, Cape Town, SAR.

[44] Http://www.fsc.org/fileadmin/web-data/public/document_center/powerpoints_graphs/facts_figures/2011-06-15-Global-FSC-Certificates-EN.pdf.

[45] Http://www.fsc.org/facts-figures.19.htm/FSC-FIG-2012-06-15-Global_FSC_Certificates-EN.pdf.

Western Europe were the first certified, and only later did some of initially targeted territories and some countries, including Russia, join the process[46].

As for certification of the 'hot spots' the situation differs depending on the country and many other reasons. For example, Africa's Congo River basin is one of these hot spots. Many FSC members are skeptical about the possibility of certification in African countries, because the majority of these countries have very corrupted governments; additionally, there are a lot of wars and illegal logging in the region.[47] In conditions of such high institutional turbulence at the sites of implementation, which are characterized not only by reforming of forest governance but also by the abrupt restructuring of socio-political systems, national legislation does not work well, and the introduction of global institutes of sustainable forest utilization is very complicated. However, even in conditions of high institutional turbulence, 'little islands' of certification become possible with support from transnational agencies. In Latin America, FSC certification is a developing actively, but not in all countries. In Bolivia, for example, state authorities included in their national legislation requirements similar to those in the FSC, which makes the process of implementation of certification rules much easier for the companies. However, due to the weakness of implementation of national legislation in Bolivia, the quality of certificates requires permanent attention from the ASI.

FSC certification is influenced by the structure of ownership. As has been mentioned above, small forest owners have problems with FSC certification. That is why in countries as Finland, where the majority of the forest owners are small, they prefer PEFC certification. In addition, NGOs are less legitimate actors in Finland than in neighboring Sweden, for example.[48]

Thus, national, regional and local contexts have an impact on certification. It has to be noted, however, that the reverse influence also exists. So, despite situations that limit certification opportunities, such as corruption, certification may 'tighten up' the practices in the certified territories to an acceptable level, creating 'little islands' of sustainable forest management in countries where, in general, forestry is not sustainable. The FSC-GGN also can stimulate development of inter-sector dialogue, democracy and civil society in the countries which have no lasting democratic traditions. This happened, for example, in some regions of Russia and Belorussia.

3.4 Dynamics of GGN interaction in transnational spaces and the spaces of places

3.4.1 Certification GGNs and their interaction: drivers-brokers, competition and cooperation

In this section I will present a closer look at certification GGNs and their interaction, involving both market competition and cooperation. Although all certification GGNs are based on a multi-stakeholder process, some of the schemes are driven by NGOs, while others are driven by TNCs (Hallstrom & Bostrom 2010). Certification GGNs can compete for the sites of implementation,

[46] Interview with the director of FSC Russia, April 2007, Moscow.

[47] Presentation of Greenpeace representative at the FSC General Assembly, November 2008, Cape Town, SAR.

[48] Interview with the Finnish NGO representative, October, 2007, Helsinki.

resources, power and legitimacy, unite their efforts by building alliances and associations based on common principles or interests, and join other networks as members or participants. The section will focus mostly on the interactions of the FSC with other certification GGNs.

Promoters, drivers and brokers of certification GGNs

By supporting certification schemes, actors become promoters of particular certification GGNs. The more influential transnational actors, including NGOs, governments, retailers, and banks, support a certification system, the more it is legitimized in both transnational spaces and the spaces of places. As part of their corporate social responsibility policies, TNCs adopt forest certification schemes and implement the standards in the space of place, e.g. transfer forest certification GGN designs into concrete practices of forest management. Almost all certification schemes require compliance with the international conventions to which a country is a signatory. In such a way, TNCs, by adopting a certification schemes, help to implement the designs of many governance networks, such as CBD, CITES, ILO, and hence become their supporters and promoters. For example, Labor standards, approved by TNCs, always correspond with the ILO requirements and are compulsory for their entire business network operating in different places in the world. In the countries that have ratified the ILO convention, but have not managed to enforce its implementation in the spaces of places, it is TNCs who change practices in localities and transfer ILO labor standards into concrete practices.

TNCs can promote one or several certifications, even when these schemes compete with each other. Although TNCs always strive for standardization and it would be more reasonable for them to use only one certification system for all their subsidiaries, sometimes a TNC simply can not afford such a unified approach. The point is that choice of a certification system is, to the large extent, defined by the local context of the site of implementation. If a corporation works in countries in which the same certification system dominates and is the most legitimate, this TNC supports that system. Thus, the TNC Mondi Business Paper works with FSC certification, as their leases are in Russia and the South African Republic, where the FSC dominates. Although generally they support the FSC system, they are forced to work with suppliers from many countries, who sell wood certified by other systems as well.

Other TNCs support several certification systems at the same time. For example, Stora Enso, which works in more than 40 countries, is forced to work with several certification schemes. Even in the countries where major offices are located, the dominant certification schemes differ: the FSC dominates in Sweden, and the PEFC in Finland. The same situation exists with UPM-Kymmene. For many TNCs, competition between certification systems is a barrier to consistency and effectiveness of operations, and they try to diminish it. For the convenience of business they try to overcome 'certification wars' between these competing systems and to obtain recognition of both systems by buyers, as the company is forced to deal with both of them.[49] Stora Enso experts designed the models, where new methods of forest management were used and both the FSC and PEFC certification systems were tested. They made an effort to prove that both systems

[49] Interview with Stora-Enso manager, April 2008, Stockholm

work well for sustainable forest management.[50] The UPM Kymmene initiated models with double certifications and made an effort to achieve mutual recognition between the FSC and the PEFC.[51] Both companies simultaneously made an effort to foster the FSC in all the countries of their operations. As a result, in October 2010, the Finnish national FSC standard was approved and probably will foster FSC certification in Finland.[52]

The IKEA company followed the path of developing their own standards of business, which are applied at all their subsidiaries. Their logging subsidiary Svedwood, which operates in several European countries, is FSC certified. Working with suppliers, they prefer FSC certified production, but purchase other certified material if the FSC certified product is impossible to obtain. In advertising its production, IKEA does not promote FSC certification, but boosts its own brand and image.[53]

NGOs usually support only one certification system. For example, Greenpeace and the WWF support the FSC, and the World Conservation Society (WCS) supports the SFI system. The PEFC originally had no supporters among the environmental NGOs.

Figure 3.2 explains how buyers, operating mainly in transnational spaces, drive certification schemes that allow changing practices in the sites of implementation – in the spaces of places. The major drivers of forest certification are public procurement programs and private buyers, often big retail stores. Both public procurement programs and retail stores accept both FSC and PEFC certification; however the PEFC is accepted only from non-tropical forests, leaving the FSC alone as being legitimate all over the world. In the case of tropical timber certified by the PEFC, buyers assume that this can serve as a verifier of legality, but not as sustainable forest management. There are many broker and consulting organizations which strive for connecting responsible buyers and sellers of wood and wooden goods. Each of them occupies their own small niche in facilitating sustainable product trade. The WWF curates Global Forest and Trade Network (GFTN), through which producers and buyer's groups are linked into a network. Most buyers and sellers join the GFTN because they feel that the 'GFTN is an important part of their corporate social responsibility policy'.[54] Consulting organizations such as Forest Trends and Proforest try to transform the markets and make commerce easier for responsible sellers and buyers. They give consultations in various regions of the world to involve companies in the markets of certified goods. The Tropical Forest Trust does risk assessment of the supply chain for companies in Europe and helps to prepare suppliers in tropical countries become certified.[55] This significantly facilitates the building up of the sustainable supply chain.

The FSC and the PEFC-GGNs have data bases that can be used by sellers and buyers of certified goods to connect with each other. Additionally, 'virtual' promoters of certification programs conduct events in localities to connect responsible sellers and buyers such as trade fairs of certified production.

[50] Interview with Stora-Enso manager, April 2008, Stockholm

[51] Interview with the manager of UPM-Kymmene, January 2009, Helsinki

[52] FSC news, www.fsc.org, Accessed 30 October, 2010.

[53] Interview with an IKEA manager, June 2005, Moscow

[54] Interview with a representative of Travis Perkins, November 2009, London, UK.

[55] Interview with the Tropical Forest Trust representative, November 2008, Siktifkar, Russia.

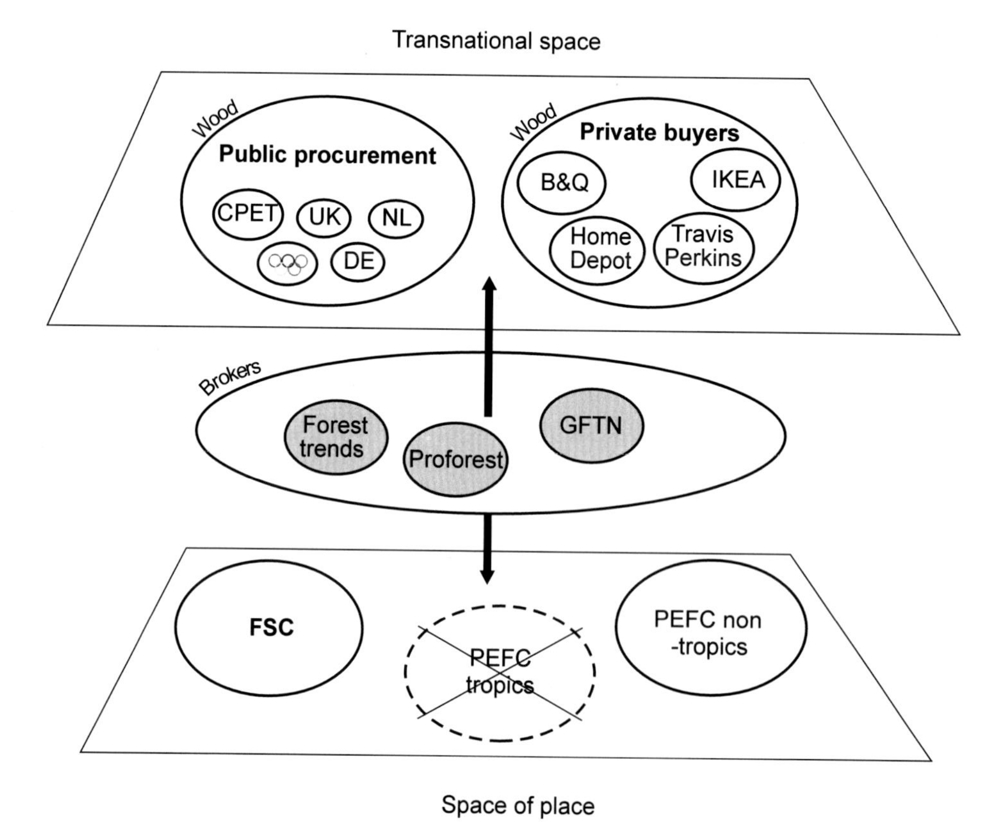

Figure 3.2. Connecting buyers and sellers: certification brokers and consultants. Symbols and arrows are explained in the text.

Competing relationships between forest certification schemes

The GGNs developing and promoting different certification schemes for the forest sector compete with each other. It can be explained by the fact that they are using the forces of the market and are, in fact, market players themselves, hence, they need buyers and sellers of their trademark, spheres of influence and other market attributes. The FSC was an absolute leader in the market, however, when the PEFC united under its own umbrella 31 endorsed national certification systems (PEFC, 2012), and together with its members it became a leader in the number of hectares of certified forest territories. In 2012, to FSC system 150 million hectares are certified (FSC, 2012a,b)[56], while to PEFC 240 million hectares, which constitutes one third of certified forests globally (PEFC, 2012).[57] The FSC logo is winning the market demand, but it is not supplying enough certified raw material to fulfill the demand. The PEFC is winning in the forest and it is supplying enough

[56] FSC involvement in Rio+20, Issue 4-20 June 2012, http://www.fsc.org/fsc-involvement-in-rio20.240.htm.
[57] Http://www.pefc.org/images/stories/documents/Brochures/PEFC_Profile_2012.pdf.

of certified raw material for the market.[58] The FSC is striving to certify more territories to out compete the PEFC.

In general, such competition between different certification systems stimulates their development, hence, it contributes to both strengthening the standards and enforcement of sustainable practices in sites of implementation (Overdevest, 2010). Certification GGNs and their supporters periodically contest the legitimacy of their competitors, which forces them to strengthen their own standards. The main tensions initially took place between the strict FSC system, which was initiated, supported and driven by NGOs, and other, business-driven systems, such as the PEFC and the Sustainable Forestry Initiative (SFI) (Cashore *et al.*, 2004).

Figure 3.3 explains the dynamics over time of the certification scheme and the roles of NGOs in strengthening the PEFC standards. In the diagram, the triangles symbolize differences in practices of forest management on the ground. Below, I explain why these practices are becoming more and more similar in both the FSC and the PEFC (triangle in the middle demonstrates the existing tendency). NGOs strongly support the FSC, while they criticize the PEFC for the weaknesses of their social and environmental standards. They have made efforts to exclude the schemes aligned

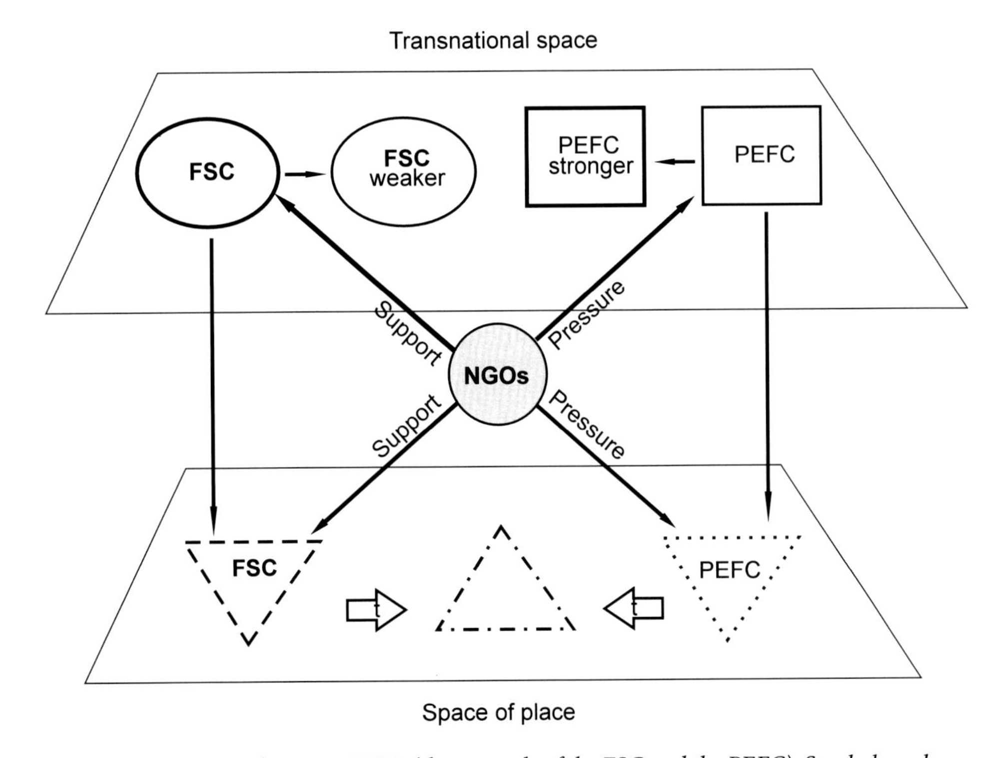

Figure 3.3. Competition between GGNs (the example of the FSC and the PEFC). Symbols and arrows are explained in the text.

[58] Interview with the Raw Material Manager at the Confederation of European Paper Industries, December 2010, Brussels, Belgium.

to the PEFC from responsible purchasing programs. In the USA, the Forest Action Network organized a marketing campaign entitled 'Do not buy SFI!'[59] NGO pressure stimulates stronger PEFC coalition member standards and their implementation. With its effort to accommodate more and more customers and compete with the PEFC, some of the FSC standards have become weaker. In addition, with rapid growth it was hard for the FSC to ensure quality of implementation on the ground. In the situation of permanent stakeholder pressures, certification systems have to improve their quality to be competitive at the markets. On the other hand, in order to win the markets, the FSC has to accommodate their standards to many institutional contexts and audiences, therefore, sometimes having to compromise and lowering the standards more than intended by the FSC's founders. The FSC scheme often does not suit small forest owners, who are often suppliers of the TNCs. Thus, the PEFC is more convenient for businesses, and suits small forest owners. The Timber Trade Federations support all certification schemes. The Timber Trade Federations lobby their governments to ensure that the PEFC, along with other certification schemes, is included in public procurement programs, including those for the Green Olympics.[60] The Confederation of European Paper Industries (CEPI) is not taking a stand on any of the certification schemes and supports all certification schemes.[61] The majority of distributors and merchants acknowledge that the FSC is a 'Golden standard', however, they support the acceptance of multiple certification systems in the markets, otherwise it is hard for them to provide their customers with the products they need.[62] For example, the Lathams family, which owns half of a distribution company, supports the FSC whenever possible. However, a family member serves on the board of the PEFC. He thinks that competition between the systems is good, as the PEFC becomes stronger 'trying to catch up with the FSC', while the FSC becomes more accommodating to businesses. For example, the FSC certified the UK's national forest management system, which is new for the FSC.[63] Previously, national schemes have only been certified by the PEFC.

Competition forces GGNs to change their policies and standards. Comparative analysis of the FSC and PEFC standards showed that over time they are becoming more and more similar (Overdervest 2010), however still the FSC standards remain stronger and more prescriptive (Auld, Gulbrandsen & McDermott, 2008b).

These changes directly influence the concrete practices at the sites of implementation. These practices in certified territories become more and more similar, although some differences, especially concerning social standards, which are stronger in FSC, continue to persist.

[59] Interview with an activist of the Forest Action Network, 2005, Washington DC, and interview with an activist, November 2008, Cape Town, SAR.

[60] Interview with the representative of the UK Timber Trade Federation, November 2009, London, UK.

[61] Interview with the Forest and Research director at Confederation of European Paper Industries, December 2010, Brussels, Belgium.

[62] Interview with a Travis Perkins environmental manager, November 2009, London, UK.

[63] Interview with Peter Lathams, November, 2009, London, UK.

Cooperative relationships between certification schemes

The GGNs developing certification schemes in different production areas do not compete, they cooperate with each other. They build alliances to solve their common problems. For example, the International Social and Environmental Accreditation Labeling Alliance (ISEAL), unites the certification GGNs, based on multi-stakeholder processes of standard development. As was described in the Chapter 2, the demand for standards and certification systems appeared, on one hand, because of the negative impact of industries on the environmental and social wellbeing of local people, and on the other hand, the appearance of ecologically and socially responsible buyers and the niche of 'sensitive' markets. In 1999, the certification GGNs united into the ISEAL alliance to share their experiences, develop common approaches to standardization, and contribute to the solving of common global trends (Bernstein, 2011). The ISEAL was founded by the FSC together with 8 other founding members, including the Fair-trade Labeling Organization (FLO), the International Federation of Organic Agriculture Movement (IFOAM), the International Organic Accreditation Service (IOAS), the Marine Stewardship Council (MSC), and the Rainforest Alliance.[64] In 2009, the sixteen GGNs developing certification systems were all members of ISEAL. Therefore, ISEAL became a kind of mega GGN, uniting certification schemes that are contributing to ecologically and socially 'sensitive' markets.

The ISEAL participates and negotiates in large forums, where the separate voices of its members would not otherwise be heard. The certification GGNs needed to have such an alliance, to put pressure on the world's trade policies for promoting 'sensitive' market niches. At the WTO meetings and forums, the ISEAL decides on issues on behalf of all its members. It also represents the interests of its members in the forums, organized by other GGNs, such as the Organization for Economic Cooperation and Development (OECD), particularly in its working group on social responsibility, or in the UN Committee on Trade and Development (UNCTAD). The Alliance also co-operates with other GGNs, which solve similar problems, such as those of sustainable development. For example, the ISEAL took part in the development of a document concerning the involvement of businesses in biodiversity conservation, organized by the Convention on Biodiversity (CBD).

Besides its representation activities, the ISEAL develops standardized approaches to standards' development. For this purpose, it issued the 'Code of good practice' for setting social and environmental standards in every area of production. The ISEAL's Code states that one of the main goals of creating certification systems is to create 'a new generation of profitable products and business processes underpinned by rules that support societies' broader social and environmental aims'.[65] ISEA also coordinates an Impact Research Network, where researchers from all over the world share their work on the impacts of private standards in both markets and the sites of implementation.[66]

The ISEAL alliance, by uniting different certification GGNs, helps to promote stronger sustainable practices simultaneously in several production areas, such as forests, fishing, fair trade, tourism and organic agriculture (Figure 3.4).

[64] http://www.isealalliance.org/content/about, Accessed 8 August 2011.

[65] http://www.isealalliance.org/code, reviewed May 2011.

[66] Observation in the Impact research network list serve, 2010-2011.

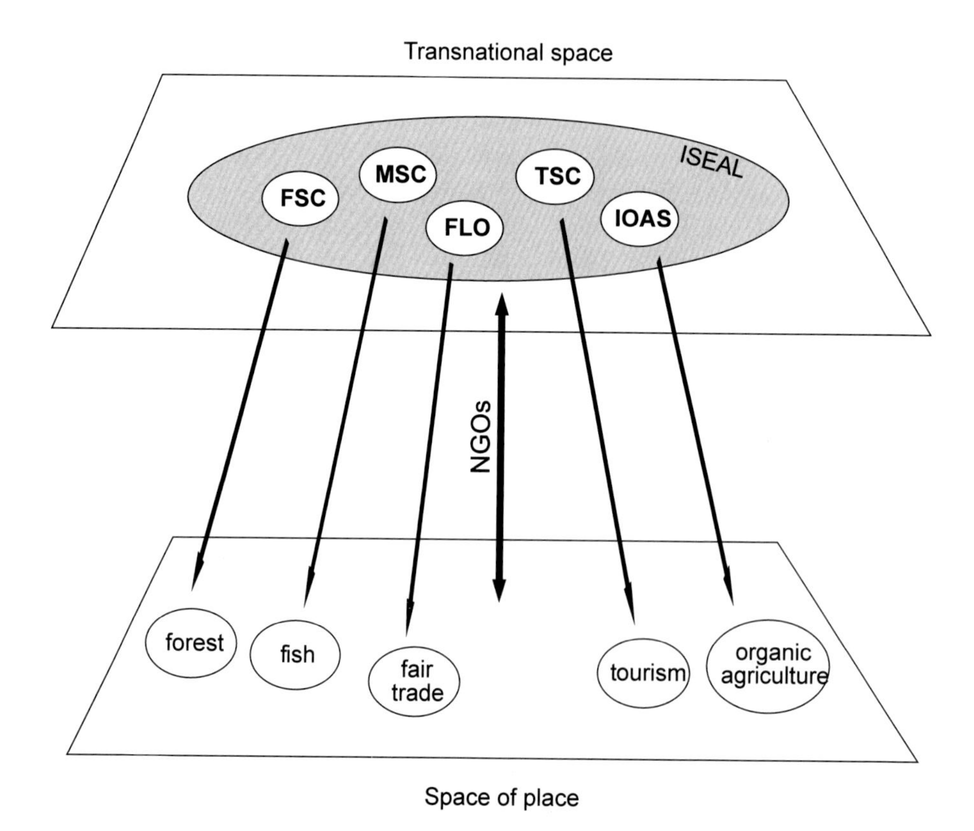

Figure 3.4. Cooperation between governance generating networks. Symbols and arrows are explained in the text.

3.4.2 Facilitating relationships between GGNs

In the organizational field of the forest sector there are many GGNs, which in a sense mirror each other in solving the same problems. In the case of one GGN's implementing a design created by another GGN, I consider such relationship as reinforcing and facilitating, although they may or may not have a common goal. This section will explain how GGNs are facilitating other networks simultaneously and their cooperation is being realized on various levels. Facilitation can be realized in transnational spaces, when GGNs contribute to the development of the global standards, or in the sites of implementation, contributing to the changing of local practices. On the one hand, such a situation creates some fragmentation of activities; on the other hand it provides a larger coverage. In all accounts, it turns out that many GGNs work on the same issues. That creates additional points of contact between the standards and policy designers, processes and implementation practices.

Figure 3.5 demonstrates that facilitating relationships occur between similar or different types of GGNs. The new practices on the ground are reinforced as several GGNs are contributing their agency to the site of implementation. For example, as will be explained below, NGO-driven GGNs, such as the WWF, the HCV-network, and the Rainforest Alliance can engage in familial

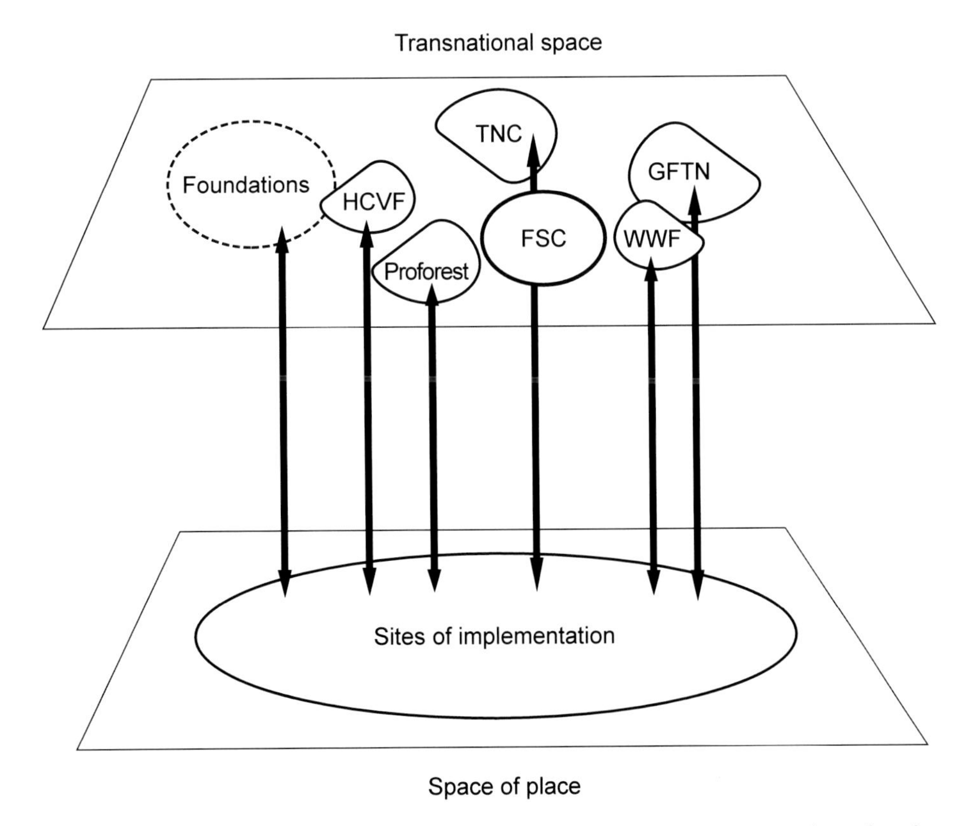

Figure 3.5. Facilitating relationships between GGNs. Symbols and arrows are explained in the text.

relationships with the FSC. Implementation of their designs, goals and interests facilitates implementation of the FSC-GGN. TNCs, when certifying their forest operations using the FSC scheme, promote the FSC-GGN by doing so, as they allow the network to expand to new sites of implementation and become more competitive on the market. Foundations that support sustainable forest management projects or the FSC institutional infrastructure on the ground also contribute to the development of the FSC-GGN.

Familial facilitators

Some of the GGN's facilitators, such as the WWF, GFTN, HCV-network and the Rainforest Alliance have familial relationships to the FSC. For example, the WWF historically was one of the founders and promoters of the FSC. As a GGN, the WWF designs preservation policies for eco-regions. The WWF node of design has a dispersed structure with multiple sub-nodes, geographically and functionally spread all over the world. Eco-regions, called 'Global 200', are the sites of implementation. The WWF designs are connected with the sites of implementation through their projects, programs and partnerships. All the WWF programs, including its Forest Program, are focused on these eco-regions. It has to be noted that the sites of implementation of forest certification do not coincide with the eco-regions, or only partly coincide, because they are determined by the companies owning or leasing territories. A certified company has to first check if their territories of operation are situated in the territories of the'Global 200'. In the event they find their territories located in the 'Global 200', they may designate them as High Conservation Value Forests (HCVF).

In many countries the FSC, in the early stages, operated under the WWF. The FSC offices gradually were separated from the WWF and became independent organizations in those places, where there were appropriate conditions for such a separation. I term WWF-FSC relationships as familial as they interact as 'relatives', where the WWF is a 'mother'-GGN, and the FSC is a 'daughter'-GGN. The WWF created the necessary expert community and institutional infrastructure for the FSC, such as certification centers, consultants, and the Global Forest and Trade Network (GFTN), which from its start operates under the WWF. The WWF and the FSC continue having close interlinks, including the sharing of individuals. For example, the coordinator of certification policy in the WWF International was elected in 2008 and in 2011 reelected to the FSC Board of Directors. The WWF supports many of its staff allowing them to participate at the FSC General Assemblies. This allows the WWF to be more influential in elaborating FSC policies.[67]

As I mentioned before, the WWF created a sub-network GFTN, aimed at connecting sustainable forest producers and consumers. The GFTN, operating under the WWF, become another facilitator for the FSC, by contributing to building its markets in transnational spaces. Both the FSC and the GFTN, being facilitated by the WWF, in turn contribute to its mission: nature conservation, through sustainable forest utilization with the use of market mechanisms.

Geographic spreading of the WWF offices and programs and their diverse structure helps both to connect the WWF node with its sites of implementation, and to play the role of facilitator for the FSC. The WWF International can, by itself, coordinate the programs or operate as a donor in some partnerships, or act through its national offices and programs, or through the offices of specific programs.[68] Many of the WWF offices are strategically located in various eco-regions where there is commercial forest utilization and the need of conservation of HCVF, rare and endangered species.

The WWF-USA sub node co-ordinates projects in 19 eco-regions. They are involved in Latin American and Asian conservation initiatives, as it is convenient from a geographic point of view. The WWF sub-nodes co-operate with the local FSC infrastructure, helping them to create systems of monitoring and control over the forests, and to involve new regions into the certification processes. The forest program of the WWF-USA deals with the forests located in the Congo and Amazon river basins, with several forest territories in Laos, Burma (Mekong river basin), Borneo, Sumatra, Chile, China and the Russian Far East (Amur river basin). In many regions the WWF co-operates with its subsidiary, GFTN. They monitor which key US companies buy wood from these regions. They help those companies develop programs of responsible purchasing and convince them to certify the chain of custody. The WWF contributes to GFTN development in

[67] Participant observation at the FSC General Assemblies, 2008, 2011.

[68] Interview with the WWF-US officer, November 2008, Cape Town, SAR.

these regions. The WWF-USA, lobbied for amendments to the Lacy Act, which helps responsible companies and punishes irresponsible ones if they purchase wood with unknown origins.[69]

Another example of a familial relationship to the FSC is the HCV network. The concept of the HCVF originally came from the FSC Principle 9, which requires designation and monitoring of forests of critical value for people and the biosphere. This value can be socio-economic, or environmental, and can be related to different scales, from small eco-systems to landscapes. The ex-FSC auditors, who decided to organize an international resource and consulting center, made it easier for the FSC to implement its standards on the ground by setting up a private company, Proforest. The HCV network was formed under Proforest and since 2005 has become a relatively independent GGN, although the connection with the 'mother' organization Proforest, as well as the FSC, remained strong.

The operation of the HCV network differs from the three-chambered FSC system, although environmental NGOs, businesses and, in a lesser degree, social NGOs are also represented here. Business is represented by such large actors as the World Bank, Mondi Business Paper, Ikea, Tetrapack, and Stora Enso. Most of the companies are FSC members and choose the FSC to certify their forest management wherever possible.

The main global design of the HCV-GGN is its Global Toolkit on HCVF designation. If compared with the FSC standards, the HVC Toolkit is more detailed and presumes more freedom in interpretations than does the FSC standard. In the majority of countries, the Toolkit was adapted to the national context by the WWF experts. Many WWF staff members are on the technical panel of the HCV network, so the relationships between the WWF and the HCV network are cooperative and facilitating. The HCV Toolkit and its national interpretations represent the major guiding document for companies certified by the FSC scheme, their experts and auditors. Therefore, Principle 9 of the FSC standard becomes 'a common territory' of the FSC and the HCV network. However, the sites of implementation of the HVC network differ from the sites of implementation of the FSC. Compared to the FSC, the HVC sites of implementation are not limited by ownership or leases of the certified companies; they are wider as the HCV network is striving to spread the process of designation of HCV to deserts, steppes, and other ecosystems. Definitions and approaches developed by the HCV-network of designation of HCV are used by other certification schemes, such as the Roundtable on Sustainable Palm Oil (RSPO).

Both GGNs, however, strive for consistency in using the HCF concept and cooperate in development of new designs. The FSC revised its Principles and Criteria in 2009-2011 and the HVC network kept playing the role of expert. Experts on the technical panel commented on drafts, and participated in a joint FSC-HCV workshop in January 2011 to ensure consistency of the new P@C interpretations of Principle 9 and HCV network approaches.[70] The HCV network organized a side event at the FSC General Assembly 2011, 'Principle 9 revision and HCV in forests and other ecosystems', where issues related to the allocation and monitoring of the HCV were discussed with wider public. On the side event, the FSC and HCV network signed a memorandum of

[69] Interview with coordinator of the WWF-USA forest program, November 2008, Cape Town, SAR.

[70] Participant observation at the FSC-HCV-network meeting, January 24-26, Bonn, Germany.

understanding on issues related to HCV.[71] Representatives of both GGNs are jointly producing the guidance document on HCV with input of a wide range of stakeholders. The renewal of Principle 9 by the FSC will lead to the renewal of the global Toolkit by HCV network.

The Rainforest Alliance GGN is one of the founders of the FSC. It combines the role of a FSC supporter with its own standard making as well as auditing and facilitating the implementation of the FSC standards on the ground. The Rainforest Alliance was created in 1989 to promote sustainable forest management and forest certification all over the world. There are several programs in the Alliance which either support FSC certification by developing additional step-wise programs, or create their own certification standards in the areas adjacent to the forest areas (mainly for agricultural lands). However, it has to be noted that although while promoting the FSC, the Rainforest Alliance is primarily devoted to their own agenda in implementing sustainability programs at the sites of implementation. Should the FSC's performance in the world go badly, the Rainforest Alliance is ready to create alternatives to the FSC.[72]

SmartWood is one of the programs of the Rainforest Alliance and represents a certification body, accredited by ASI. Subsidiaries of SmartWood work in different countries: for example, Imaflora in Brazil, and NEPCon in Russia. As an FSC supporter, the Rainforest Alliance is filling the gaps and developing standards for the companies that make a commitment to become sustainable, but cannot yet comply with the FSC standards. The SmartStep is a kind of interim certificate, which prepares the companies for FSC certification and gives them rights to join 'sensitive' market niches. Additionally, SmartWood has the SmartLogging certification scheme for small forest owners. These certification schemes are popular in tropical countries, but they are not called for in Europe, because all the companies there are able to get the FSC certificate.

Another forest program of the Rainforest Alliance is the TREES program, which is a kind of analogue to the GFTN created by the WWF. In the context of TREES, there is a SmartSource program, which works with wood processors, retailers, and consumers of office paper and packages. Its goal is to improve their environmental purchasing policies. Besides these activities, the TREES program organizes training for local forest owners, indigenous people, and other stakeholders, to help them enter the market for certified goods. Designs and sites of implementation of the Rainforest Alliance are wider than those of the FSC. Besides forests, the Rainforest Alliance certifies chocolate, agricultural goods such as coffee, tea, coconuts, flowers, fruit, as well as tourist businesses.

Foundations as GGN facilitators

Some GGNs serve as 'catalysts' or 'promoters' for other networks. Charitable foundations represent the type of GGNs that often enable NGO networks to implement projects on the ground. As do other GGNs, they have nodes of design, forums of negotiations and sites of implementation, although they have some other distinctive features. The foundation itself represents the node possessing resources. Their designs are developed in a form of a general strategy, which is taken

[71] Participant observation at the HCV-network side event at the FSC General Assembly 2011, June 28, 2011, Kota Kinabaly, Malasia.

[72] Informal talk with the Rainforest Alliance staff, June 2008, Richmond, Vermont, USA.

to the sites of implementation by the grant receivers. Giving resources to the grant receiver means sponsoring the co-designer for the project that will be delivered to the site of implementation, where it is transformed into specific place-based practices. Donors and beneficiaries act in unique partnerships, influencing each other. Often the projects funded are supposed to develop and implement models with best practices that later can be reproduced in other sites of implementation. Preliminary design of the project takes place in the form of a grant application, in which the applicant proposes its project and explains how the project will contribute to social change at the site of implementation.

Both private foundations and international development agencies often prefer channeling resources to the sites of implementation through NGOs, rather than governments, especially in developing countries with high levels of corruption, as they perceive the NGOs as more creative, flexible and legitimate actors. When the NGOs receive funding for project implementation, they become part of a GGN, which develops and delivers the design from transnational space down to the sites of implementation. Foundations make an effort to fund projects that would be sustainable in the long run and would be able to continue operating after financing stops. In this respect, agents of social change that has the future potential of using market forces to ensure sustainability has a high potential value for foundations. Therefore, the FSC has been supported by many granting agencies.

For NGOs involved in the FSC, the grant money was essential in the early stages of its development when additional resources were necessary, as the democratic process is lengthy, standards have to be tested on the ground, and market mechanisms start working only when a critical mass of responsible producers and consumers is achieved (Bartley, 2007a). The FSC certification scheme is based on market mechanisms. But in the early stages of certification, until environmentally sensitive markets were formed, the foundations took the main role in promoting certification: accordingly, they played a role as facilitators of the FSC-GGN.

In 1992, the MacArthur Foundation opened a series of grants to the FSC and other organizations related to forest certification. The Ford Foundation along with other foundations, which later united their efforts in the newly created 'Network of Charity Foundations for Sustainable Forest Management', followed the MacArthur's lead (Bartley, 2007a). Sustainable Forestry Founders Network is aimed for the creation of a private sector alternative of forest management, with the participation of NGOs and business. Financing the first attempts of forest certification, the foundations indirectly participated in the creation of 'sensitive' markets and created the infrastructure of a new organizational field. The MacArthur and Ford foundations, along with the Rockefeller Brothers Fund, all have shown special interest in this process. Between 1993 and 2001, they spent about 40 million US$ on the projects related to forest certification. The grants were given to the FSC and its National initiatives, for the training of auditors and for testing audits (Bartley, 2007a).

Usually, foundations finance specific issues within a defined time period, until that time when the topic loses its urgency, or more acute topics appear to change the donors' priorities and strategies. Thus, at a certain point, climate change problems replaced certification issues.

When the FSC-GGN became self sufficient, the foundations continued financing only innovative programs, which were not yet self-supporting, for example, the activities of the carbon standard development group. The problem of tropical forest management is still important, and it is related, in a sense, to the donor's new priority of climate change issues, which is why donors are still interested in solving tropical forest's problems. For example, in the African region the market mechanisms have not yet started working in the FSC's model, and external financing is still required. The Global Environmental Facility (GEF) grant enabled extending the FSC program to tropical forests of Congo River basin.

Facilitating relationships between state and non-state driven GGNs

Governmental agreements, ministerial processes, European and national legislation all interlace and approach the same issues from different sides (McDermott *et al.*, 2011b; Sabel & Zeitlin, 2012). In this way they facilitate the implementation of NGO-driven GGNs and *vice versa*. This facilitation can take place both through the purposeful cooperation of the founding networks, or without 'conscious' interaction between the networks, just because the designs of both are directed on solving the same problem on the ground.

This can be illustrated by the global effort of combating illegal logging. Figure 3.6 illustrates that several regulatory instruments have an impact on the logging sites and facilitate institutionalization of the designs of each other; all of them facilitate the implementation of a forest certification scheme. The Forest Law Enforcement and Governance (FLEG) ministerial process focuses, first

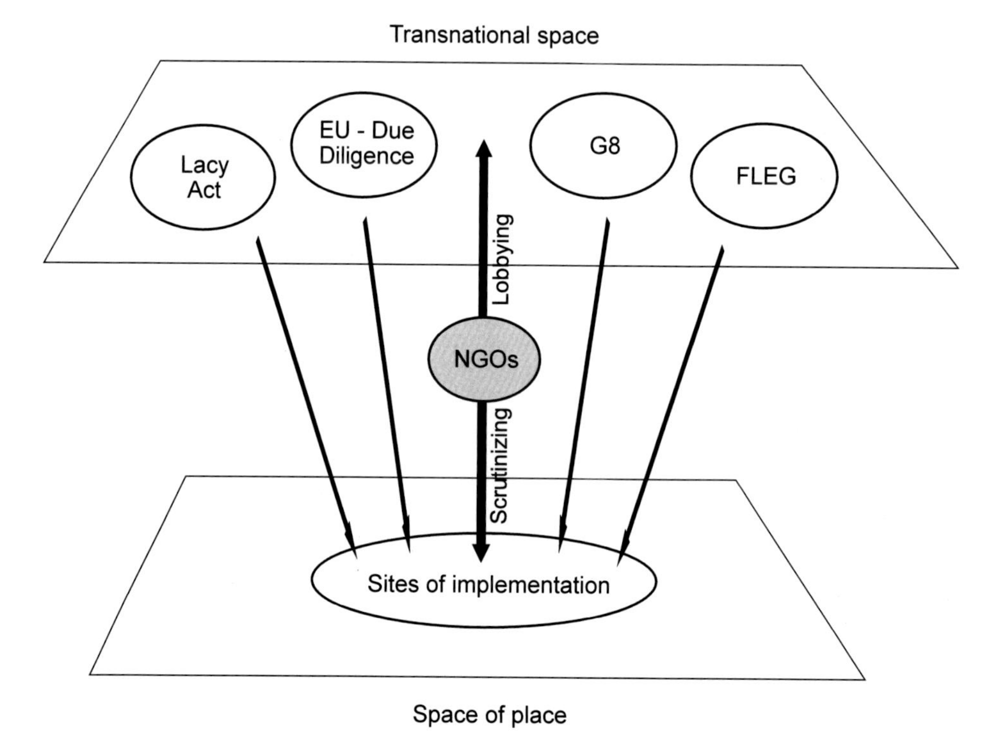

Figure 3.6. Facilitating FSC implementation: excluding trade in illegal wood. Symbols and arrows are explained in the text.

of all, on the prevention of illegal logging by the exclusion of illegal wood and the products made from it from trade. The FLEG process was initiated by such organizations as the World Bank and the US Department of State, and is widely supported by NGOs. FLEG consists of many regional processes where the problems of illegal logging are acute. During FLEG conferences, the ministers develop a Declaration to be the base of specific actions on a national level in countries that are participants in the process. The Declaration is non-binding, but it is supposed that each country which has voluntarily joined the process will try to put it into force in their localities.

The Forest Law Enforcement Governance and Trade (FLEG-T) agreement represents the action plan of the EU countries, which was adopted in 2003. The FLEG-T process concerns several regions, in particular Central Africa, Russia, Tropical South America and Southeast Asia. FLEG-T differs from other the FLEG processes, as it does not struggle against illegal logging in its own region, but instead contributes to the prevention of such practices in other countries, from which the wood is imported into the EU. The action plan involves bilateral Voluntary Partnership Agreements (VPA), which allow control of the legality of the supply chain coming into the EU and the introduction of a license scheme to verify legality. When VPA are signed they become binding for the signatories and the countries involved are obliged to introduce traceability systems into their wood trade. In VPAs the whole country has to be involved in the process. Civil society, industry and governments get involved in forums of negotiations on the definition of legality, and on the process of creating a legality assurance system. Companies that comply with the legality assurance system receive the FLEG-T export license; compliance is then verified by an independent auditor.[73] The VPAs only partly solves the issue of illegal wood in EU trade, as not all developing countries sign them and because the problems of illegal logging are acute not only outside the EU, but in such countries as certain member states, such as Hungary, or countries that are planning to become EU members, such as the Balkans.[74]

The requirements to prove the legal origin of any imported wood boosts certification of the chain of custody, because it is the best way to prove legality.[75] Cases where FLEG is implemented successfully also facilitate certification of forest management, as it is easier to certify the territories where problems with illegal logging are resolved (Overdevest & Zeitlin, 2012). Hence, legislation and other national initiatives undertaken in the frame of the FLEG and beyond stimulate and facilitate forest certification, e.g. foster development of the FSC-GGN, the PEFC-GGN and other voluntary standard-developing GGNs.

Because the FLEG-T was not effective enough, the European Union introduced legislation in which importers have to prove the legality of the origin of wood when buying and selling timber products in the EU. The 'Illegal Timber Law', or in other words, the 'due diligence' legislation, was voted by the European Council in July 2010, adopted in October 2010, and will become fully effective across Europe by the end of 2012. The due-diligence legislation was lobbied for several years by Greenpeace, WWF, Friends of the Earth, Global Witness and other NGOs. The legislation requires the first operator of the supply chain in the EU to act with due diligence and

[73] Interview with the FERN staff responsible for fostering VPAs, December 2010, Brussels.

[74] Interview with the Greenpeace policy officer in Brussels, December, 2010, Brussels.

[75] Report of the director of the FSC Russia on the Meeting of Rosleskhoz, June 2009, Moscow.

introduce honest and responsible policies of verification of legality in its trade.[76] The due-diligence legislation is supplementary to VPAs and both mechanisms reinforce each other. Moreover, the due-diligence legislation is supposed to encourage countries into signing VPA agreements as VPA export licenses are an appropriate mechanism for proving legality.

Similar legislation, the Lacy Act, was already adopted in the USA. Until May 2008, the Lacy Act had forbidden importation of plants listed in the CITES Convention. In May 2008, an amendment extended this prohibition to all timber products, including furniture, cellulose, paper, and other wood materials that were produced with illegal wood or wood of unknown origin (Ptichnikov, 2009). If illegal wood is imported into the USA, criminal prosecution can be instituted against the purchaser. This means that goods produced in China, from wood coming from Russia or any other country, and imported into the USA must have all necessary certificates of legality for the whole supply chain;[77] without the certificates the purchaser becomes subject to criminal liability. As a result, the US national legislation regulates not only their own countries' processes, but also influences the practices in the spaces of places in other countries. The environmental NGO networks and the FSC offices place their great hopes on both the Lacy Act and EU due-diligence legislation's effectiveness.[78] Both European and US legislation combating illegal logging creates a favorable context for forest certification GGNs in the regions in which the problem of illegal logging persists.

3.5 NGO-driven global surveillance: watching the operations of GGNs

There are no strong international mechanisms for monitoring the 'quality' of implementation of the international conventions that countries have ratified. Therefore, on the global scale a surveillance system in the organizational field of the forest sector is maintained predominantly by NGO networks. Thus, for example, the Global Forest Coalition, which unites NGOs and indigenous people's organizations, monitors the effectiveness of international policy and legislation. For example, they track the implementation of governmental responsibilities stipulated by the CBD. They also support and boost sustainable forest utilization, but do so in a more radical way than the certification GGNs. They comprise more radical organizations that conflict with the FSC on issues concerning tree plantations, which in their opinion have to be judged as agriculture and can not be certified as forests.[79]

The TNC's activities on the ground are monitored more strongly than the implementation of international conventions. There are many NGOs which track the companies' activities, and in case of irresponsibility in the sites of implementation, call into play the threats of boycott and mobilize other NGOs and consumers of the products. The banks investing in such irresponsible companies are also under scrutiny of these NGO networks. A whole range of NGOs are interested

[76] Http://www.illegal-logging.info/item_single.php_news; interview with Greenpeace representative in Brussels, December 2010.

[77] Report of director of FSC Russia at the Meeting in Rosleskhoz, Moscow, June 2009, http://www.fsc.ru.

[78] Interview with the head of the SmartWood Program (Rainforest Alliance), June, 2009, Vermont, USA.

[79] Interview with a representative of the NGO World Forest Coalition, and an auditor in NEPCon, June 2008, Richmond, Vermont, USA.

in the state of forests. Thus, the NGO Taiga Rescue Network monitors boreal forest management in Europe, and the Global Forest Watch, created by the World Resource Institute, monitors all kinds of forests all over the world, and tracks the legality of timber harvesting.

Forest certification GGNs also have their own surveillance systems. For example, the PEFC-Watch is taking a close look at the PEFC in Finland. The PEFC-Watch was organized by Finnish environmental NGOs, in particular Greenpeace Finland, Finnish Nature League and others, when the PEFC certified companies started logging old-growth forests in Lapland in 2008-2009. The PEFC-Watch raised this issue and made it public.

Within the FSC, surveillance is achieved mainly by certification bodies that in FSC system are verifying companies' compliance with FSC P@Cs. The ASI in turn monitors the quality of a certification body's work on the ground. The ASI and the certification bodies represent surveillance introduced by the FSC-GGN. In addition, the FSC's own member organizations, such as Greenpeace and FERN (FERN, 2008),[80] monitor the FSC's performance (Ozinga, 2005).

However, besides its 'in-house' observers, the FSC has external surveillance agents. For example, the Swedish Society for Nature Protection resigned its membership in the FSC and is criticizing the GGN for certifying forest management in primary forests.[81] The FSC membership believes that the exclusion of primary forests from the FSC certification processes would foster the conversion of forest land into agriculture in Latin America and will be impossible to implement in Russia and Canada, where primary forests occupy significant territories.[82] The Ecological Internet criticizes the FSC for the same issue.[83] Another virtual surveillance is realized through the website FSC-Watch. This website was created by one of the FSC's founders, a former FSC member who resigned his membership. The founder of the website is part of the NGO Rainforest Foundation, but he operates '*pro se*', as an individual, in order not to jeopardize its organization. Explaining the reason for creating such a website, its founder said that in his opinion the FSC's explanation of their activity did not show the whole picture: they are focused only on progress and achievements, avoiding any self-criticism.[84] On the website, the weak FSC certificates given in Thailand, Laos, Brazil, USA, New Zealand, SAR, Uganda and other countries are analyzed. The site basically co-operates with local organizations and individuals which criticize the FSC. For instance, together with the NGO Friends of the Earth UK, which is not a FSC member, the FSC-Watch conducted research in 2002 involving local NGOs and forest experts and prepared a report available to public (Counsell & Loraas, 2002). The administration of the website does not check the validity of the NGO's claims and complaints, and this has had negative results. One of these is that the website attracts radical NGOs who often pursue their goals with destructive criticism. Thus, the website published serious charges stating that the FSC was involved in green washing in a case of

---

[80] FERN, Greenpeace, Inter-African Forest Industry Association, Precious Woods & Tropical Forest Trust (2008). Regaining Credibility and Rebuilding Support: Changes in the FSC Needs to Make to Ensure It Regains and Maintains Its Credibility. A Joint Statement. October 30 2008. http://www.fern.org/node/4297.

[81] Primary forest are those that were not logged before, the FSC protects intact forest landscapes as HCVF, but not all primary forests.

[82] Participant observation at the FSC-Russia Board meeting, October 28, 2010, Zvenigorod.

[83] Http://www.forests.org/shared/alerts/sendsm.aspx?id=fsc_logging.

[84] Interview with the FSC Watch founder, November 2008, London.

forest plantation certification in Brazil, which turned out not to be true. Absence of constructive analytical criticism on the FSC-Watch website makes it difficult to achieve the goal stated by its creators: improvement of the work of the FSC certification system. The founder of the website personally has major disagreements with the FSC's policy which touch upon general principles, such as the tripartite chamber structure. These claims exclude the possibility of his work within the GGN. For example, he thinks that the economic chamber should not have equal rights with the environmental and social chambers. The FSC-Watch leader criticizes the FSC for paying more attention to creating a political space for dialogue with stakeholders than to the implementation of new practices on the ground. As a result, the forests's condition does not improve.[85] In response to certain points of criticism, the FSC members send motions to the General Assembly requesting changes, while ignoring the most radical critiques.

3.6 Conclusions

All state and private-driven GGNs link transnational spaces with the spaces of places and all are part and parcel of the forest sector institutional field, which unites actors operating in transnational spaces and spaces of places. I examined the organizational environment of the GGNs, which extends from transnational space to the space of place. Core attention was paid to the interplay of the mutually dependent networks in the forest sector, representing an institutional field.

The intercommunications within the organizational fields in transnational space have many common characteristics with the organizational fields existing in the spaces of place, although they have considerable differences, as well. In transnational spaces, the role and proportion of the non-state actors is larger than in the spaces of place, where the states' governance still prevails. There is no obvious single transnational regulating agent, which would apparently regulate the GGNs, as exists in the space of place within the nation states. There is no 'global government' and no laws that would be compulsory all over the world. However, there are mega-GGNs with mega networks, such as the Rio de Janeiro process, uniting, coordinating and to a certain extent regulating other GGNs. Despite the fact that in transnational space we see a lot of nodes of design belonging to government-driven GGNs, the NGO-driven and the TNC-driven GGNs are not subordinated to government-driven GGNs. The chapter demonstrated significant interdependence and interlacing between different GGNs, irrespective of their governmental or private origin.

The aim of the GGNs is to create the design that would solve global problems, which hardly can be addressed on national levels, and to ensure that the problems of one country are not solved at the expense of another. Governments strive for developing strategies and policies for their own countries that foster national interests. They pay less attention to sustainability and global risks. When standards, developed by the GGN's node of design, reach the sites of implementation, freedom from governmental regulations, which is part and parcel of operation in transnational spaces, disappears, and any implementation of new rules is influenced by national, regional and local legislation. This means that at the space of place governments of all levels inevitably appear as actors and stakeholders. Therefore, GGN actors operating in transnational spaces are

[85] Interview with the founder of FSC Watch, November 2008, London.

less dependent upon nation states than are actors implementing the standards at the sites of implementation.

Interactions between GGNs, as was demonstrated, have a different character. They engage in complicated relationships and interplay with each other. The certification GGNs can be competitors and/or cooperate by forming alliances for promoting joint policies in the institutional field of forest governance. Similar or different types of GGNs can enter into facilitating relationships with each other, can give birth to other GGNs and become members of bigger alliances. I examined GGNs that gave rise to the FSC, as well as those created by the FSC. Competition may exist in apparent or implicit forms. In implicit form, it exists for drawing on resources, for ensuring legitimacy in a stakeholder community, and for attention to their designs in the institutional field. Apparent competition is taking place, for example, between several certification systems in a forest sector. It does not differ much from the competition of goods, services or brands in any industry or market, except for the fact that the competition is for business's engagement in a scheme and for the 'license' to operate in the sites of implementation. Therefore, the competition in transnational spaces is in fact for power and legitimacy to operate in the sites of implementation.

Cooperation between GGNs is very frequent, for example when design of one GGN contributes to the promotion of another's design, or when various GGNs solve the same global problem. Cooperation can be between different types of GGNs, such as NGO-driven, Business-driven or Governmental-driven, as well as within these types. Within cooperating relationships it is usually possible to determine the primary GGN; therefore, the other GGNs play a role of facilitating agents.

The structure of surveillance systems, existing in transnational spaces for the monitoring and controlling of the GGNs, differ from the surveillance systems in the spaces of places. The role of a surveillance agency in transnational space is taken up by the members of the GGN, operating from within the network, as well as by external actors. As a rule, surveillance agents are represented by private actors, mostly NGOs, which are often specific for each of the GGNs.

As was described in this chapter, the same spaces of places can be sites of implementation by many different GGNs. They can coincide within political borders, but more often they do not. The place can be 'a hot spot' that attracts attention, resources and efforts of many GGNs. In such places one can see a high concentration of actors of transnational spaces. There can also be places with very few or even no designs implemented.

Chapter 4. Fostering FSC forest certification in Russia: interplay of state and non-state actors

4.1 Introduction

Since the early 1990s, the Russian forest sector has been undergoing profound change determined both by national reforms and patterns of internationalization. Although the newly emerged market economy in Russia has brought danger to Russian forests, the cross-border influence of market forces has also encouraged the introduction of responsible forestry practices into Russia. The environmental movement of the West has begun to infiltrate Russia, cooperating with national initiatives and greatly affecting the protection of nature in the country. With the NGO's efforts, FSC certification has entered into Russian regulatory processes.

Although in Russia non-governmental actors are engaged in international networks and operate independently, they have to take into account governmental policies because Russia is a state power country and all land, including forests, is federal property. All certification initiatives must to a certain extent involve the Russian government as a landowner and stakeholder. This chapter shows how the NGOs have engaged the Russian government, as well as industry and the public, in FSC certification. It also illustrates the barriers they face in persuading stakeholders in the forest and other sectors of Russian society of the desirability of certification and how they have overcome those barriers.

The FSC appears to represent a way of bringing the Russian forest industry into European markets and simultaneously of bringing the global practices of sustainable forest management into Russia. In general, if well implemented, certification seeks to increase forest profit, and to improve management and control functions. Certification is a mechanism for developing relevant trade policies, supporting environmentally responsible business, and instituting investment safeguards.

The chapter is based on field research conducted on the federal level in 2004-2012, involving interviews (#49) and participant observations. It starts with the characteristics of forestry and forest management in Russia and highlights the role of the state authority in governance (Section 4.2). The following section describes the emergence of non-state actors, both business and NGOs, in governance in the context of economic transition and institutional turbulence (Section 4.3). Next, the chapter focuses on forest certification initiatives existing in Russia (Section 4.4) with major focus on the FSC, its emergence, institutional design, operation of the National FSC office, and peculiarities of development of the national FSC standards. The current state of FSC certification with its regional distribution is described (Section 4.5). The chapter then brings light onto the interplay of state and non-state actors in forest governance in Russia on the national level. It proceeds with an analysis of how FSC global rules coincide with Russian national legislation and on how state and non-state actors interact in the FSC process (Section 4.6). Attention is given to the description of major drivers of the FSC on the internal market (Section 4.7). The chapter concludes with the FSC potential for improving forest management in Russia (Section 4.8).

4.2 Forestry and forest management in Russia

4.2.1 Forestry in Russia

Forests cover 18 billion hectares of the Russian Federation, which amounts to 69% of the entire territory of the country and for 21% of the world's forests. 1.14 billion hectares are designated for logging. There are 204 specially protected forests at the federal level on 58 million hectares and thousands at the regional level.[86] 26% of the world's virgin forests, covering 288 million hectares are situated in Russia (FAO, 2011). Most of them, 153.9 million hectares are in Siberia, 31.8 million hectares are in the European part of the country. The average usage of annual allowable cut is around 30% as only one third of the forests are economically accessible. Considerable amount of forests, especially in Siberia, remain undeveloped. As a consequence many forest landscapes remain intact, while accessible forests are over harvested (Efimova *et al.*, 2010: 5). It is hard for stakeholders to find information about forests and forestry in Russia, State Agencies have the Internet sites, however the information is lacking or poorly organized (Grigoriev, Pakhorukova, Shmatkov & Ryeïting, 2012).

Governments are concerned about 'underdevelopment' of existing forests, while NGO representatives are concerned about HCVF conservation and argue that the amount of forests harvested should not be a measure of effective forestry, as logging of certain plots can not be done in a responsible way and it would be better to keep them untouched.[87]

Russia is the leader in the world for its wood stock, which accounts for 82 billion cubic meters of wood. In 2010, the country became fourth in the world in wood production after the USA, Brazil and Canada. However, Russian exports of pulp and paper products accounted for only 2.3% of the world's exports. In 2010, the annual allowable cut was 633 million cubic meters, while only 176 million cubic meters were harvested.[88]

Approximately half of the wood is processed within the country, 7-8% goes for firewood and building houses in rural settlements, and one third of predominantly round wood is exported (Efimova *et al.*, 2010: 6).

Until 2007, Russia exported more unprocessed round wood than any other country. The Russian government tried to cope with the issue by introducing export duties on round wood. Since 2006, the state authorities started to gradually increase the export taxes on round wood in order to foster development of the processing industry in Russia. In 2009, it was planned that export taxes would grow to 80% of value (up to 50 Euros per m^3), which would make round wood

[86] http: priroda.ru/regions/forests, Accessed 6 April, 2011, Moscow.

[87] Interview with the WWF representative, April 2011, Moscow.

[88] Presentation of the head of Rosleshoz V.N. Masliakov, 'Major outcomes of forestry of Russian Federation in 2010 and goals for 2011' 21 March 2011, St. Petersburg, http://www.rosleshoz.gov.ru/media/appearance/57.

exports unprofitable. However, because of the economic crisis in 2008-2009, the tax increase was suspended until the end of 2011 and remains at 30 Euros per cubic meter. The Russian government will be obliged to reduce the export tax when Russia joins the WTO, therefore the reform has achieved less than was expected.

Despite the efforts of the government, timber processing in Russia remains poorly developed: 31.7% of income from forest resources comes from round wood exports and 25% from sawn timber (Efimova *et al.*, 2010: 6). The main income comes from pulp and paper production. However, there is a trend of decreasing income from round wood.[89]

Although the reform was not fully implemented, it entailed large-scale reorganization of a forest industry formerly focused on the export of round wood, and resulted in the closing of logging enterprises located in Russian borderlands and the closing of large pulp and paper mills in Finland. It created incentives for reorientation of both domestic and foreign partners towards investing in integrated wood processing within the country.

The lack of effective state forest policy and the permanent restructuring of the forest management system are the primary barriers to sustainable forest management in Russia (Lehmbruch, 2012; Shmatkov & Kulikova, 2012). NGOs and companies constantly criticize forest legislation, as well as federal and regional policies, as being inadequate (Tysiachniouk, 2006a,b, 2010a,c,e,g). Many high conservation value forests (HCVF) are in danger of being heavily logged – areas near roads and transportation arteries are being degraded (Yaroshenko *et al.*, 2001).

Lingering effects of the Soviet past are still affecting state forest policy. The state is not taking enough measures to increase the resource potential of the forests, and short term planning dominates in the state's decision making.[90] The key problem is that forest policy is built upon an economic foundation left over from a planned socialist economy. All the standards for planning, reforestation, harvesting, and the protection of forests, which were developed in the 1960-1980 years, take into account only the planned costs and do not take into account the real costs of a market economy. As a result, many forest management activities that were possible at the regulated prices are impossible in a market economy. The railway monopoly, which imposes costs on the transportation of wood and makes transportation of more than 150 km unprofitable for businesses is an example. In this situation, logging companies are doing only what is profitable for them. Foresters try to force their requirements on loggers, but do not understand that there is a need to change the requirements themselves.[91]

Traditionally, socialist forestry was extensive and forestry operations moved quickly from place to place allowing relatively large clear cuts, although they were typically small in comparison with unharvested areas. Since the forest code of 1997, the size of allowable clear cuts has been reduced. However, the enforcement of forest management is extremely weak, compared with Finland, Sweden, and Canada (Knize & Romaniuk, 2010: 12). This leads to a rapid depletion of the remaining intact forest landscapes and an overall inefficiency of forestry operations. Regulations of the Forest Code of 2006 do not allow for intensive forest management. Companies argue that

[89] Interview with the head of the FSC office in Russia, March 2011, Moscow.

[90] Presentation of T. Yanitskaya, materials of the seminar 'HCVF and biodiversity' 11-13 November, 2009, participant observation.

[91] Interview with the head of the FSC-Russia office, March 2011, St. Petersburg.

intensifying forest management operations will increase the competitiveness of the forest sector,[92] while NGOs argue that it will allow preserving intact forest landscapes.[93]

In 2008, the state adopted the Strategy of Developing Forest Complex up to 2020 with the declared priority of modernizing the forest sector. In 2010-2011, the problem of extensive forest management started to be recognized by state authorities. Rosleshoz organized a Council on Forestry Intensification[94] with the participation of scientists, NGOs (Silver Taiga Foundation) and major holding companies (Ilim Group, Titan and Investlesprom) to discuss ways which would allow introducing the Scandinavian intensive forest management model into Russia. Amendments to regulations were suggested. The intensification of forest management was perceived as a major innovation that has to be introduced to the forest sector in the process of its modernization.[95]

In post-soviet times, illegal logging became one of the roadblocks to sustainable forest management in certain regions of Russia. During socialism, illegal logging was extremely rare due to the strict enforcement of the law and severe punishment for stealing from the government. After perestroika's privatization laws, a criminal element quickly entered into the country's commerce, including the forest sector. The volume of illegal logging began to rise, often with the cooperation of corrupt government officials. This was especially apparent when Russia experienced an economic downturn in 1980s-1990s. The 'wild privatization' of the early 1990s saw the rise of organized crime in forestry. It involved former ruling elites of the Communist Party, as well as those from regional governments, administrators, law enforcement agencies, and police forces. Although illegal logging reduces government revenues, it serves the interests of the corrupt elite. After the government's forest production failed, its former employees found a new lucrative niche in illegal logging, especially in the Russian Far East. After Russia's borders with China were opened, illegal logging in the Russian Far East became a profitable option (Tysiachniouk & Reisman, 2004; Tysiachniouk, 2006a).

It is very hard to estimate the scale of illegal logging, as data provided by government differs significantly from that from NGOs. According to the calculations of the FSC-Russia office, logging without permits by 'black woodcutters' is 10-15% of the total volume of all illegal logging. The majority is conducted by forest tenants who use inaccurate forest inventory plans and otherwise make many violations on the leased territories.[96]

Expert estimations are that 10-35% of all timber harvested in Russia is illegal, and in certain regions, especially those bordering China, it can reach 80%.[97] This can be partly explained by the lack of the state's accounting of logged timber (Efimova *et al.*, 2010: 8). The practice of illegal logging is spread not only among organized crime networks, but also among villagers, who can

[92] Interview with an Ilim Pulp representative, St.Peterrsburg March 2011, interview with the Investlesprom top manager, May 2010, interview with Mondi Business Paper, top manager, January 2011, Siktifkar.

[93] Interview with the WWF representative, March 2011, St. Petersburg.

[94] The first meeting took place in January 2011, interview with an Ilim Group representative, March 2011, St. Petersburg.

[95] Speeches of state officials, materials of the roundtables, participant observation at the international conference 'Innovations and Technologies in Forestry, 22-23 March, 2011, St. Petersburg.

[96] Interview with the head of the FSC-Russia office, March 201, St.Petersburg.

[97] According to Mr. Safonov, the Authorized representative of the President of Russian Federation in the Far East (2009) http://www.rg.ru/2009/04/07/safonov.html.

make quick money to help them survive in a poor and unstable economy (Sun & Tysiachniouk, 2008). The flow of wood across the border skyrocketed throughout the 1990s and still remains high in 2000s. The transition to a market economy, coupled with the government's collapse and economic depression, has caused this rapid rise in commercial crime (Tysiachniouk, 2006a).

Since 2005, Russia is part of the ENA-FLEG process, and the state has made an effort to cope with illegal activities by introducing a satellite monitoring system. However, the problem remains unresolved on non-certified forest territories.

4.2.2 State authority in forest management

Since the 1990s, Russia's system of forest management has been in a state of constant restructuring. Russia took a path of de-institutionalization of environmental policies, hoping to foster economic growth (Mol, 2009). This path became apparent in all resource extracting industries and especially in the forest sector. The interactions between different divisions of government were further complicated by shifting jurisdictions. Forest management structures in the period of 2000-2012 survived through six reorganizations.

In 1997, the forest code was enacted, but already by 2002 there was a plan to change it completely and to enact a new code in 2003. The Ministry of Natural Resources, in conjunction with the Ministry of Economics, developed the new code,Radical changes, such as private property ownership on forest lands was suggested but opposed by NGOs, foresters and the public at large. The Russian Society of Foresters wrote a myriad of articles to the media opposing developments of the new Forest Code (Shutov, 2003: 136). The new code was enacted only at the end of 2006 because of lengthy delays and disagreements between governmental agencies and civil society institutions. When the new code came into force, the disagreements between stakeholders became even more apparent. Many amendments to the code were then introduced (Papilla, 2009). The Forest Code was largely implemented by 2009, however interested parties, e.g. governments, NGOs and businesses, recognized the failure to create effective forest legislation and insisted on continuing reforms (Shmatkov, 2012: 25).

The history of reforms shows patterns of de-institutionalization in forest management authority. In 1998, the Concept of Sustainable Forest Management in the Russian Federation, which declared criteria and indicators of sustainability, was approved by the Federal Forest Service. However, the concept turned out to be a declaration which was neither introduced into Russian forest legislation, nor implemented (Knize & Romaniuk, 2010: 12).

In 2000, President Putin closed the Federal Forest Service and gave its responsibilityies to the Ministry of Natural Resources. The Ministry of Natural Resources thus became responsible for both the protecting and harvesting of forests. The Ministry failed in achieving both economic efficiencies in the sector and in the appropriate protection of forests; forest operations continued to be extensive, the revenue to the governments low, and levels of illegal logging high (Pisarenko & Strakhov, 2004: 249; Pisarenko & Strakhov, 2006).

The Federal Ministry of Natural Resources policies were carried out through numerous regional branches. Local administration was carried out by forest management units (leskhozes), the traditional forest management agencies carrying over from socialist times. The leshkozes were guided by ten year plans developed by the Forest Inventory Agencies, an engineering and planning institutions usually situated in the region, and subordinated to the Ministry of Natural Resources. Although the leskhozes had little input into the formulation of the long-range plans, their authority included leasing tracts of forest to private timber companies as well as performing rudimentary maintenance (such as thinning, replanting) and protecting the forest from thieves and natural disasters. A central role of the leskhozes was to ensure that the operations of the private timber companies are consistent with laws and regulations. The leases paid by the timber companies were shared by federal, regional and local agencies. The leskhozes were mostly funded from federal government budgets, however, they could also gain money from selling timber harvested during forest management operations, such as thinning. The actual funding level, however, was often below that appropriated in the budget. Therefore, leskhozes in many regions were cutting the best trees to provide profits from sales instead of thinning, and by doing so they contributed to forest degradation (Tysiachniouk, 2006b).

Under the new Forest Code of 2006, the structure of state bodies was constantly changing. Forests on municipal lands were administered by municipalities, as it was before, but forests on agricultural lands and forests allocated for logging since 2008 were administered by Regional governments. This became a significant shift in decentralization in forest management authority. Regional governments implemented reforms in different ways. As a result, Regional Administrations related to forest management belong to different jurisdictions and have different organizational structures depending on the region (Figure 4.1).

In 2008, the Federal Forestry Agency (Rosleshoz) became subordinated to the Ministry of Agriculture of the Russian Federation, which became responsible for development of forest policy and forest legislation. The major implementing agency, the Russian Federal Forest Agency (Rosleskhoz), subordinated to the Ministry of Agriculture, controls the Regional Forest Administrations and manages forests in the Moscow district. The Ministry of Natural Resources is responsible for managing the natural reserves at the Federal Level (strictly protected reserves (Zapovedniks) and National Parks), while other specially protected areas are administered by Regional Administrations.

With the new Forest Code, Forest management units (leskhozes) were transfered to units with other functions (lestnichestvo). They continue to control management of the forest lands, but all forestry activities became the responsibility of leaseholders on their leased land, and in cases where the territories are not leased, by contracted organizations. The new forest code obliged companies that use the forests to do all the forest management operations, including replanting and thinning. This jeopardized economic stability of small and medium size companies (Kuzminov, 2010: 33). Companies started to contract workers from the forest management units to do this work, which created a conflict of interest with state authorities. The length of the leases became 10-49 years, and leases could be acquired through auctions.

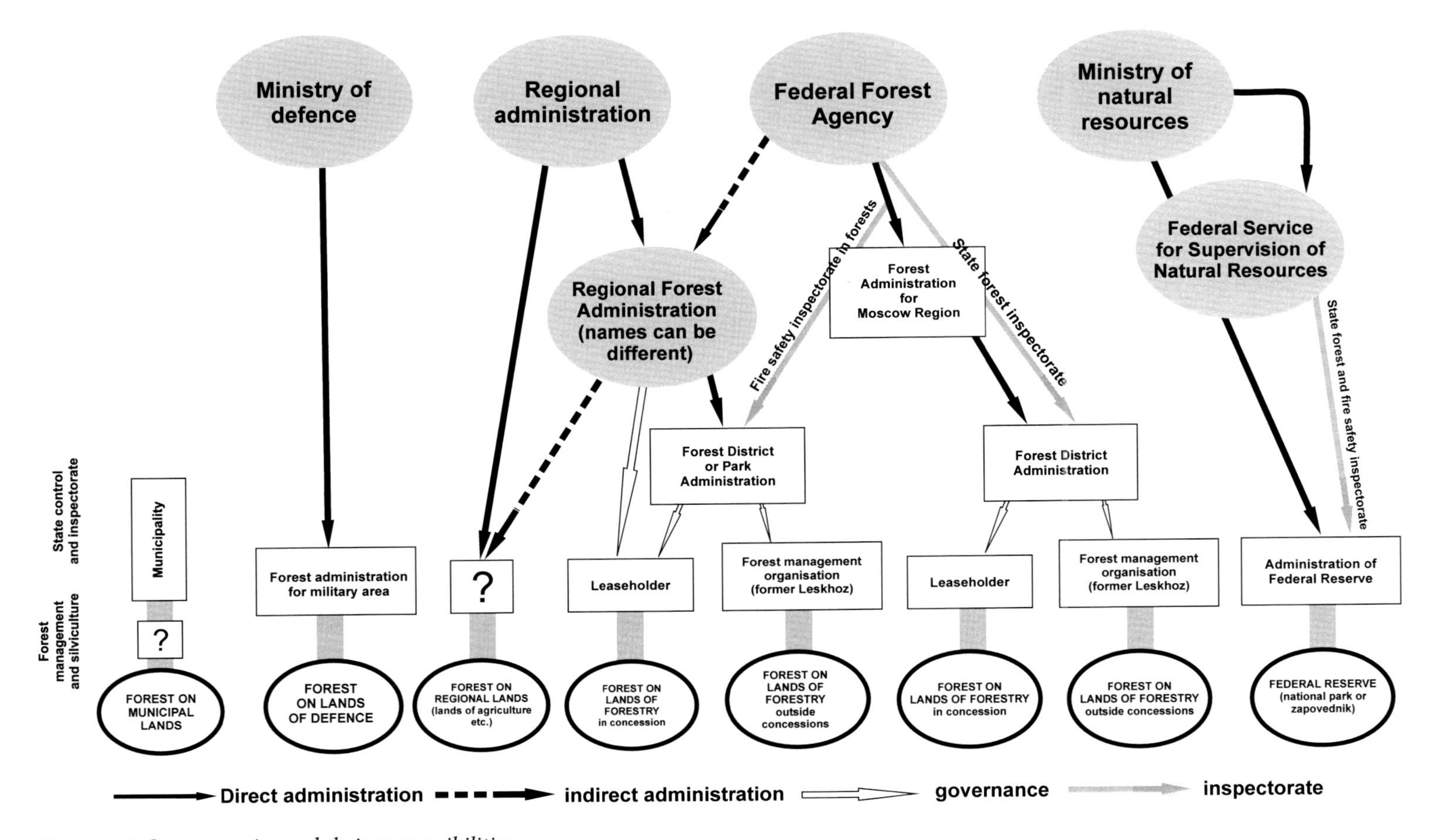

Figure 4.1. State agencies and their responsibilities.

In 2007, when the process of decentralization in the managerial systems of the forest branch began, forest management, planning, leasing, and protection was delegated to the Administrations of Subjects of the Russian Federation. Therefore, much of management responsibility was turned over to the regions, while financial management remained centralized and subject to the federal authorities' prerogatives. Regions became responsible for elaborating a general 10-year plan of forest development for their whole territory. They also have to develop regulations for the forest management units (lesnichestvo).

The Ministry of Agriculture had not previously paid much attention to the problems of forest management and it did not have enough professional resources for resolving them. The Federal Service for Veterinary and Plant Sanitation became responsible for forest protection, it was assigned to supervise and control forest fires in areas of the Russian Federation. They also were lacking in necessary manpower and technical resources. The controls over forest exploitation were eased up to a large degree, and by the time of delegating the powers it was non-existant.. The Federal Service for Natural Resources (Rosprirodnadzor), which was responsible for natural resources protection, was made subordinate to the Ministry of Nature Resources, whose task was – quite the contrary – to organize exploitation of these same resources (Papilla, 2009).

In the summer of 2010, Russia experienced intense yet poorly controlled forest fires[98] which drove the Russian government to increase the status of the Federal Forest Agency (Lehmbruch, 2012). In 2008-2010, the Ministry of Agriculture had the responsibility for introducing changes into forest legislation, and the Federal Forest Agency had executive power and was responsible for the implementation of the federal forest policies. According to the Decree of the president of Russian Federation #1074 (August 27, 2010), the Federal Forest Agency received the status of an independent agency with legislative power. The new wave of state forest management reforms started. However, the State forest policy continues to be poor (Shmatkov & Kulikova, 2012: 2). Figure 4.1 visualizes the simplified structure of forest management in the Russian Federation, developed according to the Forest Code of 2006 and the President's Decree of 2010.[99]

Constant reorganizations and restructuring caused inefficiencies in governmental policies, and created an institutional void in the governmental regulatory system. To some extent, this opened up space for private actors and voluntary regulatory mechanisms to fill the gap. In this situation, FSC certification becomes essential for contributing to the sustainability of forest management in Russia.

4.3 Non-state actors in the Russian forest sector

4.3.1 Forest industry in transformation: historical perspective

The Soviet regime attempted to expand the forest sector by launching extensive programs for building pulp and paper mills. The majority of these pulp and paper mills were build in 1960s-1970s,

[98] 33,5 thousands fires on 22,1 million hectares, with 192,2 thousands hectares of forests totally burned, http://www.rosleshoz.gov.ru/media/appearance/57, reviewed 6 April, 2011.

[99] The scheme has been developed by Greenpeace Russia in August 2010, questions remain in places where management mechanisms are not yet developed by the state authorities.

the last were constructed in the middle of the 1980s. Most of the regional forest complexes of the Soviet Union were located in the territory of present-day Russia, and the role of the mills was to satisfy the demand for forest products in other regions of the Soviet Union and in many other socialist countries. There is a disproportion in the territorial distribution of wood enterprises. The major wood processing factories and pulp and paper mills are located in the European part of Russia, while considerable wood resources are in Siberia and Far East, on the other side of the Ural Mountains (Knize & Romaniuk, 2010: 15). When the raw materials are delivered from more than a 1.5 thousand km distance, the transportation costs become critical for further production profitability (Knize & Romaniuk, 2010).

The privatization of the Russian forest industry started in 1992, and as in other sectors of the Russian economy, representatives of the former socialist state became key personnel for the newly formed Ministry of Industry, which facilitated privatization. They were also in the forefront of newly formed business associations related to forestry and forest trade, and finally became the owners of the majority of the enterprises.

During the transition period, enterprises continued to behave as social organizations, preserving wages and jobs and not prioritizing the maximization of profits. There was a danger during the transition to a market economy for these companies to collapse. However, most of them did not collapse, because a so-called 'informal economy' developed. The informal economy concept explains how the patterns of the Soviet economy partly substituted for the market system and helped Russian enterprises to avoid bankruptcy. Usually this is done through barter trade, with established and negotiated non-market prices (Olsson, 2006, 2008).[100]

Actors not related to the Socialist government structures also rapidly emerged, many in opposition to the former apparatchiks. Several industrial and quasi-state-industrial groups made an effort to build empires and a new monopoly and several other holdings were formed (Lehumbruch, 1999). By the end of 1990's, a large fraction of the large and middle-size timber enterprises had became part of holding companies and investment groups. Thus, a few dozen large and middle-size forest holdings appeared in the forest industry and united enterprises of the pulp and paper industry, wood-processing, and logging enterprises. Most of the forest holdings were created under the aegis of a large pulp and paper or wood-processing mill, and consolidated the round timber suppliers, middle-sized and small timber enterprises. With the new regulations of lease contracts through auction coming from the Forest Code of 2007, it becames harder for the small leaseholders to survive as independent units. Most of them became subsidiaries of large holdings. Logging companies were joining the holdings either voluntarily or by force, when a holding company bought a controlling amount of shares.

In the 2000's, large holding companies became Transnational Corporations (TNCs) of Russian or foreign origin. The Investlesprom holding company, of Russian origin, purchased many other holdings in Russia and abroad. Conversely, many Russian companies were purchased by foreign TNCs; for example, Siktivkar Pulp and Paper Mill became part of Mondi Business Paper, Ilim Pulp was transformed into the Ilim Group belonging to International Paper.

As concerns certification, such a state of affairs is rather favorable, as TNCs are for the most part interested in certification of their leased territories and can foster top-down certification

[100] Olsson calls informal economy 'virtual economy'.

in a centralized way. This results in an expansion of certified territories and leads to a situation when a decision by one actor can begin the large-scale process of certification. However, newly integrated subsidiaries of the holdings are affected by the restructuring of their businesses and the desire of the new owner to increase efficiency and minimize costs – therefore, they do not see the advantages from certification and envision it as an additional requirement imposed by the company owner (Tysiachniouk 2010a,e).

4.3.2 Environmental NGOs: history and current situation

During socialism, the non-governmental sector in Russia was limited to professional societies, such as the Society of Naturalists, the All-Russian Geographical Society, the All-Russian Societies of Nature Protection, and the All-Russian Society of Foresters. The state itself initiated these organizations in the 1920's, and controlled their operations through the Soviet era. In Khrushev's time in the 1960's, Nature Protection Corps were formed in the biology departments of many universities under the Young Communist Organization, Komsomol. Despite their official affiliation with Komsomol, the Nature Protection Corps united many dissidents and free thinkers (Yanitsky, 1996).

In the late 1980's, the social and political structure of Russia and the USSR experienced profound changes. Gorbachev initiated Perestroika ('reconstruction') and glasnost (free speech) in an attempt to open up the tightly closed and controlled society while at the same time maintaining Soviet power. With these initiatives, room for social mobilization soon emerged and people throughout the country began gathering and criticizing the government. The Chernobyl disaster of 1986 helped to awaken concern over nuclear energy and further legitimized the already growing environmental movement. In addition, this new movement became closely linked with liberation movements in the Baltic States. Riding on the wave of Perestroika, existing environmental organizations received new supporters, and many new grassroots NGOs were formed throughout the territory of the Soviet Union. The period between 1990 and 1991 saw a change in the Soviet Union's political landscape, following the government's enactment of many measures favorable to NGO activity. On the basis of the student Nature Protection Corps, the Socio-Ecological Union (SEU) was formed, with its center of coordination and information in Moscow. The SEU became an umbrella for many local NGOs specializing in different issues, including forestry. The Nature Protection Corps continued to exist in universities throughout the country. After the collapse of the Soviet Union, the SEU automatically became international with its NGO-members in the Newly Independent States (NIS). Fast economic reforms in 1992 killed many of the grassroots NGOs in Russia, however, the more professional ones and those formed in the territories contaminated by the Chernobyl disaster, as well as the radioactively contaminated territories in Siberia (Cheliabinsk district) managed to survive (Tysiachniouk, Mironova & Reisman, 2004). By the end of the 1990s however, only those NGOs soliciting Western funds continued to exist. NGOs focusing on environmental education began a period of rapid development throughout Russia. International NGOs, such as Greenpeace and the WWF, established their subsidiaries in Russia. International NGOs employed environmentalists who had grown up in the Nature Protection Corps, and who maintained close ties with their peers at the SEU. In the 2000s, the environmental NGOs in Russia became professionalized and quite diverse. NGOs have formed multiple networks, both within the country and with international partners (Tysiachniouk, 2003).

The Biodiversity Conservation Center (BCC), the SEU, the Nature Protection Corps and Greenpeace formed the Forest Club, which joined the transnational Taiga Rescue Network (TRN). The network focused on all kind of issues related to forests, but their major focus was the conservation of intact forest landscapes containing old growth forests (Tysiachniouk, 2003; Tysiachniouk & Reisman, 2006).

In the 1990s, Greenpeace International organized several direct actions against companies that were harvesting High Conservation Value Forests (HCVF) in the Karelia and Arghangelsk regions. In partnership with other NGOs, they created maps of all the old growth forest landscapes in Northwestern Russia and distributed them to both Russian forest producers and western forest consumers (Tysaichniouk & Reisman, 2002; Tysiachniouk, 2009). The first case study in this book focuses on a detailed analysis of the consumer boycott in Karelia and on creation of the Kalevala National Park that includes old growth forests. As I will show, the boycott was a turning point in the interactions among stakeholders. Both companies and governments began to consider the NGOs as stakeholders. NGO trans-boundary campaigns can be considered the pre-history of Russian certification.

NGOs also made considerable efforts to work with governments on forest legislation. NGOs, especially the Forest Club and the WWF, have taken an active role in the development of both the forest code of 1997 and of 2006, yet they did not achieve any significant results. Only a few suggestions have been taken into account by governmental agencies. In addition, the NGOs promoted sustainable forest management through their own programs and projects, such as Model Forests (Elbakidze, Angelstam, Sandström & Axelsson, 2010; Tysiachniouk, 2006b, 2010e). Since the early 1990s, the WWF and the Forest Club have promoted forest certification. The case studies on NGOs, Building Model Forests, in which forest certification is a part, will be described later in this book (see Chapters 6-7).

Russian NGOs became fully integrated into transboundary NGO networks. These NGO networks have tried to influence government policy, industry, and the environmental awareness of Russian citizens. The expansion of Western environmentalism into Russia since the early 1990's has brought with it ideas and concepts of nature conservation and techniques of natural resource exploitation developed by science, industry, and the 'third sector' of the USA, Canada, and European countries. Therefore, international stakeholders with different value systems, different understandings of land use issues and opposing interests came into the Russian political arena and build networks with different kinds of Russian actors. Since 2006, Western funding for Russian NGO has diminished as Russia is perceived as a country with a developed market economy, although its democratic development was still questionable. The NGO community divided into those close to the state, such as government affiliates like the All Russia Society for Nature Protection (VOOP) and the Societies of Hunters and Fisheries, (Henry, 2010a: 154) and those who were funded by western grants that provided services to business.

The most negative relationships were with the watch dog NGOs, especially those dealing with nuclear issues (Henry, 2010a: 214). The forestry-related NGOs fall into the category of lacking support from the state. The NGO activists started to serve as experts in the FSC system

(Tysiachniouk, 2010f,g,h, 2012a,b). They became actors involved in opening the Russian forest system for governance by private actors.

4.4 Forest certification initiatives

There were four different efforts to promote forest certification in Russia. First was compulsory forest certification, but that was never implemented. Next came two initiatives devoted to promoting nationally-based systems with the aim of having them accredited by the PEFS, and the forth initiative promotes the FSC system.

4.4.1 Compulsory national forest certification initiative

Article 73 of Russia's 1997 Forest Code calls for a compulsory national forest certification program to be implemented by the (now-defunct) Federal Forest Service (Rosleskhoz). In 1997, the federal government perceived the FSC as an intrusion on Russian sovereignty, while at the same time observing that many other European countries were developing national systems of forest certification.[101] Rosleskhoz gave the Forest Inventory Agencies responsibility for the development of standards and auditing procedures. The Union of Timber Merchants and Timber Exporters of Russia and the Ministry of Industry were strongly against this initiative. They perceived trade with Europe as private business. Other state agencies were also not strongly motivated to move forward with the compulsory certification program.[102] Rosleskhoz's primary mission was to create an additional law enforcement structure to generate additional annual revenues from the companies to augment the governmental budget (Tysiachniouk 2006a, 2012a,b). Compulsory national certification was never implemented and has been effectively abandoned by the government.

4.4.2 The national system of voluntary forest management certification in Russia

The Union of Timber Merchants and Timber Exporters of Russia, which consists predominantly of exporters of round wood to Finland and China, was the source of the first national initiative of voluntary forest certification. In 2001, they formed an initiative called 'The National Council of Voluntary Forest Management Certification in Russia (NCVFCR)'. The Central Research and Development Project and Design Institute of Mechanization and Energy of the Timber Industry, with participation of the All-Russia Research and Development Institute of Forestry and Mechanization of Forest Industry and the Moscow State Forest University, developed and tested a set of national forest standards. The developers of this system drew on Finnish experience of developing a Forest Management Certification System, and the system was close to the Helsinki criteria. 'The Concept of Sustainable Forest Management in the Russian Federation', approved by the Federal Forestry Service in 1998, was also used.

In August 2002, the system was tested at two enterprises in the Vladimir region. The developers claim that the system was efficient and that its criteria almost completely reflect the activities of

[101] Interview with WWF staff, who at that time worked in Rosleskhoz, March, 2004, Moscow.

[102] Interview with the representative of the Union of Forest Producers, March 2004,Moscow.

timber industry enterprises with respect to the certification requirements. The system was also discussed by timber exporters of the Russian Federation, whose recommendations were taken into account when the final standard was developed (2003). Final testing took place in January 2004, in Voziagales. The initiative is oriented solely towards the program of Endorsement of Forest Certification (PEFC). The Union of Forest Owners of Land and the Ministry of Industrial Science[103] financed it. NGOs and the forest processing industry were not involved in this process and do not support this initiative.[104] Environmental NGOs criticized the standards for weak biodiversity conservation measures, and a lack of transparency.[105]

The agencies and institutions that promoted this initiative collapsed during the restructuring of the state forest management bureaucracy. However, the initiative itself did not collapse and accreditation of its standards by the PEFC was sought. In October 2004, at the PEFC General Assembly, the NCVFCR became a member of the PEFC family, with the right to represent Russia in the PEFC Council.[106] According to the PEFC rules, several national systems can be accredited. Therefore, in 2006, after several years of negotiation, the NCVFCR signed the Umbrella Agreement with another initiative, the Russian National Council on Forest Certification (RNCFC), to jointly represent Russia in the PEFC. The NCVFCR efforts in accrediting their standards by the PEFC Council failed. They also failed to pay PEFC membership dues and in March, 2011, membership of the Umbrella Organization was suspended.

4.4.3 PEFC forest certification in Russia

The second national voluntary effort was supervised and supported by Rosleskhoz after the state initiative with compulsory forest certification failed. Part of the funding for the national system of forest certification came from the World Bank's pilot project on sustainable forest utilization. Additional funding was provided by a grant from the Finnish government.[107] The World Bank lent US$ 60 million to the Russian Ministry of Natural Resources in order to promote sustainable forestry, of which US$ 400,000-450,000 was allocated to promote forest certification and to create a 'certification climate' and infrastructure.[108] Funding available from the World Bank was an incentive for the government to participate in this initiative, although the certification component was weakly implemented and the money spent inefficiently.[109]

On May 14, 2003, the Russian National Council of Forest Certification (RNCFC) was established and officially registered.[110] The Council involved people mostly affiliated with Rosleskhoz

[103] The Ministry of Industrial Science was closed by Putin in March, 2004. At the time of my interview in February, 2004 it was still functioning normally.

[104] Interview with WWF staff, February 2004 (later this person become the head of the FSC national office).

[105] Interview with the representative of Biodiversity Conservation Center, March 2011, Moscow.

[106] Http://www.ruslescert.ru.

[107] Interview with the World Bank consultant, March 2004, Moscow.

[108] Interview with the World Bank consultant, March 2004, Moscow.

[109] Interview with the head of FSC-Russia office, March 2011, Moscow.

[110] Interview with the head of the Department of Forest Use of the Ministry for Natural Resources, March 2004, Moscow.

institutions, such as the All Russia Scientific Research Institute of Forestry and Mechanization of Forest Utilization, Moscow State University of the Forest and its affiliate, the International Institute of Forestry. It also engaged a World Bank Russian representative, and an Ilim Pulp Company representative. Representatives of the WWF and of the World Conservation Union (IUCN) were observers. Members of the Council had varying attitudes toward certification. The national standards were supposed to be 'national in content and international in form'.[111] In the early stages, the standards were intended to be similar to those required by both the FSC and the PEFC. The development of national standards was started in the International Institute of Forestry under the supervision of the academician A. Isaev, who was chairing the National Council. As with the NCVFCR, the plan was to accredit the national standard at the PEFC and simultaneously make it compatible with the FSC. The RNCFC was relying on FSC certification centers, which were set up by the WWF.[112] Therefore, in the early stages the RNCFC initiative was developed consistent with the the FSC, drawing upon the FSC's institutions and experts.

On May 22, 2005, the Council signed an agreement on cooperation with the Russian FSC National Initiative and in 2009 renewed this agreement. The FSC International supported this cooperation and started providing the Council with materials on certification and other logistical support free of charge.[113] The purpose of the agreement was to harmonize the RNCFC standards with the FSC standards. This would facilitate dual certifications that companies might undertake. It was also important to harmonize Russian national forest laws and regulations with the FSC standards and to develop a common strategy for FLEG implementation and EU due-diligence legislation. The prototype for this partnership was an agreement between the UK Woodland Assurance Standard and the FSC, which served the same purposes.[114]

A working group was established in 2006 to harmonize the standards. This group worked intensively and harmonized 85% of the standards. The 15% of the standards that were not harmonized were mostly related to High Conservation Value Forests (HCVF). Many of the FSC requirements contradict Russian legislation and the RNCFC was not making an effort to foster changes in Russian legislation.[115] In 2008-2011, differences in the standards diverged further with the processes of accreditation of national standards by the FSC and the PEFC schemes. Both national initiatives adjusted their standards according to corrective action requests by the FSC and the PEFC. The FSC was rapidly developing while the PEFC was stagnating, and the standard harmonization working group was inactive for several years. Although meetings of the harmonization group are planned for 2011, both schemes do not expect the standards to be completely harmonized and reduced their cooperation to working with FLEG and the EU due-diligence legislation, e.g. to verification of legality by both schemes.[116]

The RNCFC initiative, as does other certification schemes, involves forest management certification (FM), chain of custody certification (CoC) and joint FM-CoC certification. Certification

[111] Interview with the academician A. S. Isaev, May 22, 2005, Moscow.

[112] Interview with the WWF staff, March 2004, Moscow.

[113] Participant observation, May 22, 2005, Zvenigorod.

[114] Interview with the head of the FSC Russia office, March 2011, St.Petersburg.

[115] Interivew with the director of the PEFC-FCR, March 2011, St. Petersburg.

[116] Interview with the head of the FSC office, March 2011, Moscow.

can be individual, by a company, or territorial by a forest management unit (lesnichestvo). In territorial certifications, the lestnichestvo is supposed to be the certificate holder and supervises compliance of all leases to the standards. Territorial certification is allowed only if more than 50% of the lease holders working in the territory of a lestnichestvo agree to certify (Ryzhkov & Prokazin, 2011: 54-55).

State money was used for testing the RNCFC scheme in 12 forest areas in different parts of Russia.[117] In March 2009, the RNCFC standards were accredited by the PEFC. In August 2010, the Scientific Technical Centre on Industrial Safety signed an agreement with the PEFC International Accreditation Forum and was allowed to accredit Russian auditors pursuant with PEFC certification.[118] In March 2011, when membership of the Umbrella Organization at PEFC was suspended, the RNCFC reorganized itself into PEFC-Russia and made a new application for membership to the PEFC Council, hoping that in May 2011 they will again become members.[119] In March 2011, PEFC-Russia started a new round of promoting its certification scheme. The first seminar with 32 stakeholders was held in St. Petersburg. The representative of the Ilim Group[120], belonging to the TNC International Paper, expressed its support to the PEFC, stating that customers in Japan ask for PEFC certification. Environmental and social NGOs were not present at the meeting. Unfortunately, the dominant discourses expressed at the presentations were related to overcoming non-tariff ecological barriers with the help of the PEFC and not on promoting sustainable forest management.[121]

In July 2012, three CBs were operating in Russia and three in the process of accreditation[122] Up to date, subsidiaries of the Finnish holding company Mesaliitto, operating in Russia, received the PEFC FM/CoC certificate. Three companies held PEFC-CoC certificates.

In 2007, the Metsaliitto company initiated a pilot project of PEFC certification for its subsidiary Matsaliitto Podporozhie in the Leniongrad oblast. The Finnish company Inspecta served as a certification body. The holding company Metsaliitto belongs to more than 130,000 Finnish forest owners. For the most of them, PEFC certification is part of their corporate ethic toward the environment. Metsaliitto owners wanted their subsidiaries in Russia to be managed in the same way as in Finland, so it was logical for the company to have its operations in Russia certified by the PEFC. In addition, the company in general was interested in promoting the PEFC in Russia and in creating a competitive environment for certification schemes.[123] The company was preparing for certification very seriously, allocating HCVF and involving stakeholders of different levels in the process.[124] In March 2010, Metsaliitto Podporozie received the PEFC certificate and another

[117] Interview with the director of the FSC office, March 2011, Moscow.

[118] http://www.fcr.ru.

[119] Interview with the director of the PEFC-Forest Certification Russia, March 2011, St. Petersburg.

[120] All operations of the Ilim Group in Russia are certified by the FSC.

[121] Participant observation at the seminar on PEFC certification for stakeholders of North Western Russia, 30 March, 2011, St.Petersburg.

[122] Http://www.pefc.ru/roc_01.html.

[123] Interview with the manager in ecology, Metsaliitto, March 2011, St. Petersburg.

[124] Interview with the researcher from the Centre for Independent Social Research, who conducted longitudinal study of Mesaliitto's PEFC certification, January 2011, St. Petersburg.

subsidiary, Petrov Pasha, started its preparation toward certification. In November 2010, the company made a decision to certify its operation using the FSC scheme and to maintain both FSC and PEFC certificates as there is market demand for the FSC certified wood from Russia. In January 2011, both subsidiaries of Mesaliitto operating in Russia received both the PEFC and FSC certificates for a total 327,761 hectares.[125]

4.5 FSC certification

4.5.1 Emergence of the FSC in Russia

The first FSC certifications in Russia came via market relationships. Three enterprises: (1) Kosikhinski Forest, Altai Region with their processing enterprise Timber Production Pricebatch Ltd.[126], (2) Koverninskiy Leskhoz, Nizniy Novgorod oblast[127], and (3) Holz Dammers GmbH in Arghangelsk oblast[128] – received their certificates without any help from the WWF or forest certification centers. The enterprise Kozikhinsky Leshoz in Altay Region started preparing for FSC certification in 1997 and received the certificate in 2000. Another first company was the Paper Mill Volga, which started working on FSC certification of Koverninski Leskhoz in 1996 and received it in 2002. All three enterprises were certified privately in response to requests for FSC certification from their western co-owners and partners. Only after they had received forest management and chain of custody certificates did they begin to share their experiences, interact with the FSC institutional designers in Russia, and participate in conferences on certification (Tysiachniouk, 2006a, 2012a,b).

4.5.2 Initial NGO effort to start the FSC in Russia

Russian NGOs started promoting the FSC in the late 1990s. In 1997, at a meeting in Finland, environmental NGO representatives decided to start promoting the FSC system in Russia, but nothing concrete was done.

The Finnish Ministry of Agriculture initiated a feasibility study on the need for certification in Russia in the framework of the Russian-Finnish commission on sustainable forest management. Andrey Ptichnikov (currently director of the national FSC office) worked at that time as the national project manager at the TACIS project 'Sustainable forest management in the North-West Russia: the Karelian project' and was selected as a main expert for the feasibility study together with Finnish Agricultural consultant Pasi Miettinen, who is currently the FSC Finland director.

However, when Mr. Ptichnikov tried to report on the results of his study to the Russian-Finnish commission on forest use, the Russian representatives to the commission did not allow him to present his findings. At that time, the Russian Forest Service was concerned about Russia's international image and did not allow disclosure of information that would show the international

[125] Interview with the manager in ecology, Metsaliitto, March 2011, St. Petersburg.

[126] Trading with the Body Shop UK; received CoC certificate in 1997.

[127] Russian mother company Pulp and Paper Mill Volga; received the certificate in 2002.

[128] Co-owned by Dammers Moers, Germany, received its certificate in 2000.

community what was going on in the Russian forest sector. Some of the decision makers in Federal Forest Agency were interested in keeping full control of forest certification in their own hands.

Mr. Ptichnikov resigned from the TACIS project and took a new position he was offered by the WWF Russia and started promoting forest certification on behalf of the WWF.[129]

In 1998, the environmental organizations WWF, Greenpeace, Social SEU, and the Biodiversity Conservation Center (BCC) began to promote FSC certification in Russia. The initiative belonged to the WWF, which invited other NGO's to work on forest certification. The first meeting of the initiative group was organized in May 1998, in the WWF office and included only NGO's. Vladimir Chouprov from Greenpeace was elected as the coordinator of the first FSC initiative group in Russia and became a contact person for the FSC International (Tysiachniouk, 2006a).

Each of these organizations worked with European partners and were familiar with the FSC process in Europe. In 1998, Greenpeace sent information regarding FSC forest certification to 5,000 forest producers and forest enterprises. At that time the interest of forest companies in certification was still very low. Only 10 of the 5,000 companies requested more information.[130] Still, in 1998, the WWF organized the first multi-stakeholder based conference on FSC certification in Petrozavodsk, Karelia Republic. The goal was to start a dialogue with business and show the government that Russia needed voluntary certification. The conference was sponsored by the MacArthur Foundation. It was the first time that business representatives were invited to discuss issues with NGO representatives. Only a few forest companies attended the conference, which was dominated by scientists and NGO representatives. This can be explained by the existence of large conflicts between the forest industry in Karelia and environmental NGOs, such as those involved in the Forest Club, due to Greenpeace's direct action and consumer boycotts. In addition, forest companies in Karelia were interested predominantly in exports to Finland, where interest in the FSC was low. At that time, the Russian government was still committed to compulsory certification and opposed non-state voluntary approaches to forest governance, while environmental NGOs opposed compulsory systems and promoted the FSC. As a result of the conference, the Federal Forest Service became informed about the FSC and started to pay attention to it. Within the government the first respondents were scientists, typically the most progressive people, and they started to educate governmental officials.[131] Despite conflicts, governmental representatives participated, and the conference can be considered the first intersectoral dialogue on forest certification in Russia.

In 1999, a second conference took place in Pushkino, Moscow oblast, where a working group was created comprised of participants from business, and representatives of social and environmental NGOs. Later, the WWF and Greenpeace formed an organization that eventually became the FSC National Working Group to promote the FSC system; it used a Coordination Council as a governance body. At that time the forest companies did not feel comfortable enough to openly work with NGOs, but rather preferred to interact informally. They participated in the events as private individuals and not as representatives of their companies. The interest of the

[129] In 2004 Mr. Ptichnikov worked for INDUFOR on an assessment of FSC potential in Russia and in February 2005 become a director of the Russian National FSC office.

[130] Interview with WWF staff, February 2004, Moscow.

[131] Interview with WWF staff, February 2004, Moscow.

majority of the forest companies in certification at that time remained very low. The Federal Forest Service was still promoting compulsory certification and created a regional center for compulsory certification within the Novgorod Center for Forest Protection. The WWF awarded a grant to this center to develop the FSC model in parallel with the governmental compulsory forest certification.[132]

4.5.3 Institutional design and institutional infrastructure for the FSC

As was described in the previous chapter, the FSC connects with the sites of implementation through the contact person,[133] or the FSC national offices that represent national initiatives. All these institutions are part of the FSC-GGN. FSC institutional infrastructure also involves the certification bodies, expert community and stakeholders. This section explains how this institutional infrastructure was gradually built up in Russia. The section focus on how NGOs 'moved' companies toward certification and how they made an effort to construct supply and demand for certified products.

FSC forest certification has been promoted through a series of WWF and Greenpeace initiatives. The WWF disseminated information through the series of conferences. It first promoted intersectoral dialogue among governments, forest users, and environmental NGOs. It also initiated the national and regional working groups on standards development and model-demonstration projects. The WWF started the Association of Ecologically Responsible Forest Companies in 2000, called a Producer group; such groups were formed at that time only in Russia and Brazil. In 2002, the WWF, together with Greenpeace, IUCN, BCC, and SEU developed criteria for ecologically responsible forest businesses. The requirements of membership in the Russian Producers Group were developed by its first members: Ilim Pulp, Archngelsk PPM, Volga PPM, Kartontara PPM, Solombala LDK, and Onega LDK (Ptichnikov, 2009: 5). In 2003, the Russian Association of Ecologically Responsible Forest Companies became part of the Global Forest Trade Network (GFTN), which served as a conduit through which WWF connected forest producers with responsible buyers groups, predominantly in Western Europe.

Promotion of FSC certification started through the WWF –World Bank Alliance project and later through WWF partnerships with IKEA and cooperation with regional forest business associations (the forest companies Pomorie in Arghangelsk and PALEX in the Russian Far East) (Tysiachniouk, 2006a). The WWF-Model demonstration projects served as educational sites upon which to show how intensive and/or sustainable forest management schemes can work. The Model Forest Priluzie was essential for involving state agencies in overcoming the barriers and trends related to forest certification in Russia.[134] The Pskov Model Forest, implemented through the WWF-Stora Enso partnership was important for development of nature conservation planning combined with intensive forest management.[135]

---

[132] Same source.

[133] In the case of a country not having a FSC office.

[134] See Chapter 7.

[135] See Chapter 6.

In the early stages of certification development one mechanism for promoting responsible forest management was eco-rating, based on self evaluation by the companies, first conducted by the BCC and later coordinated by the WWF. The results were disseminated to buyers around the world and posted on the internet.[136]

To help companies make the often difficult changes necessary to achieve FSC certification, the WWF, with its partners, has developed a 'step-wise' approach for Russian companies, which guides them through this process. The first step involves adoption of an environmental policy and preparation of an eco-action plan. The second step requires the company to control wood legality, establish a chain of custody system, and conduct an internal audit. The third step involves landscape planning and High Conservation Value Forests (HCVF) protection. The last step involves achieving good forest management and certification. The WWF publishes materials with examples of good environmental policy done by companies. They also publish examples of environmental policies of international companies operating in Russia, such as Stora Enso, UPM-Kymmene, Metsaliitto, and IKEA, and explain how appropriate environmental policies facilitate the process of certification (Ptichnikov, 2009).

The WWF-IKEA project (covering Russia, China, Romania, Bulgaria, Lithuania, Latvia and Estonia) also contributed to FSC's institutional design. IKEA had step-by-step requirements for their suppliers and through a partnership with the WWF they tried to support greening processes for forest businesses. The last step was equivalent to FSC standards. The project had three major phases, 2002-2005, 2005-2008, and 2008-2011[137] and its geographical scope gradually extended from North Western Russia to the Far East. One of the components of all phases was on HCVF and included the development of methodologies, adaptation of the concept to Russia, and fostering mechanisms of implementation into the FSC framework. In the early stages, there was collaboration between WWF-IKEA and the WWF-World Bank Alliance on this issue. The WWF-World Bank Alliance project helped to develop a national interpretation of the HCVF Global tool-kit, developed by the HCV network. Later, in 2008, in the framework of the WWF-IKEA project, a handbook was developed on HCVF definitions, designation and monitoring adjusted to Russia. As was explained in the previous chapter, the HCVF element is tied to FSC Principle 9, and focuses on designating such forests and supporting them. By working on HCVF for almost a decade, the WWF adopted Principle 9 to the district and local levels. They worked with scientific institutions in an effort to create a methodology for designating HCVF, tested this methodology, and suggested amendments to federal and regional legislation, that take into account the HCVF.

WWF-IKEA created a working group on HCVF in Archangelsk with all stakeholders involved, such as administration, forest industry, science, a forest inventory team, representatives of Model Forests, and NGOs including Greenpeace. The development of mechanisms for their use and conservation was started in the Archangelsk region, because, on one hand, there are big plots of HCVF and, on the other hand, forestry is intense and export-oriented. A technical group tested

[136] Interview with WWF staff, March 2004, Moscow; with the BCC representative, March 2011, Moscow.

[137] The phase for the 2012-2014 WWF-IKEA project was approved in 2011, interview with a WWF representative, April 2011.

the methodology in the field, and reported to the working group. Later on, the project worked in Krasnoyarsk, but achieved poorer results than in Arhangelsk.[138]

Another component involved the strengthening of the Association of Responsible Forest Producers by involving new members, including IKEA suppliers, helping companies to formulate environmental policies, and strengthening contacts with buyers in the GFTN. WWF-IKEA worked with current and potential members of the Association. Their efforts included education of top company managers, connecting them with Swedish and Canadian producers, as well as organizing study tours to Sweden, Canada, USA, Brazil, and to Segezha in Russian Karelia. In the third phase, an emphasis was made on the promotion of sustainable wood trade between Russia and China, with WWF-IKEA Russia and WWF-IKEA China collaborating on this issue. The WWF-Amur office fostered and monitored the FSC certification as well as revitalized the regional working group on the FSC standards.[139]

Another initiative of the WWF-IKEA project was the creation of certification centers for education and training. They prepared and trained staff for existing and newly established certification centers. Trainees could be qualified as auditors or could work as consultants for leskhozes or the forest industry. A total of five certification centers were enhanced and supported in North Western Russia and Siberia. The WWF-IKEA project took trainees not only from their priority regions, but from others as well. They educated not only representatives of certification centers, but staff from universities and Forest Inventory Agencies. They conducted a series of seminars and workshops, some of which took place in the Model Forests. The WWF-IKEA project hosts the journal Sustainable Forest Management, where issues related to certification, HCVF, and state forest policy are highlighted.

In addition, the WWF-IKEA worked on educating forest industry staff about certification. They conducted seminars for different enterprises throughout Russia. In the 2000s, there was a huge interest in studying FSC forest management and chain of custody certification.[140] The project contributed to a university course on sustainable forest management and a textbook on forest certification published in 2011.

The next component focused on illegal logging. The WWF-IKEA project prepared an in-depth analysis and made recommendations to regional administrations on what can be done to stop illegal logging. Since the start of ENA-FLEG process in 2005, the WWF-IKEA project has helped governments to develop regional plans for fighting illegal logging. WWF-IKEA Russia and WWF-IKEA China made a joint effort to cope with illegal logging in the boarder regions.

The WWF made an effort to work on Russian federal and regional legislation. They tried to incorporate the concept of HCVF into Russian legislation. They lobbied for creating additional policy instruments for establishing an official status for HCVFs and for creating mechanisms for

---

[138] Interview with a WWF representative, April 2011, Moscow.

[139] Presentation of the director of the WWF forest program, Elena Kulikova 'WWF nature protection strategy and responsible forest use', 30 March, 5th Conference of the National Working Group, Zvenigorod, http://www.fsc.ru.

[140] Interview with a WWF-IKEA project coordinator, March 2004, Moscow.

biodiversity conservation in the territories of industrial forest utilization.[141] However, they did not achieve significant results because of institutional turbulence and the restructuring of the state system of forest management.

4.5.4 FSC national office

As was explained in the previous chapter, FSC national office (National Initiatives before reorganization) are responsible for setting up a national standard, which involves development of indicators and verifiers complying with the global FSC standard, which helps to adjust the standards to the context of a specific country. In Russia, as in the case of other countries, development of the National Standard represents a forum of negotiations, on which the actors interpret the general international standard and adjust it to specific Russian circumstances.

Prior its accreditation by the FSC, the National Initiative in Russia existed in the form of a Working Group on forest certification, which was created in May 1998. The process of forming of the group was difficult. In the beginning of its work there was a lot of friction[142] between the participants, because both more radical NGOs (like Greenpeace) and consensus NGOs (like WWF) were represented.

In parallel with the National group, four regional FSC certification working groups were organized over different time periods: in the Komi republic, Arhangelsk, Krasnoyarsk and the Far East. The working group in the Komi Republic was shown to be the most efficient and productive as it had been developed in parallel with the Model Forest Priluzie (Tysiachniouk, 2006b, 2008c), and in addition supported by the MacArthur Foundation grant (see the case study on the Model Forest Priluzie, Chapter 7). The group in Krasnoiarsk was less efficient, and in the Far East, the group stopped its activity and dissociated.

The process of accreditation of the Russian National Initiative was drawn out, and its founders faced a lot of complications over the course of it. These difficulties were similar to those existing on a transnational level in the FSC system. They were mainly related to equal representation in the economic, environmental and social chambers of the FSC office (National Initiative). It was difficult to involve indigenous people and social organizations in participation. The final difficulty was caused by the nature of Russian social organizations, which mainly specialize in specific social problems concerning single population groups, such as homeless people, the disabled, HIV-infected, drug addicted and others. These NGOs are not usually concerned with the issues related to sustainable forest management. So, it turned out that business representatives, as well as representatives of consulting and research organizations, became part of the social chamber[143] of the National Initiative. In the beginning, it was difficult to involve the business representatives. Later, when many companies had started the certification process, their participation in the work of FSC-Russia became more active. Complications were, in part, caused by the traditional lack of an inter-sectoral dialogue in Russia.

[141] Presentation of T. Yanitskaya, Materials of the seminar 'HCVF and biodiversity' 11-13 November, 2009, http://www.fsc.ru.

[142] Interview with an activist of the Forest Club, March 2004, Moscow.

[143] Participant observation, the author is part of the National Initiative.

Additionaly, the work of the FSC-Russia was not financially supported, and the participants showed differing levels of enthusiasm for volunteer engagement. The National Working Group was accredited by the FSC in June, 2006, and became the National Initiative (FSC office in 2011). In 2006, its membership consisted of more than thirty people with a coordination council (FSC Board of Directors) of nine people. The National Working Group has had, since its start, five conferences; the last, and biggest one, was held in Zvenigorod, 29-31 March, 2010. The conference elected the new Board of Directors, approved new by-laws and more than 20 new members, and held several roundtables and training workshops. The major focus of the conference was on the quality of FSC forest management and controlled wood certification in Russia, harmonization of the FSC standards with national legislation, and promotion of the FSC in the internal national market. An effort was made to conduct the conference of the National Working Group in a same manner as the FSC General Assembly, and the same voting system (each organization gets 9 votes versus 1 each for individuals) was introduced and tested.[144] The Russian FSC Board created two committees: one on conflict resolution and a technical committee to resolve all other up-coming issues, and started actively working on improving the quality of the FSC in Russia.[145] The technical committee was formed in spring 2012 to work on national standards development.

In 2010, the FSC-Russia began to receive funding from the FSC International, as the decision was made to share funding from certification fees with the National FSC offices. Money sharing between the FSC International and the FSC Russia allowed the intensifying of activities of the National office and the Russian FSC Board.

4.5.5 Russian national FSC standard

The process of creating and accreditation of the FSC National standard is also a very long process, as it must be conducted with the participation of all stakeholders and should be thoroughly discussed by them. It must be composed according to the rules of the FSC International. National standard developers had problems in keeping a balance in these two processes: changing of the standard to adapt it to the conditions of the country, and keeping it in the framework of the global standard. Due to this, the draft standard was returned by FSC for revisions several times. The system itself of the FSC international coordination center works rather slowly. Due to all this, in Russia the National Standard has taken more than 10 years to develop and accredit. In October 2003, the first draft of the FSC Russian National standard was finished, but at that same stage the National Initiative was not yet accredited by Accreditation Service International (ASI).

The FSC national standards include stronger protection for the rights of indigenous people than does the Russian governmental policy. The national FSC standard has been tested in five different places. The regional Komi republic standard was coordinated with the national draft standard. In the 6th version, both environmental and social indicators became clearer and stronger. The work on the Russia National Standard continued until November, 2008, when the 6th version was at last accredited with some corrective action requests. The 7th version was developed with

[144] Participant observation at the conference, the author been elected to the FSC-Russia Board of Directors, March 29-31, Zvenigorod.

[145] Participant observation at the meeting of FSC-Russia Board, 28 October, 2010, Moscow.

improvements to the 6th which went beyond the ASI requirements. Stakeholders were widely involved in commenting on the 6th version and suggesting changes for the 7th. In addition, guidelines on public reports for companies were developed.[146]

From the time of accreditation, CBs must be guided by the National Standard, but official validity of the norm came 12 months after the date of accreditation. As was described in the previous chapter, before the national standard is accredited by the FSC, CBs themselves adjust the FSC standard to local conditions. It has to be noted that when various CBs work in a country, the interpretation of the standard changes with a particular CB, because despite the common approach to certification, different experts focus on different issues and have their own priorities in adaptation of the global FSC Standard (Malets and Tysiachniouk 2009). After the National Standard is accredited, all CBs have to bring their activities to conformity with the national standard. As the approaches to certification were different, challenges of adaptation to the national standard were also different for the different companies. Some companies have to enhance social aspects of certification, while others also have to reorient their approach to HCVF. The 6th version of the National Standard had certain weaknesses noticed by CBs, companies and stakeholders. Therefore, changes have been implemented and in 2010, the 7th version, with many clarifications to the 6th version, was submitted to the FSC for accreditation.[147] It was several times revised since then and approved in Ocober 2012.

4.5.6 National FSC office

As was described in the previous chapter, the National FSC office is part of the FSC international system. The representatives of the FSC node of design in each country are mainly engaged in coordinating FSC activities, linking clients with auditors, maintaining data bases on FSC products and facilitating certification processes for the companies.

The national FSC office in Russia was established in February, 2005, with initial funding provided by the European Union grant program. The FSC office in Russia is mainly engaged in the coordination of the FSC's activities in Russia and the Commonwealth Independent States (CIS) countries, yet most of its work is related to Russia and work in the CIS became apparent only in 2009-2010. Their responsibilities include coordination of all the work related to FSC certification within Russia, namely: interplay with the National Initiative, the FSC-Russia Board of Directors, certification bodies, and stakeholders. The office conducts major informational work, it creates the database of certified companies, and spreads the news from the FSC International Coordination Center among all interested parties in Russia. In the office, all stakeholders can get information concerning new certification guidelines and methodology manuals on various certification aspects, information about training and other organized events. To make the process of institutionalization of the new rules easier, the FSC national office in 2008-2010 intensified its work with the companies and CBs, organizing trainings, distributing manuals and guides, etc. (Tysiachniouk, 2012a,b).

[146] Participant observation, 2008-2011.

[147] Participant observation in the process of the development of the 7th version of FSC standards, 2010.

Stakeholders are invited to discuss the information on the FSC website. It has to be mentioned that there is not a lot of discussion on the website because of low involvement from the interested parties. The National office is also forced to take part in solving various questions and problems which arise in the course of certification. In general, the FSC office should not interfere in conflicts related to certification; it can only give necessary information to the parties and reply to their questions. Despite this, the Russian FSC office was often forced to participate in discussing conflicts.[148] That happened because the stakeholders, including governmental structures, often addressed all their questions, claims and complaints from the sites of implementation to the FSC national office. Since late 2010, conflict resolutions has become the responsibility of the special committee, initiated by the FSC Board.

With reorganization in 2010, the FSC office and the National Initiative became one organization, managed by the FSC-Russia Board of directors with funding provided by FSC International and membership dues, which will start in the fall of 2011.[149]

4.5.7 Current state of FSC certification

Russia is in second place after Canada organized in the world's amount of certified territories, of ten major holdings, nine are certified, and many smaller companies are in the process of certification. For the most part, FSC certification has been achieved by companies already operating in the European market. Certification helped to increase their contacts in Europe and to ensure long term contracts.[150] In some cases, forest companies sought certification in response to demands made by their buyers, thereby protecting future trade with environmentally sensitive consumers.[151] Certification tends to make forest companies feel more secure about their future. This became clear during the economic downturn in 2008-2009, when certified companies were better off than non-certified companies in selling their wood abroad.

110 Forest Management certificates have been issued (see dynamics of growth, Figure 4.2) and around 30 million ha are certified.[152] To date, support for certification varies by region. It is greatest in the European part of Russia, it is currently booming in Siberia and has only recently started in the Far East, largely owing to European buyers' demands for certified wood, who themselves came under pressure from nongovernmental organizations to make those demands. High demand for non-certified wood by Asian markets, especially those in China, as well as both Russian and Chinese corrupted networks and illegal operations have prevented the fast development of certification in the Russian Far East. Of the 26 million certified ha, 70% of the certificates are issued in North Western Russia, 22% in Siberia and only 8% in the Russian Far East (Figure 4.3). The number of CoC certificated is steadily growing (Figure 4.4). A total of 222 chain of custody and 136 controlled wood certificates are issued.[153] With the rapid growth of

[148] Participant observation, 2007-2011.

[149] Participant observation at the FSC-Russian Board meetings in May and October, 2010.

[150] Interview with the FSC contact person, February 2004, Moscow.

[151] Interview with WWF staff, February 2004, Moscow.

[152] Http://www.fsc.ru, data provided for 03.07.2012.

[153] Http://www.fsc.ru, data provided for 03.07.2012.

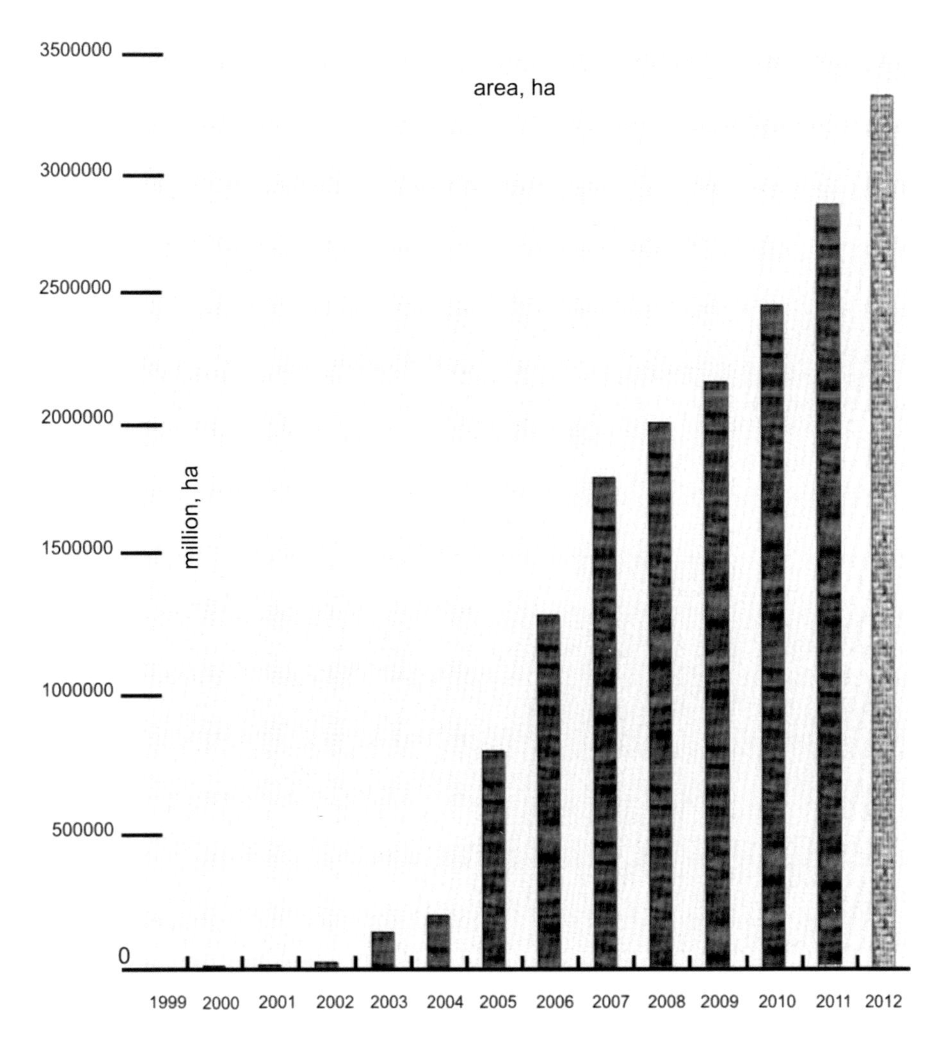

Figure 4.2. Development of forest management certification in Russia 2000-2012. 2012 is in grey because it is an estimate done by FSC office.

forest certification in Russia in the 2000s, the quality of the certificates became an issue (Malets, 2010). The FSC increased surveillance and in the territories of 1.9 million ha, certificates were temporarily suspended in 2008-2009 until forest management practices improved.

4.6 Interplay of state and non-state actors in the process of FSC certification

Certification is being advanced by non-governmental actors, i.e. environmentally responsible businesses and NGOs, and refers to the sphere of non-state governance of forest resources. Its relationship with state institutions in different countries is developing in different ways. There are countries, such as Belorussia and Bolivia, where standards for certification were immediately

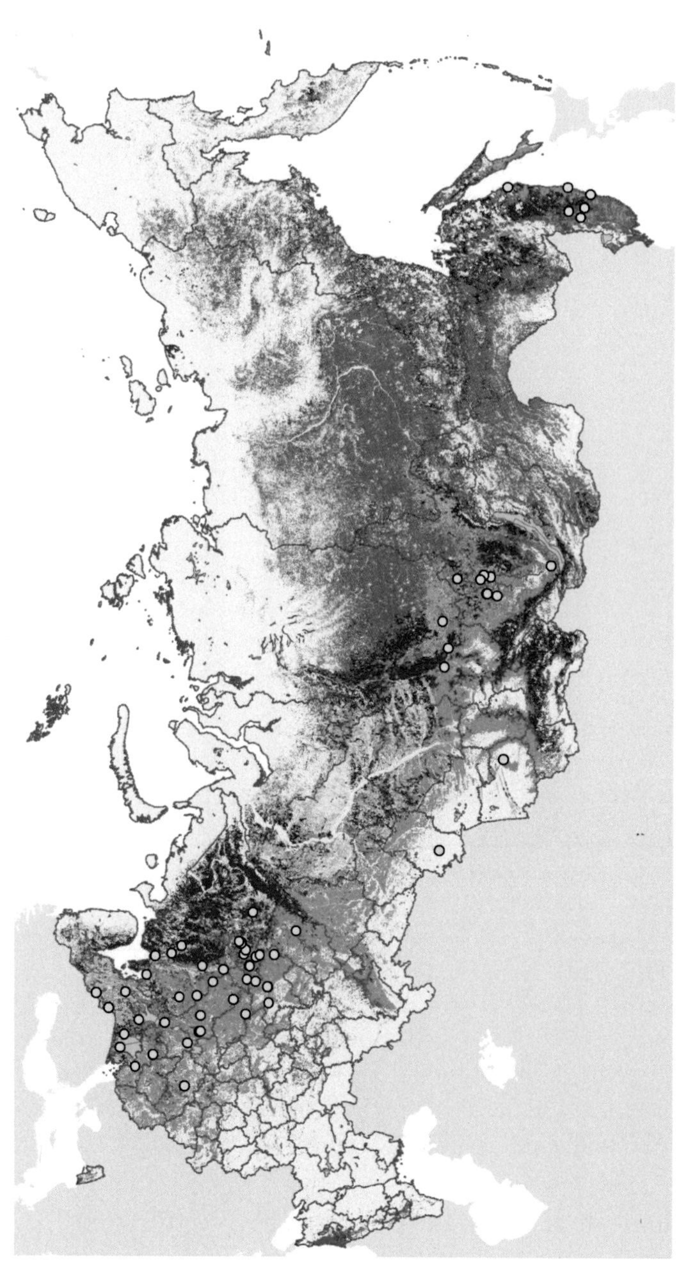

Figure 4.3. Forest management certification in Russia in 2010. Map developed and provided by the FSC-Russia, dots are forest management certificates.

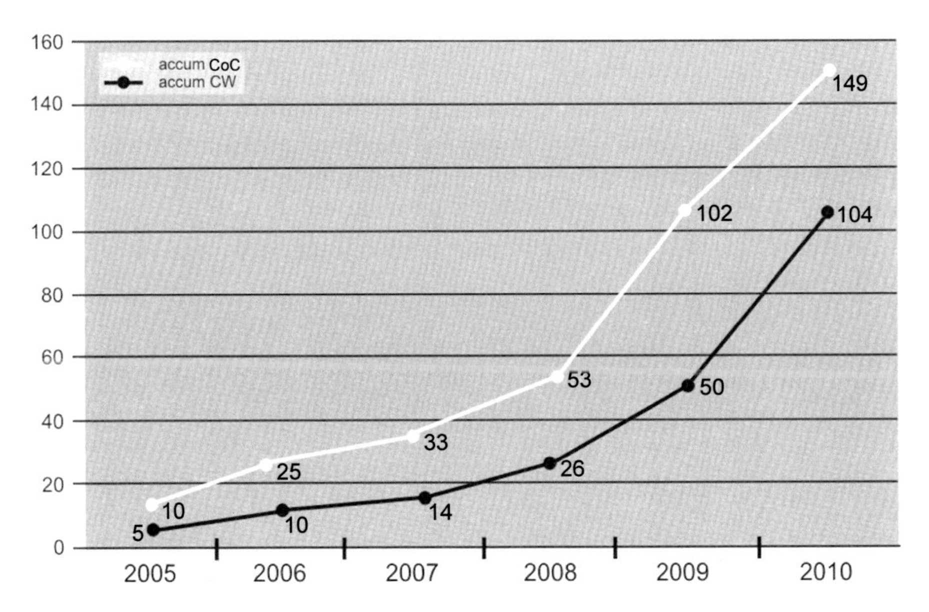

Figure 4.4. Accumulated number of CoC and controlled wood certificates in the Russian federation, 2005-2010.

adopted by the state and included in their legislation. As was described above, in Russia such integration is not the case.

4.6.1 FSC requirements and Russian legislation

In Russia, there are contradictions between the Russian legislation and the FSC requirements. The new Forest Code adopted in December, 2006, disregarded innovations developed in the process of certification; that is why the discrepancy between certification and the Russian laws continues. In 1995, Russia ratified the Convention on Biological Diversity (CBD) whose regulations are consistent with the FSC standard. Article 1 of the Forest Code of 2006 declares a commitment to the CBD requirements. However, the state has not elaborated adequate regulatory documents, which would ensure implementation. It is necessary to remark that since the Soviet times, Russia has constituently taken measures for maintaining biodiversity in wildlife and conservation. These measures, however, did not concern the sphere of commercial forest exploitation, while certification envisages regulation within this very sphere. This has entailed contradictions with the Russian legislation. For example, the concept of key biotopes, a requirement of the FSC, was not even mentioned in the Russian Forest laws. Another example is that Russian legislation provides that old growth forests should be preserved only when they belong to the first category of forests (those that are close to waterways, contain valuable species or are in specially protected areas). When forest companies lease territories for commercial forestry, these territories often contain old growth forests, forming relatively large intact forest landscapes, especially in the Arghangelsk,

Komi, Karelia, Siberia and the Russian Far East. According to the FSC certification, these old growth forest landscapes belong to the HCVF-2 category (Tysiachniouk, 2008b).

Another challenge for the FSC in Russia is the issue related to indigenous people. The reason for this is again a different understanding of the term 'indigenous people' by Russian legislation and by the FSC National Standard. Russian legislation recognizes as indigenous only Low-Numbered Populations of the North (less than 50,000 people). The Russian FSC National Standard recognizes any community consisting of one or more ethnic groups as indigenous people if they are engaged in traditional forest utilization.

Several forested regions of Russia are populated by indigenous peoples. Indigenous cultures throughout Russia – the Komi, Koryak, Itelmen, Udegeis, Chukchi in the north, and many others – have suffered greatly since the advent of Russia. In Tsarist times, the Russian Empire's eastward expansion brought Christianity, as well as marauding Cossacks demanding tributes in fur from the native peoples.

Later, the Soviet policy toward indigenous peoples brought even more far reaching changes to their cultures and ways of life. The State Committee for Numerically-Small Peoples of the North, Siberia, and the Far East oversaw this policy, operating with the primary goal of turning the native people from aboriginal semi-nomads into full place-tied citizens of a modern Soviet society. The policy of 'centralization' moved subsistence-based community clans into more centralized villages. This allowed the state to more efficiently deliver subsidies, which included bread, coffee, tea, sugar, and the other basics. Native people were put to work on collective farms, and the children of the reindeer herders were sent to boarding schools for education. After perestroika, subsidies halted abruptly, rural economies soured, and indigenous people became even more disempowered (Tysiachniouk, 2006a).

In the Far East, forest conflicts and tensions occur with the Udegeis populations. Since the early 1990s, there has been new legislation and a policy process to create 'Territories of Traditional Nature Use' for indigenous peoples, also called ethno-ecological refuges (Zaporodsky & Murashko, 2000). This policy is applicable to Indigenous Low-Numbered Populations of the North. The absence of appropriate norms inhibits the designation of such territories. Many native communities, such as Komi, Pomor, Karel, are not considered low-numbered and there is no government policy to incorporate them in the forest decision-making process.

FSC certification has the potential to clarify and protect the rights of these people. However, the issue continues to be very complicated. Tensions and conflicts concerning indigenous peoples rights occur almost on all certified territories where these groups live.

Some ethnic groups are not interested to be 'qualified' as indigenous, while others, on the contrary, strive for recognition. This issue arose in the Komi republic, because the Komi people are officially an autonomous ethnos, having their own republic. Representatives of their national movement 'Komi voiter' did not agree to be referred to as indigenous people because they were afraid to provoke conflicts between Russians and Komi who live in same places and had never before considered their ethnic affiliation. On the contrary, the Pomor people in Arhangelsk, consisting of several ethnic groups including Russians, are interested to be recognized as indigenous people and obtain their rights. They actively participate in the FSC and designate their hunting grounds as HCVF (Kulyasova, 2010).

4.6.2 FSC-state relationships

The FSC-Russia pays great attention to fence-mending with state bodies and strives for reaching several goals in this interplay. First, it tries to lobby for necessary changes in national legislation, in order to eliminate contradictions with the FSC rules. For this purpose a working group was created to resolve the contradictions between requirements of the FSC certification and the new Russian forest code, especially the issues concerning biodiversity, because this issue was the stumbling block for the companies during the process of certification. Having achieved some results, the working group, however, has not managed to resolve the problem once and for all.[154]

Since June 2003, when Rosleskhoz was still functioning under the Ministry of Natural Resources, the FSC Russia, as well as the WWF took part in the Rosleskhoz Public Ecological Council. The goal of their participation was the dissemination of information on the purpose and progress of certification, and on its requirements. Through the Council, the NGOs lobbied for harmonization of Russian legislation with the FSC requirements. With jurisdictions changing in 2008-2009, the Council met irregularly.

The breakthrough in state-non-state actor's mutual understanding took place at the Parliamentary hearings on 'The legal basis of forest certification to ensure the legality of exports and imports of timber and processed wood', which took place May 20, 2010. Parliamentarians, representatives of Rosleskhoz, the Ministry of Agriculture, the WWF, the FSC, and large holding companies used the participatory approach for development of policy recommendations.[155] Issues on contradictions of the FSC requirements and Russian legislation were addressed. The Committee of Natural Resources, Nature Use and Ecology agreed to become a platform for negotiations between the different interest groups. WWF and FSC representatives acknowledged that there was a high probability that the changes to Russian legislation would be made, as the state authorities, companies and NGOs all finally understood each other.[156]

Shortly after the Parliamentary hearings, the working group on harmonization of forest legislation with the FSC standards was formed and had its first meeting. The FSC again received an opportunity to negotiate contested issues with state authorities. However, since September 2010, the negotiations have been frozen due to a new reorganization of state agencies. As was mentioned, Resleskhoz received the status of an independent agency, combining both executive and legislative powers.

The Public Forest Council convened under the jurisdiction of Rosleskhoz in April, 2011. The Participants acknowledged the need for modernization of forest legislation. Governmental officials and NGOs once again discussed contradictions between the FSC requirements and the Forest Code of 2006. They analyzed the conflicts that arise due to these contradictions and decided to intensify work on harmonizating the Russian legislation with the FSC standards.[157] When a new

[154] Presentation of the head of the FSC office, August 2008, Moscow, participant observation at the meeting of the National Initiative with certification bodies.

[155] Http://www.duma.gov.ru/cnature/parl_conf/parlam/home.htp.

[156] Reflections of WWF and FSC representatives at the FSC-Russia Board meeting, May 2010, participant observation.

[157] Participant observation at the Public Council at Rosleshoz, April 4, 2011, Moscow.

conflict related to HCVF preservation arouse in Russian Far East in Spring 2011, Rosleshoz took the side of the FSC in the process of conflict resolution.[158]

The collaboration between the FSC and state agencies in the process of FLEG implementation has been mutually beneficial. FSC-Russia submitted to Rosleskhoz a proposal in the framework of FLEG on including FSC certification into the national plan of combating illegal logging. The main purpose of FLEG has been the struggle for the legality of wood in trade, and certification ensures such legality. Thus, the acknowledgement of forest certification by FLEG would help to avoid overlapping regulative functions of these two tools of forest management. The FSC representatives made attempts to coordinate their work with events conducted under FLEG, as obtaining recognition of certification is one of the best mechanisms of verification of legality.

4.7 Fostering FSC demand on the internal market

Both the WWF and FSC aimed to achieve internal demand for certified products through state policy. In 2007, they started negotiations with federal authorities lobbying for a governmental purchasing program based on certified timber, as is customary in most European countries. In 2008, the FSC national office, together with the WWF, started a campaign of promoting FSC certification in the internal Russian market.[159] They organized an informational campaign with businesses and representatives of governmental structures to explain the advantages of certification. Many roundtables with representatives of business and state authorities have been organized. For example, in December 2009, a roundtable focused on promotion of FSC certified packaging as an alternative to plastic.[160] The roundtable was organized by the FSC, WWF and the Investlesprom holding, and the Moscow government was the major target group. The most recent roundtables in 2010-2011 were organized with the aim of fostering green purchasing programs and policies in Russia.

Governmental agencies in Russia are generally responsive to the demands of large businesses; therefore, TNCs operating in Russia can significantly impact national policy. There are several companies in Russia that are driving FSC certification on the internal market and fostering visibility and recognition of the FSC trademark. Mondi Business paper Siktivkar Pulp and Paper Mill is producing office and printing paper called Snegurochka (Sowmaiden). Three printers certified their chain of custody, therefore publications are produced on FSC certified paper. The holding company Investlesprom produces paper packaging and one of its subsidiaries is involved in green building, based on FSC certified wood (Figure 4.5).

As in other countries, the Olympic Games in Russia are one of the major drivers of the internal FSC market, as the Olympic Committee requires the games to be green. In 2008, Rosleskhoz and the President of the Russian Federation approved an agreement with the FSC that only certified wood will be used in construction projects in the 2014 Olympic Games in Sochi.[161] This means

[158] Participant observation at the Public Council, April 4, 2011, Moscow.

[159] Presentation of a WWF-IKEA project coordinator at GFTN meeting, 2008, St. Petersburg.

[160] Materials of the Roundtable 1 December, 2009:'Ecological packaging, world experience and state in Russia' http://www.fsc.ru.

[161] Presentation of the head of FSC office at the meeting of the national initiative, August 2009, Moscow.

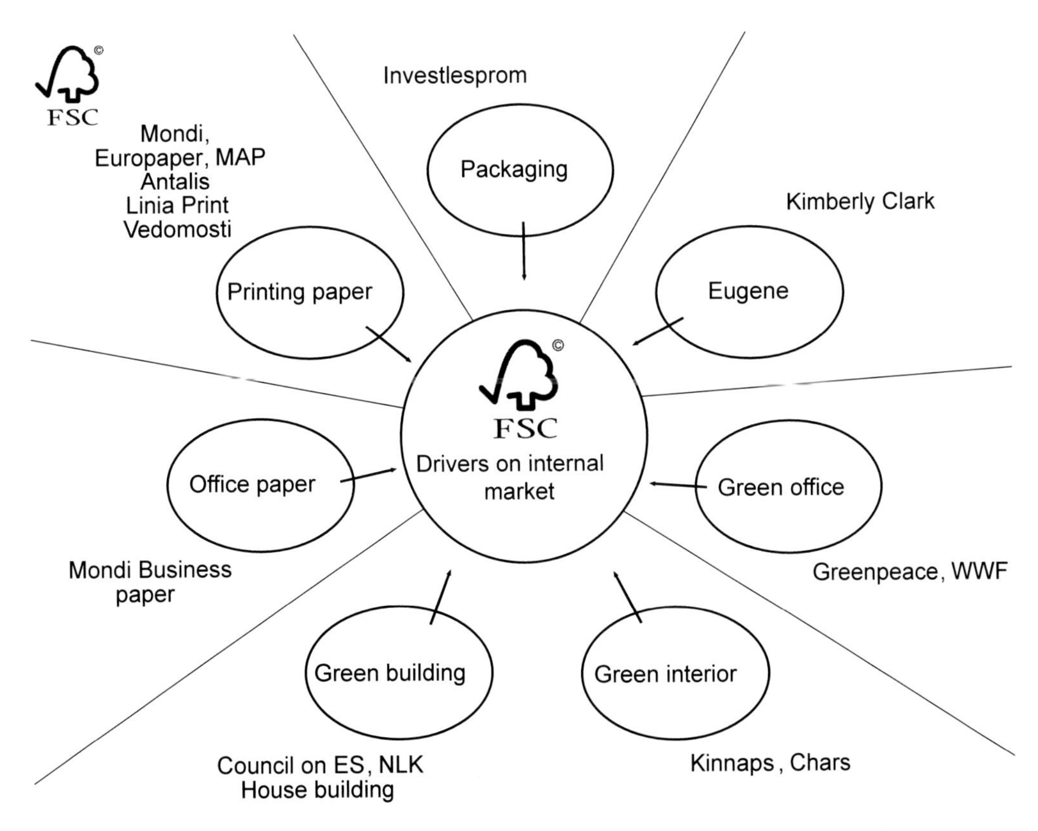

Figure 4.5. Drivers of the FSC campaign in the internal Russian market, contributing to the recognition of the FSC trademark.

that building companies will use exclusively FSC certified wood and all the suppliers will be obliged to have certificates of the chain of custody for this project. Such a decision of governmental structures became a considerable step in boosting FSC certification in Russia. In April 2011, the state agency Olympstroy (Olympic Construction) switched to FSC certified office paper and furniture. The FSC office decided to become an intermediary between the supply and demand side of Olympic construction.[162]

The national FSC office is continuing negotiations on converting all public purchases to FSC certified products. An environmental purchasing program for the government of Moscow is under way. In May 2010, the State Council of the Russian Federation made a decision on making the state purchasing program more ecology minded, that if implemented will help foster demand for FSC products on the internal market. 2011-2012 is the second phase of the FSC campaign, in which major target groups are state bodies, retail stores and consumers.[163]

---

162 Http://www.fsc.ru/?mod=news&id=219.

163 FSC office Power Point Presentation: 'The FSC campaign, second phase, http://www.fsc.ru/index.phd.

4.8 Conclusions

With its major focus on the federal level, this chapter has shown the complicated process of institutionalizing the FSC in Russia. Despite the efforts of several initiatives introducing other certification schemes in Russia, the FSC has become the only one that plays a significant role in forest management.

The FSC emerged in Russia, in part, because certain buyers in Europe requested certification from their Russian suppliers. The FSC also emerged because environmental organizations, especially the WWF, promoted it. As was described, the WWF demonstration projects, WWF-Stora Enso and the WWF-IKEA partnerships contributed to institutional design of certification. Therefore, the chapter demonstrates the essential role of environmental NGO networks, especially the WWF and Greenpeace, in promoting FSC certification. Interestingly, much of the WWF's promotion of sustainable forest management in Russia has been funded by western governmental agencies, including the World Bank, the Swedish International Development Agency, and the Swiss Agency for Development and Collaboration.

The chapter shows that despite resistance of state authorities step by step, the pressure of private authority in governance of Russian forests is increasing and making its way through the state regulatory system. As was demonstrated in this chapter and will be further demonstrated in the case studies, certification addressed many ecological, social, and economic aspects of forest management in Russia. The most contested ecological issue that was described is the consumption of wood from pristine forest landscapes. Despite contradictions with Russian legislation, certified companies are required to identify and protect HCVFs, taking into account biodiversity and adopting sustainable forest management. Moreover, certification has made it possible to protect forests in territories that are leased to forest companies, and not only those in specially protected areas. As a result, FSC certification has significantly reduced the threats to HCVF on certified lands, especially in the European part of Russia. This is especially true in the Arghangelsk region, Karelia Republic and in the Komi Republic (Tysiachniouk, 2006a, see case studies in this book).

Multiple problems occur with implementation of the social aspects of certification (see the case studies in the book), especially those related to support by companies of the forest settlement's infrastructure. Certification in Russia occurred simultaneously with the period of general post-perestroika economic reforms. In the course of these reforms, the forest settlement's infrastructure, which traditionally was supported by the forest enterprises in Soviet times, was transferred to the governmental responsibility. Thus, the local people's expectations of support from the companies turned to be much higher than the businesses were able to provide, even under the conditions required by certification (Tysiachniouk, 2008c, 2009).

Generally, FSC certification appears to have great potential as an economic instrument for the management of forests allocated to concession or rent. It can help strengthen forest governance structures, because it integrates the interests of producers, consumers, nature protection and effective participation of civil society. Internationalization of forestry and foreign investments may also help the Russian processing industry, which may in turn help address the problems of extensive forestry.

Chapter 5. Preserving old growth forests in Karelia: involving Western consumers in a market campaign

5.1 Introduction

In this chapter, I will analyse a market campaign that was initiated by NGO networks to save old growth forests in the Republic of Karelia in the Russian Federation. In the early 1990s, NGOs started monitoring the area of old growth forests in Russia, including Karelia, and were simultaneously engaged in designing preservation policies that could influence the Russian environment. ENGOs, including Greenpeace, and Taiga Rescue Network began discussing the preservation of old-growth forests located on the border between Finland and Karelia. Greenpeace led an effort to propose the Karelian forests for inclusion in the international UNESCO World Heritage List. The World Heritage Convention was adopted at the General Conference of the UN on November 16, 1972, and the Russian government was among its 168 signatories. The 'Greenbelt of Fennoscandia' would have been a joint nomination from Russia, Finland, and Norway, and would have included 20 forest massifs along 1000 km of the border region. In all, the area compromised 1.5 million hectares of forest, some of it virgin. This effort failed. Therefore, the ENGOs chose consumer boycotts as a strategy. ENGO networks already had vast experience of organizing similar campaigns. They learned to shape consumer preferences through international information campaigns and to raise international awareness of the importance of old growth forests.

The focus of this chapter is on the NGO network which set up a market campaign that stretched from the space of place in Russian Karelia to transnational space actors, the TNCs involved in buying wood from valuable old growth forests. In Chapter 2, three ideal types of governance generating networks (GGNs) have been outlined: bottom-up, top-down and two-fork. The case study described in this chapter belongs to the first type, a bottom-up network. The focus of the chapter is a deliberately initiated transnational conflict, 'a market campaign', which NGO networks use all over the world in order to force business actors to change their practices. As was described in Chapter 2, market campaigns are usually related to the behaviour of multinational companies in both producing and consuming countries and pertain to the processes of production and/ or management of natural resources. Market campaigns take the form of a social movement, however, at the same time they represent GGNs because they are involved in global governance, play the role of a global regulatory agent and generate new governance arrangements in specific territories. The campaigns are typically initiated against corporations with a highly visible brand in cases where those corporations operate in ways that violate internationally accepted rights of workers or that destroy the environment and natural ecosystems. Such campaigns are commonly organized by civil society actors to change the behaviour not only of the forest-producers but also of the companies in industries such as oil, textiles, bananas and coffee. Market forces and consumer preferences are used by the campaign organizers to influence a company's behaviour (Conroy, 2001; Tysiachniouk, 2009).

In the forest sector, the best known and most significant campaigns were conducted by Rainforest Action Network, Forest Ethic and Greenpeace to save the Amazon's forests, targeted not only corporations, but also their investors. One of the goals of the actors involved in market campaigns worldwide in the forest sector is the preservation of old growth forest landscapes. Activists, from a number of NGOs, join their efforts in identifying the problem, planning and implementing the campaign against companies involved in logging or buying wood from the intact forests. This group of actors represents the node of design of a market campaign. Territories involving old growth intact forest landscapes are the sites of implementation of these GGNs. As a result of a market campaign social changes are expected to occur in both transnational spaces and the spaces of place. Transnational actors are expected to understand the risks related to purchasing wood from the old growth forests, while in the space of place local practices are expected to be reassessed and renegotiated. Purposefully initiated conflict shapes the institutional environment in both transnational spaces and spaces of places and calls for reassessment of existing practices. Market campaigns involve coercive measures, thus negotiations take place at all stages of the conflict in both transnational spaces and the spaces of places. Forums of negotiation are part and parcel of the conflict management and the design of new institutions.

This chapter is based on field research conducted between 2002 and 2009, during the later stages of the conflict. Qualitative methods, such as semi-structural interviews (total n=273) and participant observations at public hearings and informal conversations in the villages were used during four expeditions to the area, each lasting 15-20 days, in October 2002, November 2006, June 2006 and August 2008. An additional trip to the area was done in August 2009 during the preparation of the Investlesprom subsidiary for FSC certification. The interviewees comprised NGO representatives in both Moscow and Karelia, representatives of the governments at different levels in Petrozavodsk, Kostomuksha and Kalevala, representatives of logging companies and local stakeholders of the Kalevala National Park, including local people living in the park (see Appendix B-C for details). Additional interviews have been conducted in the headquarters of the corporations, based in Stockholm and Moscow.

The following section introduces the reader to the specific regional context of the Republic of Karelia and the local contexts on the Russian and Finnish sides of the boarder (Section 5.2). Next, I describe the market campaign's stages and chronology of events (Section 5.3). In the following section attention is given to the analysis of stakeholder interactions, describing tensions at the negotiation processes leading to the preservation of the old growth forests. The role of transnational, regional and local stakeholders is highlighted (Section 5.4). The chapter proceeds with the description of the outcomes of the market campaign (Section 5.5) The chapter concludes with a model of the cross-border market campaign's GGN, oriented bottom-up from the space of place to transnational spaces and specifics of the operation of its node, forums and sites of implementation. Special attention is given to organizational isomorphism in both transnational spaces and the spaces of places (Section 5.6).

5.2 Regional and local contexts

5.2.1 Republic of Karelia

The Republic of Karelia is a heavily forested area in northwest Russia. Since imperial times, forestry has been the republic's primary industry with an orientation towards export. (Autio 2002a,b). In 2006, forest production was 55% of the total industrial production in Karelia, with 60% of all forest products being exported (Kozyreva, 2006). Karelia's border with Finland is 700 km long and forms a large part of Russia's longest land border with Western Europe. While most of the country's forest resources are far to the east in Siberia and less accessible, Karelia offered huge tracts of virgin, old growth forests with proximity to the important timber markets in the west. In addition, Karelia's extensive waterways, over 27,000 rivers and 60,000 lakes, provides an effective means to transport logs and so helped to orient the forest sector towards export (Kozyreva, 2006).

Under socialism, two zapovedniks were created in Karelia in the 1930s – Kivach and Kandalaksha (Kantalahti). These encompassed 19,300 hectares of mostly forested land. In addition, the Vodlozero and Paanajarvi National Parks and the region's many nature reserves contain some of the largest tracks of virgin forest in Europe. An area of 404,000 hectares in the Vodlozero National Park was nominated for the UNESCO World Heritage List in 1998, but was not included.

Despite Karelia's economic importance, it remained fairly undeveloped until 1917, with over 90% of the population living in rural areas, as compared with 74.1% living in cities and towns today (Autio, 2002a). The Soviets had plans to rapidly industrialize much of the country, and Karelia's forests became an important resource for efforts to enter the world's timber markets. Timber was sold at low prices in Western Europe. Many researchers see this as a strategy of the Soviet government to acquire hard currency, which was much in need at the time (Autio, 2002b). As the country became increasingly centralized, the economic planning for Karelia became disconnected from the regional level. Quotas were set high and labour was often in short supply (Autio 2002b). In addition, processing plants were overwhelmed and so increasing amounts of unprocessed logs were exported (Autio, 2002a). Very few funds were put towards the development of the forestry infrastructure in Karelia. In addition, the Soviet Union's economic approach to logging was based on the belief that increased harvesting meant increased profits. This trend accelerated in the 1950's with increased large-scale industrial logging and clear-cuts.

This somewhat one-sided economic policy for the region had important impacts at the local level that persist today. During the 1990s, Karelia experienced an economic downturn and forest production declined nearly 40% (Autio, 2002b). This downturn somewhat reduced pressure on the forests, but the government continued to promote the policy of 'more harvesting means more economy'. Thus, in the 1990s, with increasing economic difficulties, the republic's officials saw increased exploitation (Laine, 2002), especially of unexploited old-growth forests, as a necessary and urgent task.

Over 60% of Karelia's virgin forests were harvested (Yanitsky, 2000) in the course of the 20th century. The forest industry was especially interested in logging the old growth in the former military zone, where transnational NGO networks were interested in creating a new Kalevala National Park.

5.2.2 Local context, the Russian side of the border

The territory of the disputed Kalevala National Park represents a forested area that lies partly in the former Finnish border military zone and has only one small village inside it (Figure 5.1). The same families have lived in this territory for generations; but they were relocated to a bigger town during socialism, returning after the fall of the Soviet Union. Currently, there is a small farm inside the park and small-scale tourism with Finland is developing. From the very beginning local villagers were strongly against logging in the area, yet they were not interested in federal agencies governing the park and building its infrastructure. The major local stakeholders lived in the villages of Voknavolok (Vuokkiniemi), Kalevala and the small city of Kostomuksha, located across the border in Finland, outside the disputed area. There were no local NGOs participating in the issue on the Russian side of the border, but Finnish civil society groups were involved.

5.2.3 Local context, the Finnish side of the border

There was also a conflict going on in Finland that centred on saving borderland forests. The commercial logging on the Finnish side of the border was much more intense, as the Finnish military zone was much narrower than the former Soviet Union's zone, and thus only small spots of old growth forests were left. Both international and Finnish NGO networks easily mobilized the local population in support of the Kalevala National Park on the Finnish side of the border. As a result, the Finnish Kalevala National Park was formed, consisting of several small specially protected areas (Chriakenen, 2005). People who were mobilized in support of the Finnish Kalevala Park became dedicated supporters of the Kalevala National Park on the Russian side of the border, and they adopted and understood in full the value of old growth forests and participated in the actions in Russia organized by NGO networks.

The old growth forests were not the only reason for the Finnish communities' support for the Russian Kalevala National Park. The Karelian borderlands are perceived by both Karelians and Finns as culturally close; local people on both sides of the border share a common language, common traditions, common Karelian-Finnish epic runes and folklore. The *Kalevala*, the Finnish national epic, was collected in the Finnish Karelian borderlands, especially in the area close to the park, and Elias Lönnrot (1802-1884) invented the title Kalevala for the poems collected. The first edition of the *Kalevala* was published in 1835 (Lehtinen, 2006: 175). Later on, the name Kalevala was given to the park area.

Finnish people were thus eager to preserve old Karelian villages with their traditions and surrounding forests. In their perceptions, if the forests around Karelian villages were logged, the villages would change and lose their ethnic spirit.[164] Finnish ethnographers, especially those who study folklore, collaborated with Russian scientists from Petrozavodsk and Kalevala and organized expeditions to the Karelian villages to collect songs and describe traditions. These people, although they did not participate in direct actions, were strongly supportive of all kinds of preservation programmes in the area.

[164] Interview, Finnish ethnographer, April, 2006, Vocknavolock.

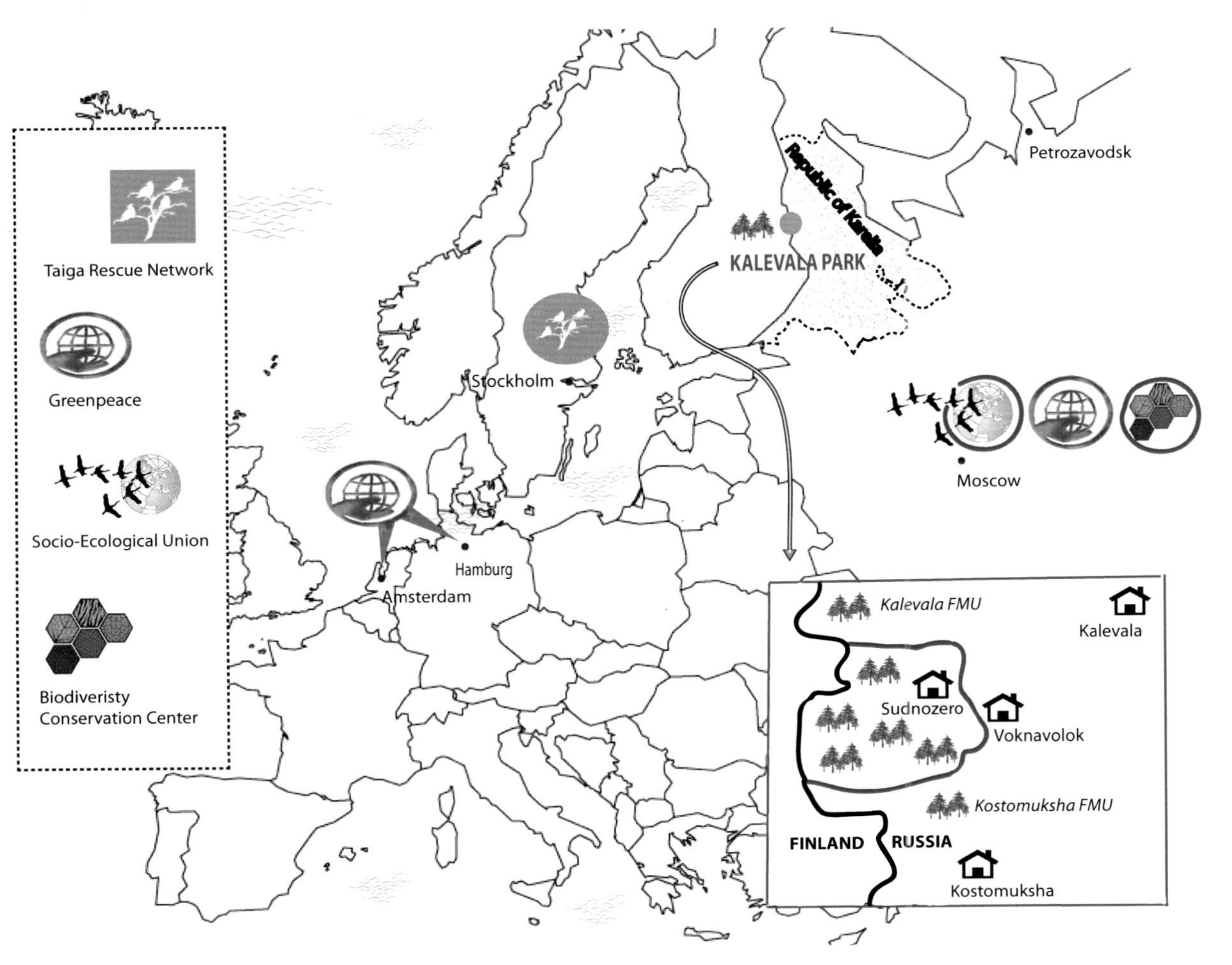

Figure 5.1. Kalevala National Park: the case study with actors involved.

Finnish tourist firms developed special trips to the Karelian villages, including areas inside the Kalevala Park, in order to introduce tourists to Karelian culture and their simple lifestyle, which is still not modernized; Finnish people reside in Karelian villages as they might have lived in the last century.[165] The village inside the Kalevala park is a special attraction as it is home to a family whose ancestors were rune singers; their house is like a open air museum, in which people live and exhibit old traditions, including animal sacrifice, that were common in remote Karelian villages in the 19th century and date from pagan times. Both historic culture and pristine nature are advertised as part of these trips. Currently, a new tourist route called 'the Blue Road'[166] is being discussed that would link ethnic places in Sweden, Finland and Karelia. The Kalevala National Park is on the route, between other stopping places in the villages of Voknavolok (Vuokkiniemi) and Kalevala.

All these projects, although not directly related to the preservation of old growth forests, constituted the context in the space of place for the success of a NGO-led market campaign organized in transnational space.

5.3 The Karelian old growth forest market campaign: stages and chronology

The international NGO-led market campaign took place during the period 1995-2006 with the aim of saving old growth forests in the Republic of Karelia, Russian Federation (Tysiachniouk, 2008b). This campaign was organized by Greenpeace (International, Germany and Russia) and the Taiga Rescue Network in conjunction with Russian Moscow based NGOs, in particular the Biodiversity Conservation Center (BCC) and the Socio-Ecological Union (SEU). It also included the student group SPOK, based in Petrozavodsk, the capital of Karelia. The campaign built on the experience of the Forest Action Network and the Forest Ethic, whose headquarters are based in the US (Figure 5.2).

Representatives of Greenpeace, Taiga Rescue Network, CBC and SEU formed the Forest Club, which designed and led the campaign. The campaign included numerous publications, videos, conferences, and protests. Using satellite images, the Forest Club inventoried and mapped virgin forests in the region. They investigated the timber pulp sources for publishing houses located in England, the Netherlands, and Germany, and requested that they boycott the logging of Karelia's old-growth forests. This culminated in 1996 in a series of publicized protests, both in the forests of Karelia and at the pulp and paper mill of the large Finnish logging company Enso (Yanitsky, 2000; Tysiachniouk, 2009). This led to Enso's announcement of a moratorium on logging on three important sites in the disputed forests in Karelia. In 1997, several companies, both Finnish and Russian, joined the moratorium.

The Forest Club's message was manifold: they listed companies logging the old-growth forests, as well as the buyers in Europe that accepted wood from these companies. They implored the European public to boycott products made with Russia's old-growth wood. Companies logging old-growth forests in Karelia were breaking no laws or norms of the Russian Federation; however, the NGOs were trying to enforce new informal global environmental rules and values beyond the control of any one state.

[165] Interview, Finnish tourist, April 2006, Kalevala.

[166] The road is described as blue due to the many lakes along the way.

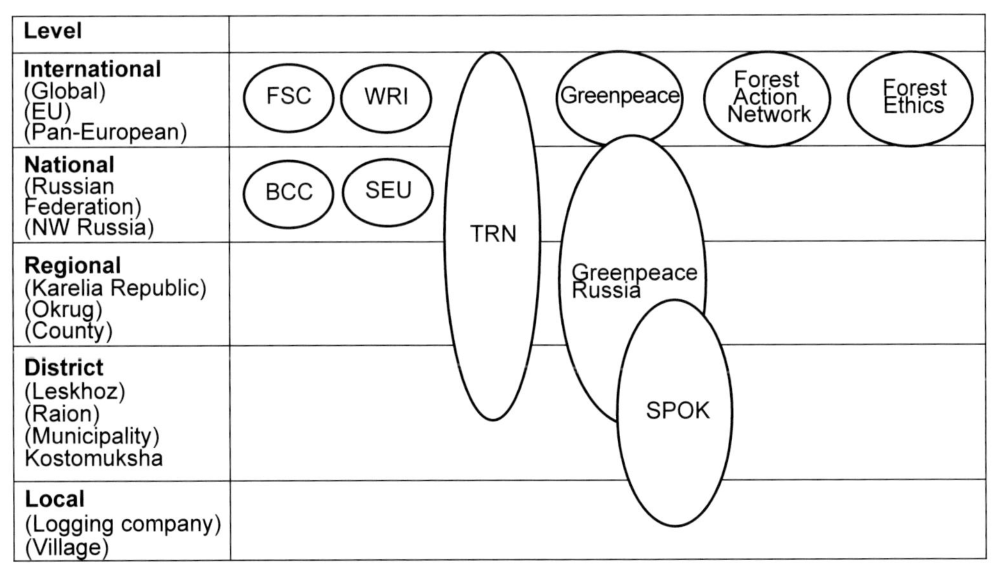

Figure 5.2. NGOs involved in the market campaign.

With the Russian government, the NGOs tried to initiate a process of creating a specially protected natural area in order to preserve the old growth forests. This last effort also witnessed the introduction of nature protection measures, created in transnational spaces, including national parks, and UNESCO World Heritage Areas created by the UN.

After eleven years, the conflict was resolved with victory for the environmental organizations. This was a transnational conflict, based on the core belief of environmental organizations, that the old growth forests all around the world need to be preserved.

5.3.1 The foundation for the Kalevala National Park: latent stages of the market campaign

In 1995, in the Kostomuksha Nature Reserve, environmental NGOs from 10 countries, including representatives from Canada, Sweden and Finland, held a meeting dedicated to the preservation of old growth forests. At this meeting the Forest Club was formed, comprising representatives from the NGOs Social Ecological Union, Greenpeace, Nature Protection Core and the Karelian student organization SPOK. The Forest Club started developing criteria for identifying old growth forests and mapping the existing old growth forest landscapes. They also started to monitor the logging and cross-border trade operations in the area.[167] In 1996, together with the Russian NGOs Socio-Ecological Union and the Centre for Biodiversity Conservation, Greenpeace created maps of the virgin forests along the Finnish border.

In 1997, in the framework of a Russian-Finnish project on biodiversity conservation and sustainable forest management in northwest Russia, money was allocated to the Kostomuksha

[167] Interview, Kostomuksha Zapovednik director, November 2006, Kostomuksha.

Natural Reserve and the Karelian Scientific Centre for field research, and for the designation of the borders of the future Kalevala Park. On the basis of a scientific report, the Karelian Committee for Natural Resources, as well as federal agencies in both Russia and Finland, were to make a decision on establishing the Kalevala National Park[168] (Figure 5.3).

5.3.2 Escalating the market campaign

In parallel with efforts to negotiate the creation of the national park, the Greenpeace network and Russian Forest Club started their market campaign. They informed companies in Great Britain and Germany about the actions of their suppliers from Sweden and Finland who were logging in Russia in the areas close to the future Kalevala park. The maps of the old growth forests were presented to publishing companies and governments, including those of Karelia and Finland, as well as to timber producers and consumers. Direct actions began in 1995, and the Finnish company Tehdaspuu was one of the first to break off its relations with a logging company working in the disputed 'green belt' in Karelia (Figure 5.3, step 2). However, in early 1996, the Karelian government supported continued logging in the area, citing economic interests (Yanitsky, 2000). The next year, the chairman of the government, V. Stepanov, issued the following statement:

> *The government of the Republic of Karelia has to evaluate the actions of various ecological NGOs as interference in the internal affairs of Russia that is aimed at undermining both the Karelian and Russian economies and that violates the basis of the boundary policy of the Russian Federation (see Yanitsky 2000: 242).*

A series of protests in August 1996 brought the area greater publicity. Members of Greenpeace blocked Finnish harvesters in the Kostomuksha district of Karelia and staged a protest a few days later in Finland at the pulp and paper mill of the Finnish company Enso (currently Stora-Enso) (Yanitsky, 2000).

5.3.3 De-escalating market campaign

This led to Enso's announcement of a moratorium beginning January 1, 1997, on logging on three important plots in the disputed forests in Karelia. Enso promoted creation of a working group with NGOs and governmental representatives to discuss conservation issues and further inventorying of old growth forests in Karelia (Lehtinen, 2006: 187). In 1997, several companies joined the moratorium, including UPM-Kymmene, MoDo, Vapo, Kuhmo, Pölkky, and Ladenso (Vorobiov, 1999). Greenpeace made agreements with a local forestry unit (lespromkhoz) stating that the latter would not lease the territories, even though it was already nearly impossible to find a willing lessee. Soon after this, a working group was formed to set guidelines for dealing with the virgin forests and for making future decisions on the issue. This group included environmental NGOs, government officials, and representatives of industry. The Karelian government joined the group, but again stated its opposition to the environmental NGOs when the chairman accused

[168] Interview, Karelian Scientific Centre, November 2006, Petrozavodsk.

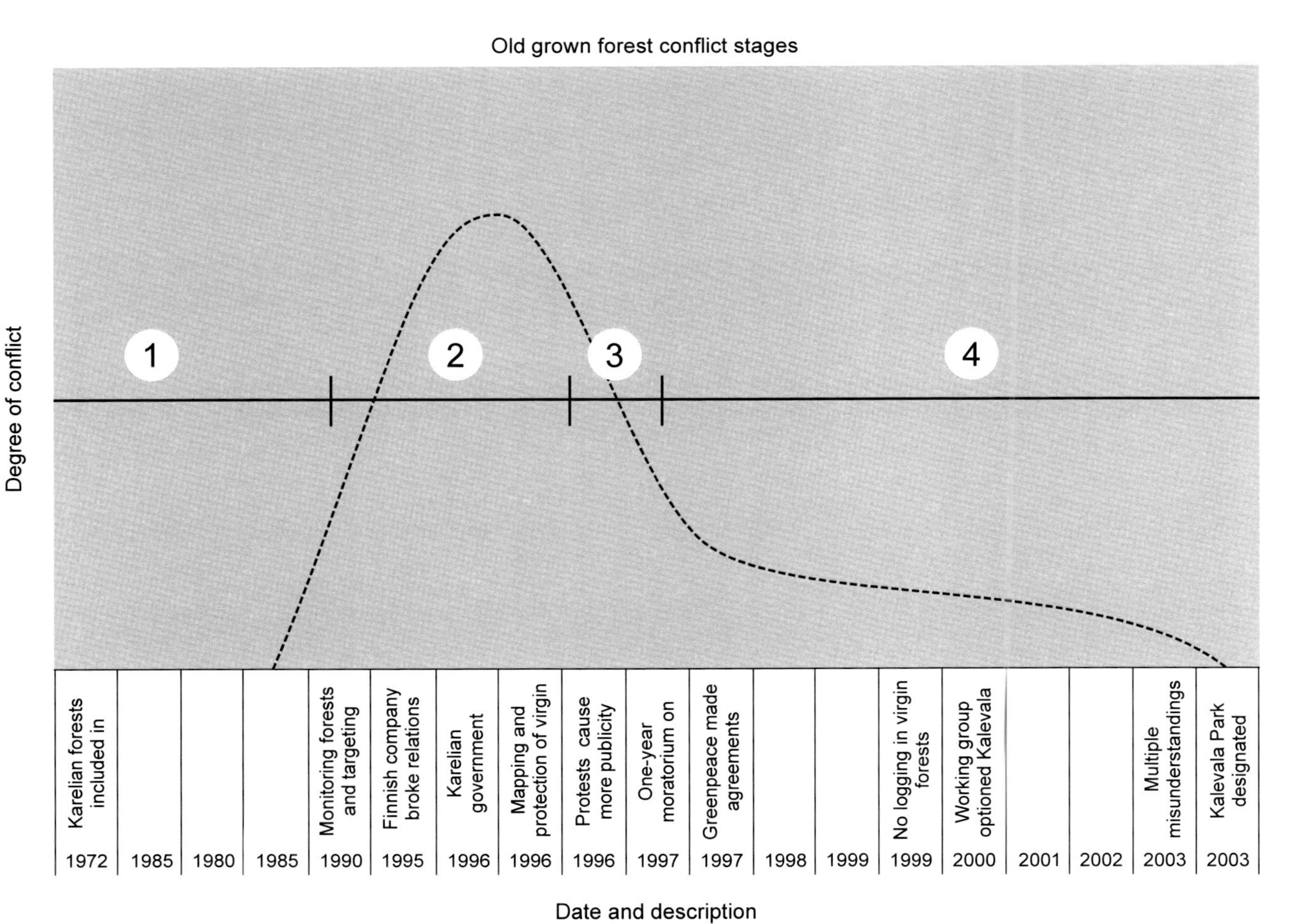

Figure 5.3. Market campaign stages. On the vertical axis is the degree of conflict, on the horizontal-time, vertical lines separate phases of the conflict.

them, in a letter to the deputy chairman, of illegal acts aimed at undermining the Karelian economy (Yanitsky, 2000) (Figure 5.3, step 3).

5.3.4 Period of stagnation and following resolution

In 1997, the Karelia Science Centre began preparing detailed justifications for the specially protected area of Kalevala. By 1999, there was no logging going on in the old growth forests and the first round of justifications was finished. In the framework of the EU TACIS grant a decision was made to establish the park on 95 thousand hectares. At the end of the project in 2001, the park had still not been designated, but the equipment for the park was purchased from grant money. A municipal agency, 'Kalevala Park', was established to manage the area during the transition period with the idea that when the park was set up the equipment would be forwarded to the park's federal administration.[169] In the framework of the Dutch-Russian programme Matra in 2002-2003, 1,500 signatures were collected from Karelian citizens in support of the future national park and a set of recommendations was developed for the park's development.

Negotiations related to the park took place in parallel on different levels of governmental structures with deadlocks and delays due to conflicts between the different state agencies. In 2001, an agreement on the park was negotiated on the level of the municipalities; however, on the level of the Republic of Karelia, the government agreed to a park with an area of only 74.4 thousand hectares (Yanitsky, 2000). On August 6, 2002 the government finally issued a decree on setting up the park and documents were submitted to the federal authorities. The period from 2002-2006 was a stagnant stage in Karelia, as the decisions had been forwarded to Moscow.

In Moscow, the Kalevala Park documents were received just as state agencies were being restructured, and began their journey from cabinet to cabinet and from one agency to another. When the documents reached the stage of environmental impact assessments, it turned out that this procedure was not in the state agency's budget, so Greenpeace contributed US$ 5,000 towards making the assessment, as designation of the national park was essential for the 'happy ending' of the market campaign.[170]

Such intervention seemed a bit ridiculous – a radical environmental organization taking financial responsibility, instead of the government (Tysaichniouk, 2009). Finally, on the federal level, it was only in November, 2006 that a positive decision was made to establish the Kalevala National Park and to allocate money for its infrastructure (Figure 5.3, step 4).

5.3.5 Starting new forest management practices: FSC certification

In 2005-2010, the logging companies that had leased the land around the Kalevala Park area completely changed their attitudes toward the old growth forests, choosing the path of FSC certification with compulsory preservation of high conservation value forests (HCVF), especially intact forest landscapes (HCVF-2). The maps of old growth forests created by the NGOs became a regulatory guideline for business operations. Swedwood Karelia (a subsidiary of IKEA) and

[169] Interview, former director of the municipal agency, November 2006, Petrozavodsk.

[170] Interview, local governmental representative, November 2006, Petrozavodsk.

Segezha Pulp and Paper Mill, belonging to the Investlesprom holding company, made an agreement with the Forest Club for a moratorium on the logging of old growth forests and stopped buying wood from their suppliers cut from old growth forests.[171] Segezha Pulp and Paper mill rejected a lease for land adjacent to the future Kalevala National Park. The informal agreements between the companies and the NGOs made the total area removed from commercial use, including Kalevala National Park, about 95,000 ha (the officially designated park, at the governmental level, is 74,500 ha). Therefore, the total amount of land now set aside is the same as originally proposed by the environmental NGOs in early 1990s, although the area was diminished by governmental decree. In addition, since 2007-2008, in the framework of FSC certification, companies started consultations with local communities identifying social value forests (HCVF-5-6) and pursuing social programs that contributed to the infrastructure of the local forest settlements. Since 2008, the dialogue between the companies and stakeholders has shifted to the forest management in the buffer zone of the Kalevala National Park, the creation of ecological corridors, preservation of hunting grounds, ethnographic tourism and the development of ethnographic trails.[172]

5.4 Stakeholders involved in the conflicts and the negotiations of the Kalevala National Park

A market campaign is a purposefully initiated conflict, thus relationships between the stakeholders can be more confrontational than those developing in day-to day interactions. This conflict is about natural resources, yet all economic, social, cultural and natural considerations come into play. The market campaign can be described as a multi-stakeholder conflict situated in both space of place and the transnational space. The stakeholders of the Karelian Old Growth Forest campaign can be subdivided into two groups: affected stakeholders of the space of place, who have a stake in the contested territory of the Kalevala Park (both in Finland and Russia), and interested stakeholders in transnational spaces, who have a stake in what is happening globally with old growth forests and for whom the particular territory of Kalevala Park is but one example where a global preservation policy has to be implemented (Table 5.1).

5.4.1 Stakeholder tensions on the level of the Karelian Republic, difficulties in negotiation

As we see from Table 5.1, the conflict involved many stakeholders. However, despite the fact that the conflict occurred mostly between NGOs and logging companies, after the moratorium on logging in the area, the major arena of negotiation was left open to only to several major stakeholders: governmental agencies, scientists and NGOs, which were constantly blaming and shaming each other.

[171] Iinterview, SPOK November, 2006, Petrozavodsk.

[172] Participant observation during the expedition to the area in summer of 2008.

Table 5.1. Stakeholders in the space of place and the transnational spaces. In the table the names of the state agencies are used as they were at the time of the events, prior to the new forest code that was enacted in 2007, after which the state jurisdictions changed. Interests of stakeholders are reconstructed from the interviews.

Spaces	Stakeholder	Type	Interest
In the space of place (Russia, Finland)	Ministry of Natural Resources	GO	At the time of the campaign they were managing forests on the federal level. They are interested in the creation of the national park as it is in federal jurisdiction. Their concern is that they will need to provide funding for park rangers and infrastructure. However, in their opinion, the old growth forests generally should be logged as forests 'ready to harvest'.
	Lesoustroistvo	Agency of regional government GO	Creates plans for harvesting in the region. Regulates a forestry cycle of 10-12 years. It does not have much stake in this issue, and argued for creation of the national park.
	Leskhoz	GO	Managing forest on the local level. Leases land to the logging companies, monitors logging operations. Interested in the fast resolution of the Kalevala Park issue as even during the moratorium it was responsible for managing the area, including fire protection. Sceptical regarding NGO's participation in the decision making process.
	Regional Karelia Government	GO	Has an interest in logging Russian old growth forest as an important economic stimulant and as a tool to decrease unemployment. Logging of the old growth forests would bring fast returns and foster economic growth.
	Russian timber companies	Commercial company	In the 1990s, economic interests, fast cash, and logging of old growth forests. In 2000s, interest in FSC certification and stable access to sensitive European markets; they accepted the idea of preservation of old growth forest landscapes because they belong to the HCVF-2 category and their preservation is required by the FSC.
	Forest Club	NGO network	Protects and monitors old growth forests and high conservation value forests in Russia, and particularly in Karelia, organizers of the market campaign. The Club was eager to create the Kalevala National Park.
	Local community	Civil society	Most communities both on the Russian and Finnish sides of the border supported the creation of the Kalevala National Park; interested in tourism development

Spaces	Stakeholder	Type	Interest
	Tourist firms	Commercial companies	All local firms supported the creation of the Kalevala National Park; interested in showing tourists both the Kalevala culture and nature in Russian pristine taiga forests.
	Russian scientists	Civil society	Interested in designating new specially protected areas.
	Finnish scientists (ethnographers, folk music archive staff, biologists)	Civil society	Interested in creating the Kalevala National Park and in preserving both Karelian culture and nature.
	Finnish NGOs	NGO	Simultaneously organized direct actions on the Finnish side of the border in order to create a Kalevala Park there; supported initiatives to create the Kalevala Park in Russian Karelia.
Transnational spaces	Multinational companies	Commercial company	Long-term economic interest in Russian forests. The distance (proximity to the border) to the resource they want to harvest is very important for these companies. Interested in reducing risks in trading with Russia; aware of the danger that the NGO networks can create to their images.
	Stores, retailers	Commercial company	Cheap products (economic interests), aware of the damage of consumer boycotts and the danger of selling wood from old growth forest landscapes (HCVF-2).
	Consumers	Not necessarily organized	Pay attention to the price and quality of timber, but also have moral issues in mind, as most do not want destructive forest practices to be implemented.
	European Union	GO	Supports creation of specially protected natural areas in Europe, allocates money in the form of grants to implement preservation.
	WWF network	NGO	Supports creation of high conservation value forests around the world, supports FSC certification around the world; supports the Kalevala National Park as it contains HCVF-2.
	Greenpeace network and Taiga Rescue Network	NGO	Assigns planetary values to all old growth forest landscapes in the world. Fights for the preservation of all possible old growth forests, including the Kalevala Park. Blocks the export of old growth timber to environmentally sensitive markets; organizes direct actions and consumer boycotts. Their stake is preventing any wood from old growth forests in getting to Western markets.

The Karelian governmental agencies and their supporters had several points of disagreement with the NGOs:

1. *Economic concerns and job loss.* The Karelian government stressed the financial importance of the republic's forest sector to the economy. One informant said, 'I worry that if a specially protected area is created in Karelia, Russians will lose jobs. That's why in preserving old growth forests we need to think about people. In Kalevala there are 150 people working in forestry. If it is closed, what will these people do? Who will create new jobs?'[173] Another official said, 'Maybe they [environmental activists] are good and want to do good, but if we do as they want, there can be negative results, especially for our economy'.[174] They argued that Karelia already has a high percentage of officially preserved territories compared to the other regions of Russia. My informants were supportive of protection in general and described it as positive, but they were reluctant, based on economic concerns, to increase the amount of protected area. One said, 'I say to the Finnish side, 'why don't you create specially protected areas in other regions of Finland. And when they have the same percentage of protected area [as in Karelia] then we can talk about new specially protected areas here'.[175]
2. *Financial concerns.* Government officials also complained that the pressure from the international community was not supported by enough financial help for creating the specially protected areas. According to one informant, the chair of Karelia's government wrote to the heads of various European countries requesting such funds.[176] He said, 'Many didn't even reply. Some replied and said 'we can't help you'. After a national park is designated and officially signed for, there are necessary expenditures, such as for tourism infrastructure. Concerning this, the respondent said, 'A huge sum is needed from the budget. In Karelia now, salaries are low and there are unemployment problems. There's no oil and no gas, unlike in Komi. We are concerned ... Who will invest all of this money?' The government claimed to be in a bind, without enough money to fulfil the requests of the international community.
   On the local level, the following statement by the town administration of Kostomuksha supports the same point: 'The town administration well recognizes the historical and cultural importance of this part of Karelia, but due to the critical situation in both Karelia and Russia, there are no funds for financing nature protection activities, or other important social expenditures'.[177] One of the science informants also agreed, saying, 'We have two functioning parks and if it were not for international projects like TACIS, the parks would sit on the budget and nothing else. The TACIS programme significantly increased the salaries for the administration of the protected areas, but such support is impossible to get permanently'.[178]
3. *Issue of sovereignty.* The government of the Karelian Republic tried to see this issue as a domestic problem, and felt that environmental activists from outside of Karelia and outside of Russia had unjustly intruded into the republic's issues. For the Russian government, the overwhelming

[173] Interview, Karelian governmental representative, October 2002, Petrozavodsk.
[174] Interview, staff in the Karelian Ministry, October 2002, Petrozavodsk.
[175] Interview, Karelian Ministry head October 2002, Petrozavodsk.
[176] Interview, head of the Karelian Ministry of Natural Resources, October 2002, Petrozavodsk.
[177] Letter from the mayor of Kostomuksha to the Taiga Rescue Network, 1996, provided to the author in 2006.
[178] Interview, scientist, October 2002, Petrozavodsk.

interest in Karelia's forests by the European NGOs and governments was not readily explicable. In my interviews with government officials, I heard various judgments: NGOs are saboteurs trying to undermine Russian forestry for the benefit of Scandinavian competitors, or NGOs are exaggerating the urgency of protecting Russia's virgin forests and biodiversity, both of which currently abound. Whatever the discourse was, a prevalent accusation made by all government officials was that Europe had logged nearly all of their own old-growth forests and thus was creating a double standard by forcing Russia to preserve those which remained.
An independent scientist agreed with this last sentiment, saying, '[Kalevala Park] will have significance for all of Europe. They cut everything in their own territories and we will preserve on our territory. In their own countries they use forests like orchards in rows, and they want to use our territory for recreation so that they can see wild nature and have fun'.[179]

4. *Ability to compromise.* The government accused the environmental activists of not compromising. One government official said, 'We always go from compromises to understandings. But our opponents also need to compromise and understand ... It doesn't always happen though, and we are blocked and not understood. We are not understood by the Greens'.[180] In addition, many officials were suspicious of the NGOs' goals. One respondent said, 'Sometimes, we think that in the bottom of their hearts they have their own interests ... Greens sent us maps coloured in green and they absolutely don't want to explain why these forests are virgin. They just say 'from satellite pictures'. We can conclude that they are acting in somebody's interest'.[181] He also said about the maps, 'Whenever an enterprise here begins to work in international markets, its territory is always coloured by green.
5. *Lack of science behind the NGO position.* Government officials claim that much of the NGO's position in this conflict was not backed up by science, or that it was based on incorrect science.
6. *Value of old growth forest.* One official argued with the science of preservation in general, 'They [NGOs] say don't touch the forests, they belong to the planet. But on the contrary, if the forest becomes older and older then tomorrow it will take all our oxygen and we will not get it back. Things happen in such a way with old-growth forests'.[182] Another official argued for the benefits of forest management over total preservation:

> *In Europe there are no virgin forests, but I don't think this is bad. You must be very careful when dealing with old growth forests. If they are not taken care of, they'll become problematic areas. Forests that are not taken care of, when you don't cut sick plants, they become problems for forestry. When forests are old it is like in life when old people are alone. Sure I don't mind if old people exist, but it is bad when they are alone. You need young people to help them. If young people don't care about the old, they will die. A forest is like a living body and if it is not revitalized it will become older and diseases and pests will spread and this is bad.*[183]

[179] Interview, scientist, October 2002, Petrozavodsk.

[180] Interview, state committee October 2002, Petrozavodsk.

[181] Interview, state consultant October 2002, Petrozavodsk.

[182] Interview, the Karelia Ministry of Natural Resources, October 2002, Petrozavodsk.

[183] Interview, state consultant, October 2002, Petrozavodsk.

The NGO network that had a stake in setting up the Kalevala National Park operates partly in transnational spaces and partly in the space of place. The Forest Club and Greenpeace participated in local direct actions, worked with the Russian government on the federal level, and took the lead in the Internet consumer campaign. Only a few NGOs in Karelia, SPOK being the most active, interacted intensely with the government on the republic level. The major disagreements were conceptual, especially regarding the preservation of old growth forests, and procedural, related to the interaction itself:

a. *Old growth forests have a planetary value.* The Forest Club, led by Greenpeace, has tried to establish the concept of a 'virgin forest' both in the legislation of the Russian Federation and in the awareness of industry and the public. The goal is to convince the forest's stakeholders that virgin forests have a value in the West and therefore must be preserved in Russia. The attempt to import this idea into Russian industry and government has not proceeded smoothly, because Russia, unlike Western Europe, contains vast stands of virgin forest and old growth forest landscapes are not that unique.
b. *Soviet mentality of the Russian government.* Environmental NGOs see the mentality of the Karelian government as harmful to the population and the republic's economy. Our informant said that government officials still have the mentality from Soviet era forestry that 'more logging = more money'. The NGOs criticize this viewpoint and the lack of sustainable forest management in general. One informant said, 'They [Karelia's government] have never done an assessment of how many forests are left, they live under the illusion that there are still plenty, and that's why our forest resources have become weaker and weaker'.[184]
c. *Lack of effective forest policy.* Both independent scientists and environmentalists alike see a problem with the republic's forestry strategy, and therefore feel that even if the old growth forests are not preserved and logged, they will not bring economic benefits to the republic.
d. *Government is corrupt and represents industry interests.* Some NGO representatives see close connections between the government and forest producers as a cause of the government's desire to allow old growth forests to be logged.[185] In addition, one informant further criticized the republic's head and Karelian politics in general as corrupt: 'Decisions there are made by a very narrow circle of people close to power, and they make them without any logic ... the Karelian government is a museum of socialism'.[186]
e. *Government is not doing its job.* NGOs in Russia frequently must do the government's technical work. A SPOK activist said:

> *In the West ... you spread knowledge about the problem and immediately the public begins to participate. The authorities get kicked and then they understand the problem well and begin to do something ... there is no action from Russian state agencies ... To achieve something here you need first to make a big noise, and then secondly you*

[184] Interview with the NGO Biodiversity Conservation Center (BCC) representative, March 2002, Moscow.
[185] Interview with BCC coordinator, June 2006, Moscow.
[186] Interview, Greenpeace March 2002, Moscow.

> *need just to do everything yourself, instead of the government. And then you will achieve results.*[187]

In both the government and NGO narratives, it seems that there were many gaps in communication between the two parties. The impossibility of arranging a meeting with high-level government officials is one illustration mentioned by the NGOs. All of our NGO informants explained this by saying that the government is simply unwilling to cooperate with them. In the years of dispute over this issue, representatives of the environmental position were unable to meet, even once, with the governor of Karelia, Leonid Katanandov. One informant said:

> *They tried very much to arrange meetings. And when the meeting was set up, they [the Karelian government] had only assistant governors who were available at that moment. This came to no agreement. They came to the meetings and were saying 'yes yes' and shaking their heads. Or they would show up and say that they cannot have the discussion today. Sometimes they were just silent and said nothing. Sometimes they simply said 'we don't support the issue'. But a meeting was never arranged with Katanandov himself, whose status allows him to say either yes or no. This is just not understandable.*[188]

Similarly, environmental activists failed to meet with the head of the republic's Ministry of Natural Resources, another government official with many responsibilities linked to nature protection issues. One of my informants said this about him:

> *It is absolutely impossible to reach him or arrange a meeting. He does not talk with representatives of NGOs. It is his principle. He doesn't even say hello. His vice chair is polite and he always says that he cannot solve the problem because he has no power or because he does not have his own personal opinion. So it is impossible to have a normal conversation with them.*[189]
> *To convince the governor of anything is absolutely impossible ... for him power is most important. Everything is about power. Not only environmental organizations think of him in this way. Foresters themselves also think like this.*[190]

Many environmental activists at the late stage of negotiations saw miscommunication with the Karelian government as a big hindrance to the creation of the Kalevala National Park. The local administration in the two regions, scientific organizations, the local forest management unit (leskhoz), and industry had all agreed to the creation of the national park; however, Governor Katanandov did not sign the proposal. This would have been the final step in Karelia towards the

[187] Interview, NGO SPOK, November 2006, Petrozavodsk.
[188] Interview, NGO SPOK, November 2002, Petrozavodsk.
[189] Interview, Greenpeace April 2002, Moscow.
[190] Interview, SPOK, November 2002, Petrozavodsk.

park's creation. Attempts to contact the government and find out what was causing the hold-up met with delayed and vague responses. One of our informants said:

> *For each letter sent, according to the rules, they [the Karelian government] have a whole month to respond, so they wait the whole month and then send a letter ... After a month we get responses like, 'yes, we wrote to Katanandov and he will analyse the situation and answer to the government and to you what exactly is going on' ... but no one did anything. Everyone just corresponds and nothing is done.*[191]

In the late stages, the Karelian government officials worked together with representatives of the scientific and the environmental movements in a working group on this issue; however, our NGO respondents felt that this cooperation was not genuine. One of the informants explained governmental delays in decision making by saying: 'The decision was not made because their ideology is to measure things in cubic metres. Really big money is involved and what exactly they can get from the park is unclear'.[192]

5.4.2 Stakeholder's interaction on the local level

On the local level the key stakeholders on the Russian side of the border were the local administration, the local forest management unit, the local community, the Kostomuksha natural reserve, scientists from the Karelian Scientific Centre of the Russian Academy of Sciences who came periodically to do research in the area, and the logging companies themselves.

There were no local NGOs in the area. The input to the space of place by NGOs from the outside, especially Greenpeace, was strategic. They organized short, direct actions. For example, 'they brought a bus with journalists to the area, chained themselves to a tractor and put up posters. The journalists took pictures and everybody left very quickly before the police came'.[193] It was only when the central TV stations showed the report that people found out that the direct action had taken place; that the local stakeholders understood that the exposé had occured and that intense conversations and negotiations between interested parties had taken place. Following the exposé, the Forest Club representatives again came from their Moscow office and held professional negotiations with local stakeholders. 'They were very flexible, when someone behaved aggressively with them; they never responded aggressively, they were carefully explaining their position on the issue'.[194]

The local administration did not express any open support for the direct actions taken by the NGOs, but they generally argued in favour of creating the Kalevala National Park and participated in the negotiation process intensively. After the TACIS grant money ran out they helped to establish the municipal organization 'Kalevala Park' and for several years paid a salary to its director. They also argued that the former military infrastructure that was left on the Finnish border should be

[191] Interview, SPOK November 2002, Petrozavodsk.

[192] Interview, Karelia Scientific Centre November 2002, Petrozavodsk.

[193] Interview with local state authority June 2006, Kostomuksha.

[194] Interview, local administration, June 2006, Kostomuksha.

part of the park in the future.[195] The people's deputies council in Kostomuksha also supported the establishment of the Kalevala National Park and new candidates stated their support for the park during their election campaign.

Local people were supportive of the park from the onset. Most local people did not perceive logging operations as sources of possible employment, as modern technology does not need many workers. For the local people, the development of tourism, as well as using the forest for both recreation and gathering mushrooms and berries is essential. The village inside the proposed park is very small and all villagers were strictly against logging, although they were concerned about hotels that might be built that would destroy the spirit of the village; they were also worried by the federal infrastructure that might be built in the park. One local farmer was interested in developing family tourism with Finland, revitalizing old Karelian traditions and running his family business on his own without any intervention.[196]

The local state forestry unit (leskhoz) in Kostomuksha felt frustrated about the moratorium on logging in the territory of the Kalevala Park. It argued that if a geographical area continues to be in its jurisdiction, it should be leased to a logging company and logged. They did not mind the creation of the national park, however, and the ceding of the land to federal jurisdiction, but they were against the responsibility of managing the area during the stagnation of a long-term moratorium.[197] Their major frustration was related to the fact that the forest companies had signed informal agreements with the NGOs on moratoriums in the areas of old growth forest in the territories of their leases, using Greenpeace's maps as accepted legal documents.[198] They blamed both the NGOs and the companies for neglecting governmental agencies in such decisions. They viewed the NGO direct actions sceptically and believed that the NGOs were interested in conducting such actions not to save the forests, but because they were receiving money from the West.[199]

From the very beginning, the Kostomuksha Nature Reserve supported both the preservation of old growth forests and the creation of new specially protected areas. They provided support for NGOs, for example, by hosting the NGO international conference that took place in 1995. At this conference the goal of preserving old growth forests was articulated and the Forest Club was formed. The Kostomuksha Nature Reserve also participated in the research conducted in the Kalevala National Park in order to provide a scientific basis for the park's creation. The Karelian Scientific Centre of the Russian Academy of Sciences followed up on the Kostomuksha Natural Reserve's research and supported the establishment of the park in the Kalevala area. Later, in 2008, a staff member of the Kostomuksha Nature Reserve became a director of the new Kalevala National Park.

In the 1990s, logging companies were strongly against establishing new specially protected areas in the border regions in Karelia. These areas with their substantial old growth forests became accessible for logging operations after the fall of the Soviet Union; previously the land had been

[195] Interview, Kostomuksha's mayor June 2006, Kostomuksha.

[196] Interview, local community representative June 2006, Sudnozero.

[197] Interview with the director of the local forestry unit, June 2006, Kostomuksha.

[198] Interview, forest unit employee, June 2006, Kostomuksha.

[199] Interview, director of the local forestry unit, June 2006, Kostomuksha.

part of a military zone along the border. This land was desirable for leasing by Russian, Finnish and Swedish logging firms, but they were faced with a consumer boycott. This case saw virtually no interaction between NGOs and industry, beyond the consumer campaign of the 1990s. All the international companies working in the disputed old-growth forests of Karelia abandoned their leases. No company, Russian or foreign, would apply to log these forests, and so the NGOs had no further interactions with industrial stakeholders on the local level, except those who had leased land around the 'moratorium' zone and in the buffer zone of Kalevala National Park.

All European companies that are currently involved in wood trade in Russia are aware that there is a possible threat to their image if they were to buy wood harvested from old growth forests. They request transparency from Russian companies on the origin of the wood. Many companies in Northwest Russia have taken the forest certification pathway in order to gain legitimacy in the European market.

Maps of the old growth forests created by the coalition of NGOs were used as tools for land use decisions and preventive measures. They became an informal rule for logging companies, and compliance with this rule has become a kind of licence to operate in Russia. Companies that leased areas containing old growth forests made informal agreements with the NGOs and established voluntary moratoriums in these areas, even though Russian governmental agencies were pushing them to pursue logging.[200]

In 2007-2012, the forest users around Kalevala National Park have become a diverse group; some are subsidiaries of large holdings, but several relatively small enterprises have survived the economic crisis and forest sector reforms. Since the National Park buffer zone was not officially established by the government, it existed informally and was respected only by companies taking the certification path. Other enterprises have periodically violated water protection zones and requirements for land use in the buffer zones.

In 2007 Swedwood, a subsidiary of IKEA, who was leasing land in the area, received their FSC certificate. The old Soviet enterprise Kostomuksha lespromchoz, who was in the middle of the conflict in 1990s, was purchased by the holding company Investlesprom. In 2008, they started preparation for FSC certification with the plan to become certified in 2011. In 2008, the first consultations with stakeholders and public hearings were held. Compared to the other parts of the Karelia Republic, the level of stakeholder engagements were much higher, probably because of the experience gained during 15 years of negotiations around the Kalevala National Park.[201] Ecological tourism started to develop in the Kalevala National park and beyond, and the tourist firms became concerned about logging as it creates unattractive views and tourist's demand is high for pristine forests and wilderness. Representatives of the Kostomuksha Nature Reserve (Zapovednik) suggested special logging regimes in the area between the Nature reserve and the Kalevala National Park[202]. The representative of the Kalevala National Park suggested banning even small clear cuts in the Park's buffer zone. The Kostomuksha Hunting and Fishing Society was concerned about the reproduction of the moose population and also suggested special logging regimes on leased territories. Stakeholders expressed concern about the temporary bridges that

[200] Iinterview, Swedwood, November 2006, Kostomuksha.

[201] Participant observation during the expeditions and consulting work in Karelia 2006-2010.

[202] Interview with the representative of Nature Reserve, August 2010, Kostomuksha.

the company was using that could harm migratory salmon routes through the Liva River[203]. Representatives of local communities, who had built their dachas or restored their old houses (summer houses) in the village of Ladvozero adjacent to the border with Finland, suggested preserving a whole variety of places of cultural significance related to Kalevala's rune singers, including several ethnographic trails and buffer zones around them.[204] Some of the issues were taken into account by the company, but many of them remain unresolved.

In 2009-2010, the preparation for FSC certification was postponed by the company because of economic and financial crisis The company planed to continue negotiations around the above mentioned issues in 2011.[205] In 2012 it decided to abandon its lease in the area, as NGOs found there so many valuable forests that operations there became unprofitable.[206]

5.4.3 The role of transnational stakeholders

The transnational space is structured by different kinds of transnational actors and their interests. In transnational spaces, large transnational environmental NGOs with their transboundary networks were part of the NGO-led consumer boycott GGN. They tried to stop the Finnish and Swedish logging industry from destroying the old growth forests and to influence government policy in Russia, by forcing the government to establish a new national park. The case demonstrates that their interaction with multinational corporations proved to be more successful than that with Russian governmental agencies. The Taiga Rescue Network (with its major office in Sweden) and Greenpeace were the key actors in transnational spaces that brought the message from the Russian Forest Club to the stakeholders in transnational spaces. They were the key players in organizing the consumer boycott against logging of old growth forests in Karelia. By encouraging European buyers to boycott products from Russian old growth forests, Greenpeace effectively eliminated the threat of logging by forcing the business actors to change their practices. As in most Greenpeace-led campaigns, the media were essential for shaping the issue in transnational spaces. Through the media, Greenpeace used the ecological sensitivity and environmental conscience of European buyers, appealing to their values in order to mobilize consumers to participate in a boycott. Values based on transnational-space discourses of biodiversity conservation and old growth forest preservation have been used as instruments for purposefully escalating the conflict with the goal of social change.

In the beginning, the conflict was barely recognized by the business actors in transnational spaces. The transnational corporations were not aware of the problems related to the logging activities of their subsidiaries and suppliers in Russia. The Finnish and Swedish companies operated legally; their subsidiaries came into the country by invitation of the Karelian government and officially leased territory or were purchasing wood from Russian suppliers. Enso (Stora-Enso after merging), for example, was buying wood from Russian Karelia for more then 100 years. During

---

[203] Interview with the representative of the Society of Fishemen, August 2010, Kostomuksha.

[204] Participant observation during the expedition to Kostomuksha, August 2008, including public hearings August 7, 2008.

[205] Follow up interviews with company managers in Segezha, 2009-2010, Moscow.

[206] Interview with the top manager of Investlesprom, June 2012, St, Petersburg.

Perestroika all trade patterns changed, and logging operations have been dominated by small wood suppliers to Finland. Enso was finding new business partners and making contracts with them. 'It was interesting times for all of us, it was a learning process, we needed to start from zero, find how to operate under new conditions'.[207] They never expected the conflict to occur. However, after media reports and direct actions, Enso reacted relatively quickly by establishing a moratorium in the disputed area. In 1996, they made a decision that all potential old growth forests in Karelia needed to be identified by the NGOs, mapped and inventoried. My informant described the decision on establishing the moratorium on old growth forests as exceptional for Enso:

> *We made a public declaration, a public commitment of not buying wood from the disputed areas until they are inventoried. The whole society was developing, evolving, it started to work in a more similar way as in other countries, NGOs started to become active, similar kinds of forest conservation conflicts happened. We wanted to leave room for the regional processes, for national stakeholders, so that they can decide on their preservation policies.*[208]

After Enso took the lead in establishing the moratorium, other companies followed and supported it. In early 2002, many of the business partners of the Russian logging companies asked their Russian suppliers to get FSC certified. This request had a multiplier effect and was directed not only toward the companies in Karelia, but also many others in Northwest Russia.

As a result of multiple campaigns, multinational corporations around the world understood the harm that the NGO networks could cause to their businesses. Many companies have not only become sensitive to the NGO's grievances, but have started partnering with the NGOs in sustainable forest management projects. This trend has led to partnerships such as WWF-IKEA. Such partnerships have facilitated the creation of environmentally sensitive niche markets and the promotion of forest certification systems, which became an alternative to consumer boycotts. The FSC-GGN has become one of the most important regulatory forces in transnational spaces for stemming forest degradation. Therefore, in transnational spaces NGO networks created the institutions in place to convert harmful companies (Cashore, 2002) by discouraging them through market campaigns from damaging the environment and encouraging them to adopt sustainable forest management through FSC certification. The changes in transnational spaces also shaped the context in different localities around the globe, influencing governments, local NGOs and other stakeholders.

Other transnational actors have also played an important role. The European Union, with its TACIS grant programme, was an important actor in transnational spaces. In the framework of the Convention on Biodiversity, the European Union established the priorities for the TACIS programme that channelled money into countries of Eastern Europe and Central Asia. Funds for the establishment of four natural protected areas along Karelia's border with Finland came from the TACIS programme in 1999-2001. TACIS gave the Karelian government 3.5 million dollars to establish two new national parks, one of them being the Kalevala Park, and build a tourist and nature protection infrastructure for the two already in existence. There were many tensions and

[207] Stora Enso manager on certification, March 2008, Stockholm.

[208] Stora Enso manager on certification, March 2008, Stockholm.

miscommunications between the actors in transnational spaces and actors from the space of place, especially representatives of the Karelian government, in the framework of this transaction. According to our informant, Karelia has acquired a negative image in the international community from this controversial issue and their handling of it.

He said, 'In Europe, Karelia is considered a scandalous region where no one can agree on anything. They only have conflicts and scandals there.[209] There was poor communication between the TACIS programme and the Russian governmental officials and scientists about how to spend the money, and much money was wasted. By the end of the grant period none of the planned parks had been designated; only the related research had been carried out. However, money channelled to the government for designating the Kalevala National Park created a favourable context for successfully resolving the issue of preserving old growth forests after the NGO-led consumer boycott against the logging companies. The availability of external resources made negotiations between the NGOs and state agencies more peaceful.

5.5 Outcomes of the market campaign

As a result of the market campaign, institutional change occurred first in transnational space and later on in the space of place. The logging companies in Russia were forced to make changes in their operations, as their business networks requested them to do so. As a result, market campaigns to save the Karelian old growth forests had a significant overall impact on the space of place, on forest management in Russia. In the 2000's, logging companies went through a profound restructuring. Almost all the small logging enterprises became subsidiaries of large multinational holding companies that have recognized images in the international arena. These large holding companies recognized the risks related to the logging of old growth forests and contributed to the institutionalization of new practices.

For the Republic of Karelia, however, the effect was twofold. On the one hand, on the positive side, there was no negative impact on local communities. This conflict took place in areas which were not heavily populated by local people. Those who were employed by logging companies continued with their jobs in the forests close by, but not where forests were old growth. On the other hand, the market campaign resulted in long-lasting misunderstandings between regional stakeholders, especially between the NGOs and governmental agencies, as there was not effective negotiation and conflict management. This case helped to shed light on the difficulties and complications of bringing global practices of sustainable forest management to Russia. The situation was not recognized in the beginning by governmental agencies as a threat to doing business as usual. The NGO's influences were not known in Russia, where a young and newly established democracy still maintained a traditionally strong top-down approach to policymaking. Because of institutional instability, turbulence and endless reforms, it was not possible to designate the park until 2006. During the period from 1997-2006 it was not possible to carry on commercial logging, or to build tourist infrastructure; yet it was economically unfeasible to keep the territory unproductive. The local state forest management unit was forced to do maintenance work, such as fire protection, without any economic rewards.

[209] Interview SPOK November 2002, Petrozavodsk.

Even those financial benefits that were coming from the transnational spaces in the form of TACIS grants were not used properly. Instead of implementing the new park in the space of place, the funds were used only to do related scientific research, and even the equipment that had been purchased for the park was kept in storage. This income, however, was softening the tensions that occurred between the stakeholders in the space of place, moving the local governmental agencies to favour the park. The peculiar situation in this case was related to the fact that Kalevala Park represented a contested issue not only for Russian stakeholders, but for Finnish stakeholders as well. The participation of Finnish stakeholders in both the debates and protests contributed significantly to the successful outcome.

This case demonstrates asymmetry in institutional change in transnational spaces and spaces of place. When NGOs operate bottom-up and direct their grievances to the actors in transnational spaces, stakeholders in transnational space react in the way the NGOs expect them to react; companies implement changes. The change of business practice is then forwarded and implemented in the space of place, in this example Russia. However, a market campaign in transnational spaces is not intended to target national governments. Governmental agencies continue to operate according to the old path dependency and, even worse, in a situation of institutional turbulence, can become significant barriers to implementing innovative institutional change.

5.6 Discussion and conclusion

5.6.1 Time-space configurations in a market campaign

The Karelia Old Growth Forest case can be divided by time into two phases: the conflict that resulted from the consumer campaign in transnational spaces, which took place before the conflict in the space of place. In the escalating stage, the space of place was used only for short direct actions with the goal of delivering images to transnational spaces. The acute phase in the peak of the conflict was quite short (1995-1997) compared to the long, stagnant conflictive stage in Karelia afterwards. When access to the natural resource was restricted by a moratorium on logging in the Kalevala area, the conflict shifted overwhelmingly to the space of place and continued there for almost a decade (1997-2006).

The NGOs involved in the market campaign's GGN largely used modern communication technologies. The internet provided opportunities for the mobilization of transnational NGO networks and movements. Modern communication technologies allowed for obtaining rapid results with very little human or financial resources. For example, only four people from Russia participated in the organization of this campaign. Although direct actions were included in the repertoire of the NGO's collective actions, the campaign mainly took place in virtual space. The consequences of these fast results happened in clock time and dragged out through the next long 15 years in the space of place. Stakeholders at the site of implementation regarded the interference of global space in different ways, hence, a lot of conflicts and friction followed. Strong and prompt decisions made in transnational spaces resulted in continuous resistance in the spaces of places. As a result, negotiations around the Kalevala National Park lasted much longer than they might have had if a more smooth intervention been conducted by the agents of institutional change.

The information campaign in the transnational space was essential for transforming business practices. In the course of the market campaign, the NGOs in transnational space used both normative and coercive pressures to foster changes in corporate behaviour and to institutionalize new practices. Without the belief that the ancient forests are of planetary value, it would not have been possible to mobilize consumers to boycott the logging companies. Simultaneously, the NGOs applied coercive pressures by naming, shaming and blaming the companies that were responsible for destructive activities in the old growth forest in Karelia, and introduced the new values while appealing to consumers at large. Before the NGO's started the campaign, there was little international awareness about old growth forests. This awareness was fostered when organizations like Greenpeace started their informational campaigns. By doing so, they institutionalized new consumer preferences based on ecological standards and facilitated the future changes in companies' path dependence. It is important to note that the informational campaign required relatively few resources compared to its significant outcomes. Only four NGO representatives in Moscow and one in Karelia were needed to frame and deliver the message to the transnational NGO networks that conducted the mobilization in transnational space and took the lead in driving the consumer boycott.

5.6.2 Market campaign GGN: generalizations

The market campaign represents a bottom-up NGO-driven GGN, which involved consumer preferences in order to change practices at logging sites, in the spaces of production. NGO-led market campaigns, despite being conflicts, offer another form of governance that converts business practices to be more environmentally friendly. They help promote and construct environmentally friendly niche markets for forest products and by doing so, foster forest certification.

From the empirical findings described in this chapter, it is possible to make some generalizations about operations of the market campaign GGNs. The 'market campaign' is composed of two distinct conflicts between the NGOs and business networks: one in transnational spaces and another in the concrete space of place (Figure 5.4). The conflict began with the recognition that a company belonging to a certain business network in a certain place (B-p) was involved in a practice that is detrimental to the environment. The NGOs in the space of place (NGO-p) monitored the activity of company on the ground and informed their counterparts NGOs in transnational spaces (NGO-t). The NGO network mobilization took place and the information about wrongdoing was passed across borders, to the consuming countries in particular. NGOs further mobilized resources and organized a campaign involving consumers (C-t), which targeted a multinational corporation's headquarters in transnational spaces (B-t). Both the NGOs and the consumers participating in the campaign viewed the behaviour of business actors in transnational spaces (B-t) as detrimental to others. Therefore, they became involved in the conflict.

This changed the stakeholder's interaction in transnational spaces and made doing business as usual too risky for the business actors involved in the transnational network. The business actor (B-t) faced a consumer boycott, and preferred to force its counterpart (B-p) in the space of place to change its practice. Again, the conflict is shifted from a transnational space to a space of place. In the space of place many other stakeholders are involved in addition to the business and the NGOs. However, the business actor (B-p) and a watch-dog NGO (NGO-p) are essential for the

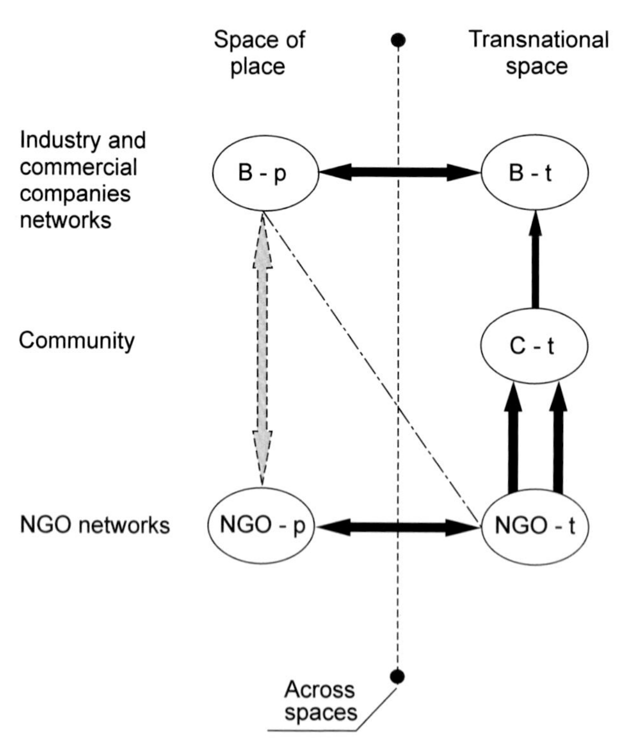

Figure 5.4. Relations between NGOs, companies and wider society in a market campaign.

conflict to happen. Other stakeholders include local, regional and/or national governments, local citizens and many others who become involved in renegotiating logging practices in the space of place. When the practices are renegotiated by stakeholders, new governance arrangements are being institutionalized.

5.6.3 Analysis of the market campaign on saving the Karelian old growth forests using the GGN concept

The case study described in this chapter falls into the category of a bottom-up NGO driven GGN. This assumes that the network stretches from the actors in the space of place towards actors in transnational spaces, e.g. network initiators are based in the local place and grievances of the activists directed mostly toward companies in transnational spaces. A closer look allows seeing the divergence of the studied network from the ideal type. The main actors of the campaign were international NGOs, namely Greenpeace and Taiga Rescue Network, and the initiators of the boycott do not originate from the locality, as the main actors of the Forest Club are Moscow based NGOs. The NGO SPOK, which helped to organize the campaign, is located in the Karelian capital, Petrozavodsk. Local civil society representatives largely supported the network, but were not part of it and did not participate in the initiation of the market campaign.

The market campaign was not the single essential network for saving the Karelian old growth. Two other networks supplemented the market campaign's efforts. First, a top-down foundation-driven GGN played the role of facilitator in providing EU grant resources (the TASIS grant program) for creation of specially protected natural areas that included the Kalevala National park. As was described above, the foundation's money was essential for the Karelian government to accept the idea of creating the Kalevala National Park. Without the TASIS money, old growth forests in Karelia might still be under a moratorium on logging, yet without the official status of a specially protected area, the same as many other plots of intact forest landscapes in Northwest Russia.

Another facilitator of creation of the Kalevala National Park was a trans-local network. The park was created in the territory of two countries, Finnish and Russian Karelia, and trans-local networks stretched from Finnish to Russian stakeholders, i.e. from one space of place to another. These were both environmental and cultural ethnographic networks interested in the preservation of both nature and the culture of villages located near the territories of the Kalevala park. Ethnographic and cultural tourism provided good economic alternatives to logging in the area. Thus, despite the domination of a bottom-up market campaign GGN, a top-down foundation GGN and trans-local activist networks were also essential for the story's success.. The main bottom-up GGN provoked the worst conflicts in the space of place, as its activity was interpreted as neocolonialism which forceably undermined the regional economy, while the above mentioned top-down foundation GGN, and trans-local ethnographic-cultural people-to-people networks smoothed the conflict between the transnational space and the space of place. In the long-term perspective, the market campaign positively affected the space of place, and, finally, the result obtained satisfied all the stakeholders. The NGOs were satisfied by preservation of the old-growth forests, since that was a component of sustainable forest management. For those who were concerned by economic losses, new windows of opportunity opened with the development of cultural tourism in the region (Tysiachniouk, 2010f).

The market campaign GGN had its specificities. The node of design of this market campaign involved very few people and was organized by activists of the more radical international and national NGOs. Activists located in different spaces formed an effectively functioning node, providing a threat to the TNC's desire cut old growth forests (Figure 5.5: nodes of design). Forums of negotiation were specific for the market campaign GGN. Initially, they had a conflictive character and followed this during settlement of the conflict instead consensually discussing a new way of governance. Such a course determined the dispersive character of the forums (Figure 5.5: forums of negotiation). In transnational spaces negotiations were carried out between NGOs and TNC and the conflict was settled pretty fast (transnational NGO-TNC). The forums of negotiations in Russia occurred irregularly for about 10 years, on all levels. On the national level, the agreements concerning the Kalevala park were very difficult to achieve because of the constant reforms, restructuring and low priority of the issue for the federal government, therefore the process went slowly (national NGO-NG). On the local level the process of negotiations went more smoothly, because for the most part the local stakeholders were against cutting and for the creation of the protected area (local NGO-SH). The most difficult and conflictive negotiations were at the level of the Republic of Karelia, because here the intervention of the transnational actors was taken as neocolonialism by the regional government, and the value of the old-growth

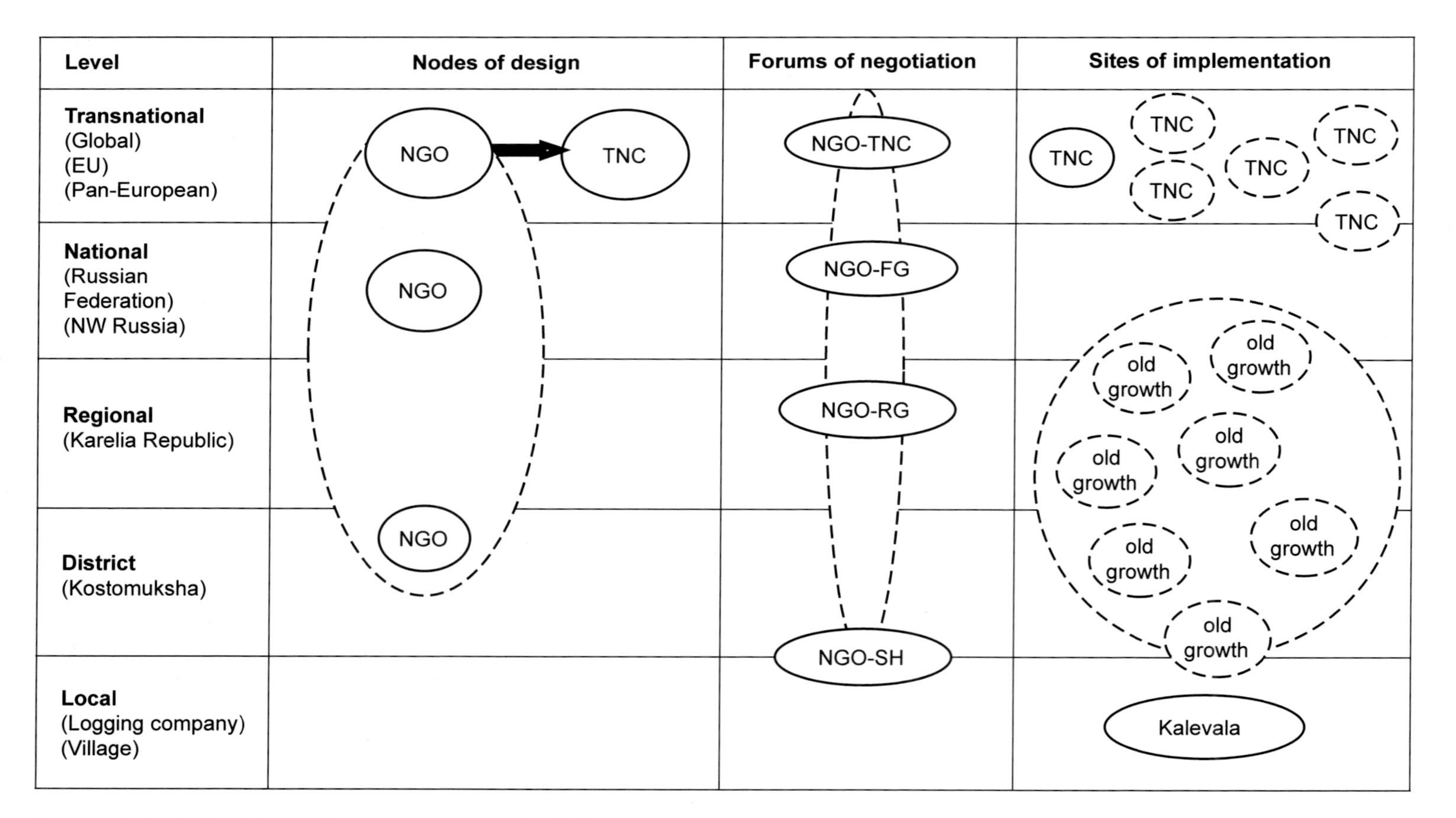

Figure 5.5. Analysis of the market campaign using the GGN concept.

forests had not been recognized (regional NGO-RG). To all appearances these problems were defined by initially conflictive stakeholder dialogue, which was difficult to transfer into more constructive negotiation.

The peculiarity of this case affecting the dialogue was that the same NGOs were making an effort for negotiations with local, regional and national actors. The stakeholder dialogue is supposed to be conducted in the space of place, where the bottom-up NGO network does not function; yet consensual organizations come into play. In the case examined, the radical NGO SPOK, a participant of the boycott, was occupied with all of these things: searching for constructive decisions, building relationships with state authorities, expert communities and with other actors. Therefore, there was none of the usual division of labor between more the more radical and consensual NGOs. The lack of appropriate negotiation skills and the conflict, that SPOK provoked by its radical actions at the beginning of the campaign prevented constructive dialogue with stakeholders. That is why decisions on the creation of the Kalevala Park and on preservation of the old-growth forests were delayed for many years.

The institutional change in both transnational spaces and the spaces of places turned out to be highly isomorphic. In transnational spaces the effect of the consumer boycott of one company resonated and changed the behaviour of the other companies. The 'old' path dependence was broken and companies looked for a new way of doing business. The pressure that was put on corporations influenced almost all the TNCs that had leases in the boreal forests of Russia (Figure 5.5: sites of implementation, TNCs). Before the boycotts focusing on the preservation of the Karelian old growth forests, the concept of old-growth forests referred only to tropical forests. After the consumer campaigns in Russia, Finland and Sweden this concept was extended to all boreal forests. This brought change into the corporate policies of TNCs.

The worldwide series of boycotts similar to Karelian had wider implications. In order to avoid future boycotts, companies improved their social and ecological policies, developed programmes that demonstrated their corporate social responsibility, joined associations of responsible producers, participated in eco-ratings, improved codes of conduct and certified their forest management practices. By doing so they created a 'new' niche market that is more environmentally friendly. The market campaigns also impacted the behaviour of investment bodies that used the toolkits and other methods to maintain responsible investment. Responsible investment provides support for the newly established path and decreases the market risks of sustainable businesses. More companies join the sustainable business 'club', and the 'new' path dependence starts to work as a supporting institutional infrastructure. To legitimize their operations in the spaces of places, companies pursue forest certification and become engaged with certification GGNs as supporters, facilitators and/or members.

The institutionalization of the norm in the described case was highly isomorphic as well at the sites of implementation in the space of place. During the consumer campaign, the site of implementation was the old-growth forests located in the territory of the future Kalevala park, but as a result, the area of implementation of the old-growth forests concept extended to all plots of old growth forests throughout northwestern Russia. The Kalevala park was the target of the boycott and the site of implementation, but in actuality, this was a struggle for the institutionalization of the old-growth forest landscapes (Figure 5: sites of implementation, national, regional, local: old growth). One of the results of the boycott was the recognition of Russian NGOs as legitimate actors, with whom one should consult about issues concerning forest utilization. Boycotts resulted in a chain of informal agreements between the NGOs and businesses on voluntary logging moratoriums in these forests all over northwestern Russia. New business networks came into the space of place and the old business actors changed their practices, new path dependence was established and both the forests and the local stakeholders ultimately benefited.

Chapter 6. WWF-Stora Enso strategic partnership for promoting sustainable forest management: Pskov Model Forest

6.1 Introduction

Interactions between NGOs and businesses may be based on different principles. The interactions vary from confrontation, in the form of market campaigns and consumers' boycotts (see Chapter 5) to 'soft inducements' through the mechanisms of standardization and certification, or even in more advanced ways by creative NGO-business partnerships. An example of cooperation aimed at building models of sustainability on the ground will be analyzed in this chapter, demonstrating an innovative type of partnership which created a network linking actors in transnational spaces and localities. Joint development of sustainable business models involving economic, environmental and social components has become one of the types of partnerships between NGOs and corporations. The development of such models helps transnational corporations adapt to the specificity of social, economic and political contexts in the countries in which they work. In different countries they have to adjust their businesses to numerous challenges all at once. On one hand, they have to take into account the legislation and established rules of the country in which they are located, and on the other hand, it is necessary to conform to international ethical and regulatory norms of business activity. In addition, they usually adhere to their own corporate policies, which they should follow irrespective of the locations of their subsidiaries. Building models that can be later transferred to multiple sites of business operation is one of the ways of developing optimal practices to adjust to both national and global contexts. If the lessons learned in the process of model building can be subsequently applied in other localities of particular countries or regions, the business operations can be standardized and transaction costs reduced.[210] At the same time, sustainability practices will involve more sites of implementation, which is beneficial for the spaces of place. NGOs within such networks contribute to the development of these models by helping enterprises contact stakeholders, and thus become 'transmitters' of ideas from business to the society and *vice versa* (Tulaeva & Tysiachniouk, 2008; Tysiachniouk, 2006a,b, 2010c,f,g).

This chapter presents a case study in which I will analyze a NGO-transnational corporation (TNC) strategic partnership. Stora Enso, in order to adapt to both transnational and local pressures and dynamics, partnered with the WWF to establish meaningful and effective relations with both transnational and local actors. FSC certification was used as a necessary component for legitimating Stora Enso operations in Russia. The partnership between these private actors ran into rather stubborn and unpredictable Russian state legislation and governance issues. The Model Forest initiative served as a platform for (re)negotiating global-local power relations in forestry management by the company, governments and all other stakeholders involved.

[210] Transactional costs, as we see it, are those not connected directly with production: as, e.g. expenses obtaining needed permits, endorsements with different actors; establishing trust relationships with local authorities, the community and stakeholders.

I will focus on relations between transnational and local actors, which have developed in the course of building a model of sustainable forest management and certifying forest management operations using the FSC scheme. Outside donor funding was used to enable the NGO's operation and reaching out to the local community and stakeholders. The chapter dwells upon advantages which strategic partnership with NGOs provides for business, and the role of partnership in transmission of global discourses on sustainability to the concrete practices in a specific locality. Furthermore, I assess the possibility of reproducing these practices in other places.

In this case study several governance generating networks (GGNs), namely the Swedish International Development Agency (SIDA), the World Wildlife Fund (WWF), TNC-Stora Enso and the FSC were involved. As I defined in Chapter 2, global actors are termed GGNs if they are engaged in global governance by developing new global norms and standards to be implemented in specific territories around the world, namely the sites of implementation. The development, adaptation and implementation of the standards takes place through forums of negotiation between the different kinds of stakeholders. In the case study, the SIDA represents a foundation-GGN, which is involved in designing strategies that foster sustainable development, poverty reduction and democratization in developing countries. The grant receivers co-design specific projects to be implemented in specific territories at sites of implementation. As I described in Chapter 3, where I introduced the reader to different kinds of GGNs and the interactions between them, foundations, being GGNs themselves, often play the role of facilitators in providing resources for the implementation of the designs of other GGNs. In this case study, the SIDA sponsored the WWF-Stora Enso partnership to build a model of sustainable forest management, the Pskov Model Forest (PMF), with FSC certification being one of the components. In this way, it became a co-designer of the PMF and facilitator of the WWF and the FSC-GGNs. As was mentioned, the WWF-GGN, in addition to designing the preservation strategies for the Global 200 ecosystems, is fostering sustainable forest management and has familial relationships with the FSC, promoting and facilitating its development worldwide. In this case study, the WWF facilitated the FSC's certification of the Stora Enso subsidiary in the PMF's territory, and by doing so, fostered development of FSC certification in Russia (Figure 6.1).

Transnational actors, involved in building the PMF, represent different GGNs with different agendas, yet in this case they engaged in a top-down network, led by the TNC-NGO partnership, and sponsored by the foundation with the aim of developing, testing and reproducing the model of sustainable forest management in Northwestern Russia.

The chapter is based on field research conducted between 2002 and 2009. Data were collected during 6 trips to the research site, in July 2002, October 2002, August-September 2002, July 2006, November 2007, and March 2008 (see Appendix A). Both semi-structured and open ended interviews were conducted (total #248). Interviews involved NGOs (#14), representatives of state agencies at different levels (#29), representatives of local administrations (#12), local state forest management unit representatives (#7), employees of Stora Enso and its subsidiaries (#9), community activists and ordinary citizens (#91), representatives of local museums (#8), teachers (#22), representatives of houses of culture (#22), and librarians (#10). Transnational actors, including representatives of SIDA, international NGOs, Stora Enso managers, buyers of Stora Enso products, including publishers, were interviewed in Stockholm, Helsinki, Imatra, New York, London, and Germany (see Appendix B-C).

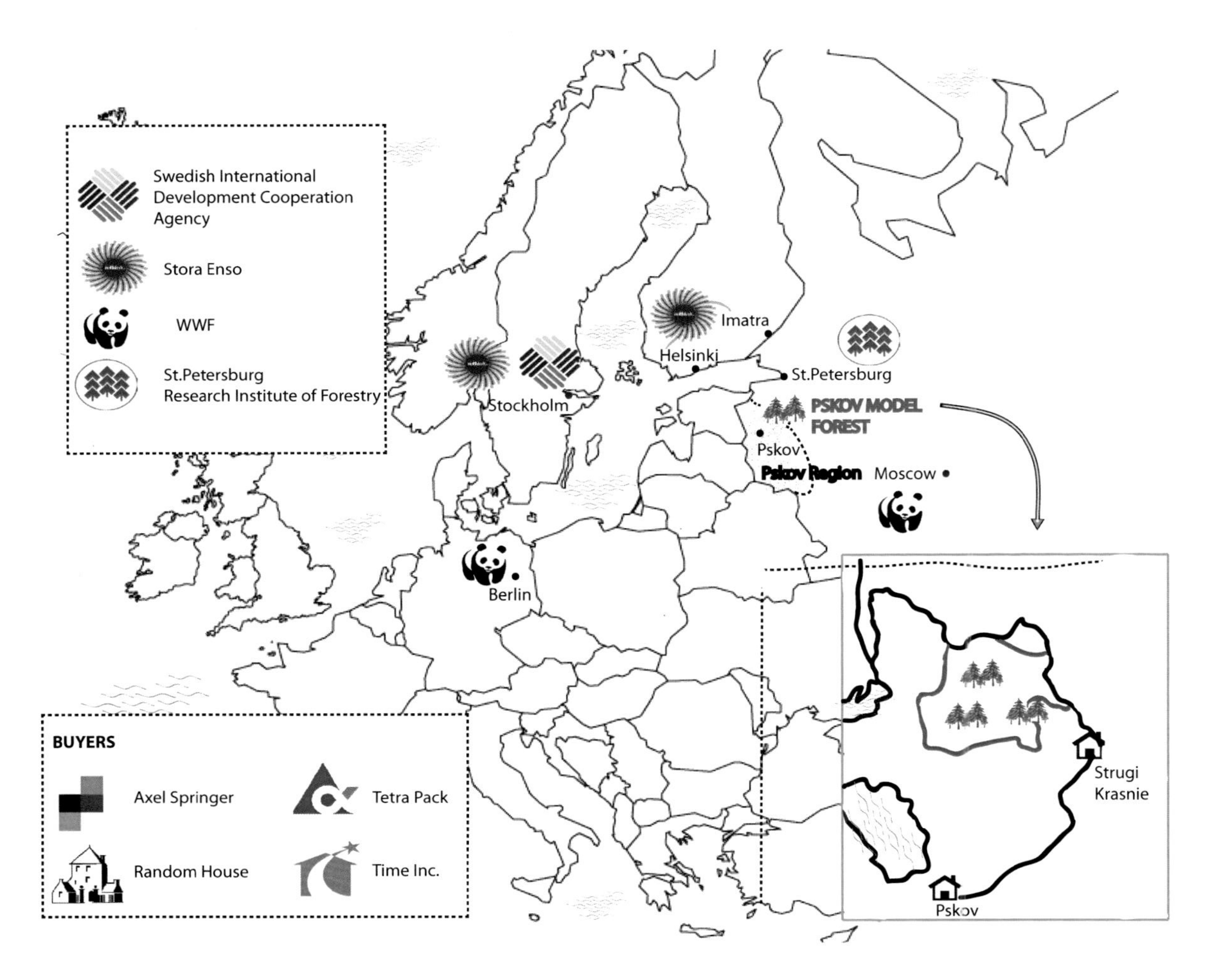

Figure 6.1. Pskov Model Forest: study area and actors involved.

The chapter starts by introducing the actors from translational spaces and the actors from the spaces of places involved in implementing the PMF project (Section 6.2). The chapter proceeds with the description of the interests and stakes of major transnational actors, namely the WWF, Stora Enso, SIDA and buyers of the final products (Section 6.3). All these actors serve as drivers and/or agents of institutional change transforming global designs, corresponding with their own goals, into local practices in the spaces of place using different channels. Next, the chapter focuses on a description of the local context (Section 6.4), a chronology of the events in the process of the PMF's development and the way new global designs were introduced in the space of place by transnational, national and local actors (Section 6.5). The chapter analyzes the strategies of adaptation of global rules to local practices, focusing on the relationships of the PMF's implementers with major national stakeholders, e.g. different levels of government and local community members (Section 6.6). Next, the major outcomes of the PMF are discussed, with emphasis on the difficulties and trends that the project met making an effort to reproduce lessons learned in other parts of Russia (Section 6.7). The chapter concludes with a generalized model of FSC certification of the TNC subsidiaries facilitated by NGO and donor funding. The specifics of the GGNs involved in building the PMF project, the role of nodes of design, forums of negotiation and organizational isomorphism at the sites of implementation are discussed (Section 6.8).

6.2 Actors involved in the PMF project network

The study examines the cooperation of the major agents of institutional change, the WWF and the Stora Enso company, in translating ideas designed in transnational spaces to concrete practices in the space of place during implementation of the PMF project. Based on empirical findings, Figure 6.2 visualizes the involvement of actors in this development.

The PMF project was financed by three transnational actors: SIDA was the major granting agency, Stora Enso and the WWF Germany (the German section of the WWF), in differing amounts. Major founders were involved in the steering committee, which was in charge of interaction with the actors at a transnational level. They also directed the project's activities into the space of place and were responsible for the project's strategy, approval of spending, general coordination and planning within the project. The project was managed by the WWF Russia. The project implementing team was located in the St. Petersburg Forestry Research Institute.

The Advisory Board was created for ensuring successful application of new practices in space of place. It was constituted by the main project stakeholders, who facilitated implementation through their personal and organizational networks. The WWF-International, WWF-Germany and the WWF-Russia were all part of the advisory board. The advisory board included representatives of regional and local administrations, the forest inventory enterprise, the WWF-Russia and the Forestry Research Institute. A representative of the federal forest agency became the Chair of the Advisory Board, which greatly facilitated promotion of the project at the federal level. Although representatives of national, regional and local state forest agencies were part of the advisory board, their role was limited and inactive. Actors from local civil society were represented by the Forest Club (Figure 6.2), created by the PMF.

Since the WWF, as an international nature protection organization, could not participate in consulting and render paid services to businesses, a separate NGO named GreenForest was

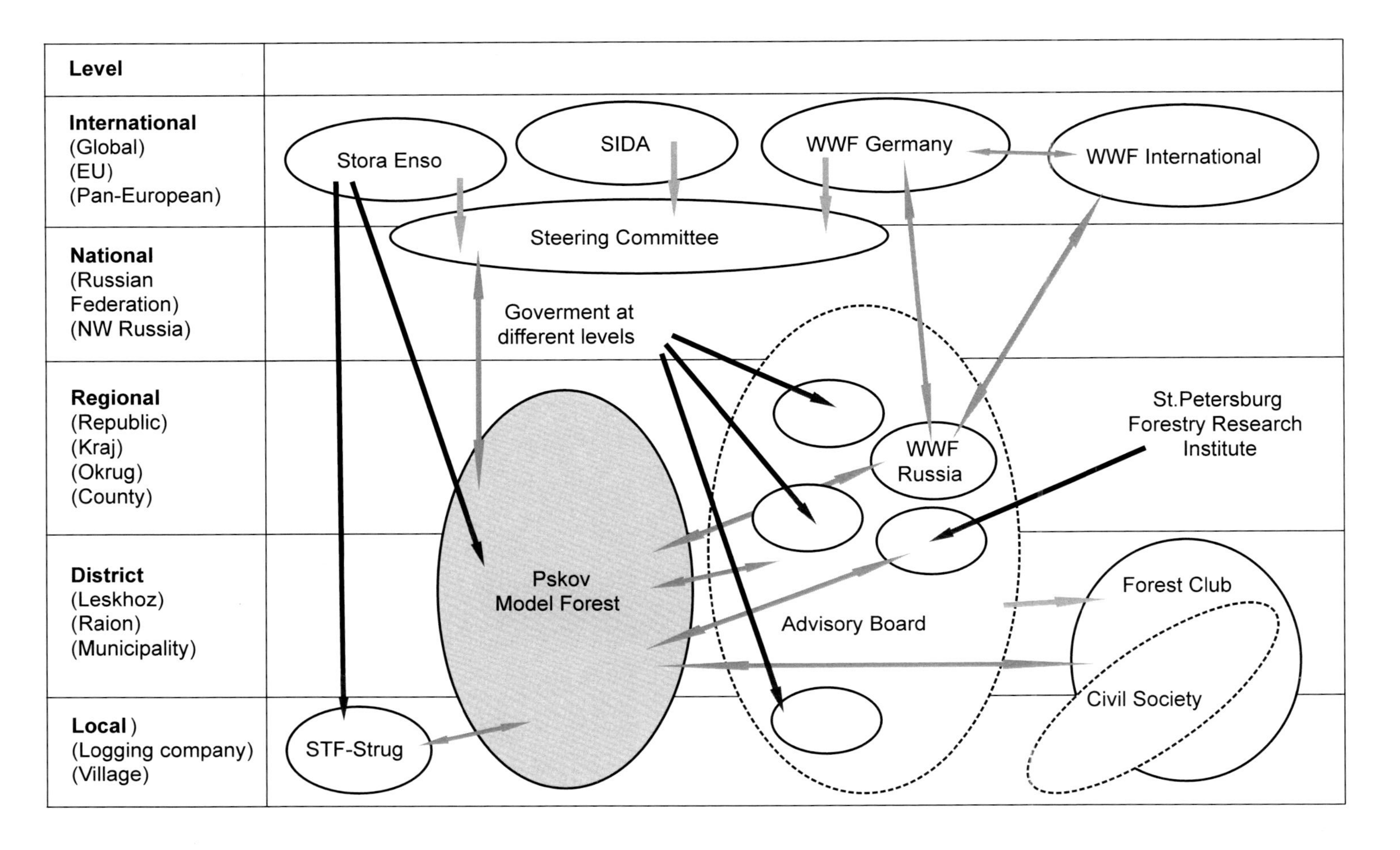

Figure 6.2. Actors involved in implementation of the PMF.

founded. This organization became responsible for disseminating the experience of the PMF and applying the acquired techniques in the other regions. Although most of the people involved in the PMF implementation and in the GreenForest were the same individuals, GreenForest was a separate channel for translating global practices to localities, as they operated under different rules from the WWF and were able to conduct consulting services (Figure 6.2).

Project implementing staff was hired especially for this project and served as major conveyors for translation of global standards and desires of the transnational actors into local practices in the space of place. Project experts, originally from the Research Institute of Forestry, developed innovative methods of felling, promoting relations with stakeholders, and coordinating implementation. The Stora Enso subsidiary STF-Strug was obliged to strictly follow the novel practices and do felling in accordance with the innovative approaches developed in the framework of the project. Although the project was being implemented by hired specialists, the project's activities were often associated with the WWF, which was not a neutral partner in the PMF project.

6.3 Transnational actors, their interests and stakes in the PMF development

6.3.1 Involvement of the WWF

The WWF is known to be one of the largest transnational NGO networks. Forest conservation and promotion of sustainable forest management are within the wide spectrum of its interests. The WWF-GGN, as was described in Chapter 3, has familial relationships with the FSC-GGN and facilitates its implementation by: (a) fostering preservation of High Conservation Value Forests (HCVF), (b) building partnerships with responsible TNCs, (c) supporting the FSC institutional infrastructure worldwide, and (d) maintaining the Global Forest and Trade network (GFTN) that links responsible forest producers with buyers. Since the 1990's, the concept of promoting sustainable forest management in the form of model forests and FSC certification has become the basis of the WWF's 'Forest for Life Campaign', and is one of the major components of its global strategy (see Chapter 4, where I explained in detail how the WWF fostered FSC certification in Russia).

In the PMF a key component of the project was FSC certification. The WWF engaged Smartwood, a program of the Rainforest Alliance, to be an FSC auditor.[211] SmartWood (and later, its subsidiary NEPCon) also contributed to the interpretation of how the sustainability standards should be implemented in a Russian locality (Figure 6.3 that visualizes in more detail the NGO involvement in the project).

Despite their similar aims, model forests and FSC certification are different mechanisms for implementation of sustainable forest management. Model forests are a wider global endeavor, although they may include certification as part of the measures undertaken, as in the case of the Pskov Model Forest. The WWF's first successful attempts of constructing model forests in the former USSR territories were made in the republic of Komi and in Latvia. The Pskov Model Forest was the second in the Northwest region of Russia.[212]

[211] Interview with researcher at Forestry Research Institute, December 2002, St.Petersburg.

[212] Interview with WWF coordinator, December, 2002, Moscow.

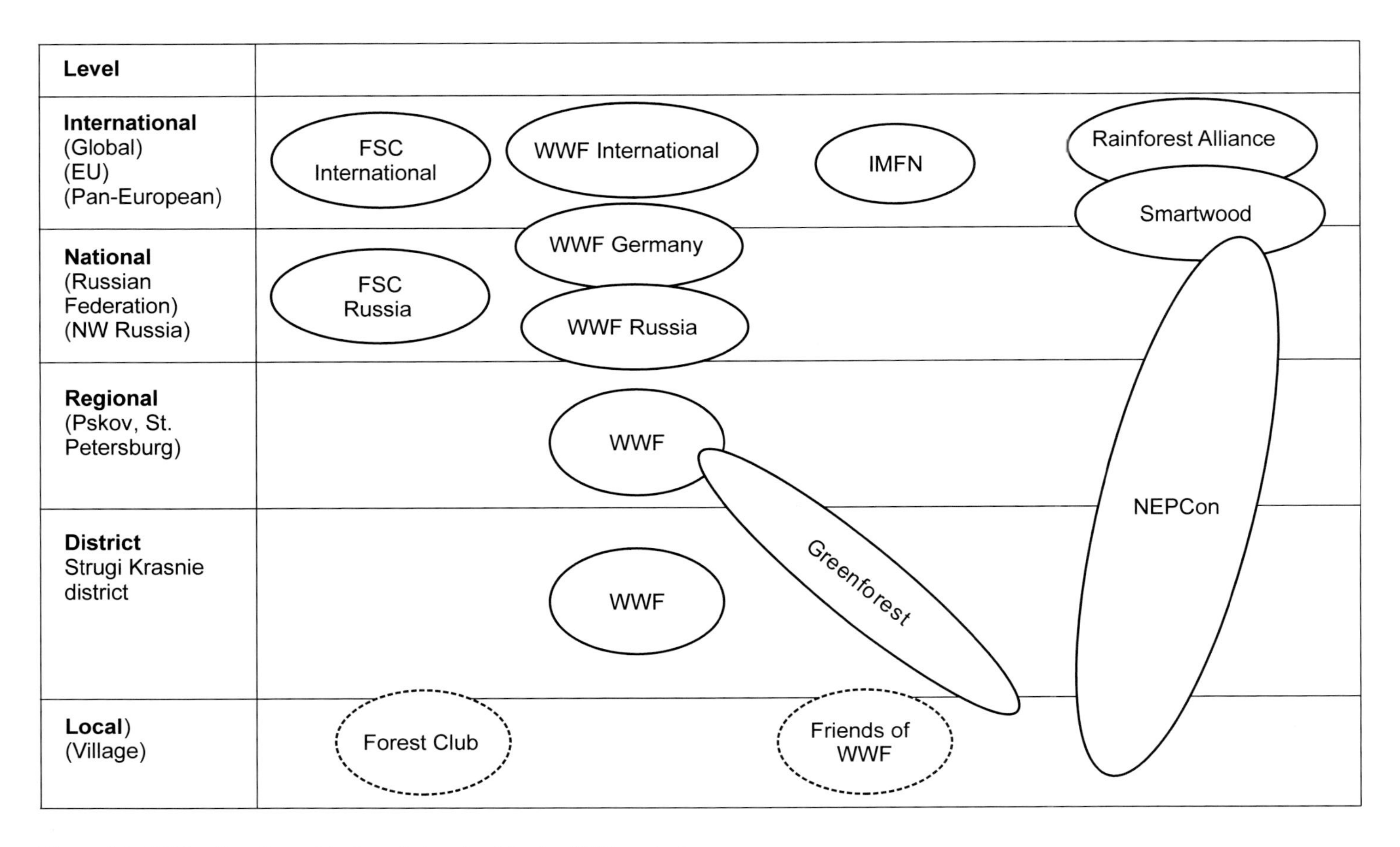

Figure 6.3. NGOs directly and indirectly involved in the PMF project.

The Russian Pskov Model Forest has been implemented into the framework described in Chapter 3; the WWF's global strategy. One of the sponsors of the project was the WWF office in Germany. The German branch of the WWF, besides promoting its own agenda, is engaged in fundraising for projects such as 'economies in transition', therefore, they supported the Russian model of sustainability. WWF international has attracted two other sponsors: the transnational Swedish-Finnish timber company Stora Enso, operating in 40 countries, and SIDA (Figure 6.3).

For the WWF, the promotion of projects in model forest development is a way to involve businesses in sustainable forest management and to encourage environmental and social responsibility. The Model forest has become a type of laboratory for development of social, environmental, and economic innovations in forest management. With this project, the WWF was following its mission, sequentially advancing its corporate strategy and its logo, while introducing innovative environmental approaches to forest management. Co-financing has allowed the WWF to position itself as a project holder and as an implementer of the project's achievements. The role of the WWF in transferring transnational concepts into local practices was essential, as they operate at all levels from global to local (Figure 6.3).

6.3.2 Stora Enso, linking transnational spaces and the spaces of places: its stake in the NGO-TNC strategic partnership

For Stora Enso, the project has became, on the one hand, a source of legitimacy in their interaction with buyers and customers in transnational spaces, and on the other hand, one of the tools for adaptation to local contexts in the space of place. Therefore, Stora Enso played an essential role in linking transnational spaces and spaces of places. Figure 6.4 represents a simplified business network of Stora Enso. In transnational spaces the network involves the buyers of its paper, publishing houses. Wood Supply Russia represents a unit with offices in both Finland and Russia that coordinates the functioning of logging enterprises in Russian localities and the purchase of wood from Russian suppliers.[213]

One should note that establishing strategic partnerships with NGOs and developing models for facilitating the company's activity is distinctive of the management system at Stora Enso. Participation of Stora Enso in the PMF project was an innovative strategy of business integration in another country of operation. Stora Enso in Russia had to solve the problems encountered in the post-socialist transition period: the constant reforming of state governing bodies, new forest legislation, and institutional turbulence. It was important for the company to find models for commercial forestry in Russia: 'It was a development project; our business was developing together with Russian social development, its legislative system and law enforcement ... We needed the holistic forest management model with information on forest resources, inventory, planning, participatory methods and effective implementation.[214] In this situation the partnership with a powerful organization like the WWF helped the company adapt their business to Russian conditions. The PMF would also become a proper place for demonstrating the advantages of Scandinavian harvesting methods, with the goal of their further dissemination in Russia. The

[213] The Figure 6.4 serves the case study, the Stora Enso logging subsidiaries were sold in 2008.

[214] Interview with a Stora Enso manager on certification March 2008, Stockholm.

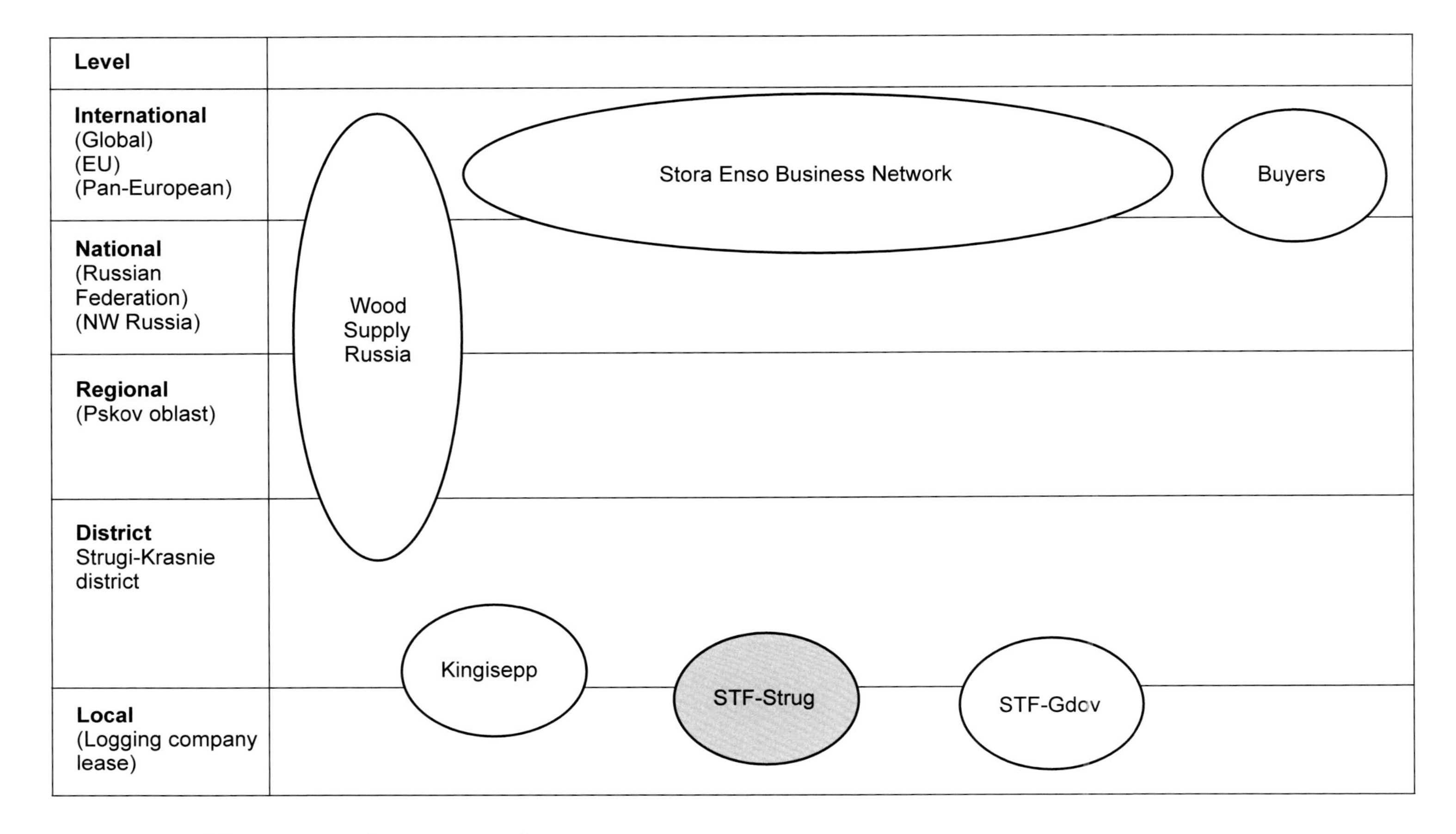

Figure 6.4. Simplified Stora-Enso business network.

WWF, as Stora Enso's strategic partner, assisted in resolving problems coming from state structures, local stakeholders and the population; it also contributed to 'legitimizing' Stora Enso in the eyes of international stakeholders (Table 6.1). The brief history of Stora Enso's operation in Russia prior the project can bring light to the motivation for the company to get engaged.

Stora Enso was instituted in 1998 as a result of a merger of the Swedish company Stora and the Finnish company Enso. Before the merger, both companies had worked in Russia. In 1988 Enso started preparations for establishing the Russian joint venture in Karelia, and in 1990 the enterprise Ladenso was put into operation. In the course of creating this joint venture, the company met with many difficulties, taking into account the Russian context which strongly differs from the West-European. From the very beginning of its operation, the enterprise tried to introduce new harvesting technologies which were unknown in Russia at that time. It was necessary for the company to adapt technologies for the new social and environmental conditions as well as to make the use of technologies consistent with Russian legislation. The enterprise attempted to achieve a special status for its leased territory where it could show the advantages of the Scandinavian methods of timber cutting. However, it failed, although the interest of the state structures and experts was evident. Despite the existing difficulties, the enterprise made efforts to introduce new technologies of road construction and assorted timber cutting in Russia. They worked within existing regulations, although they tried to adapt their own technological innovations. In the 2000s, Ladenso successfully carried out their activities and has become one of the advanced timber enterprises of Stora Enso.

In the Stora company (the other enterprise which merged into the Stora Enso company) the decision to create a joint venture in Russia was made in 1994. The Stora company was aware of possible difficulties connected with launching the business in a country with a transition economy.

One of the problems in the 1990s was the sharply negative attitude of the local population and other stakeholders towards the penetration of western logging companies into Russian territories. At that time, local stakeholders were expecting the foreign companies to deplete the Russian forest resources and export revenues out of the country. Lastly, the foreign companies were not prepared to work in the Russian wild, uncultivated forests. The company collided with the impossibility of directly using Scandinavian harvesting technologies in Russia because of the existing forest management standards and forest ecosystems. The Swedish forestry education did not provide the necessary knowledge for such unaccustomed conditions, and the middle range managers faced difficulties while arranging forestry-related actions. It was necessary to adapt both practices and standards of timber cutting, and the production control systems, to the conditions of the local forests as well as for the turbulent institutional environment.[215]

Therefore, Stora Enso was interested in participation in the PMF project and in partnership with the WWF because this would help in solving many of the problems encountered in managing the company in Russia. At the time of research in 2002-2008, Stora Enso had logging subsidiaries located in Pskov (Figure 6.4: STF-Strug), Leningrad (Figure 6.4: Kingisep STF-Gdov), and Novgorod regions and in the republic of Karelia, and was eager to standardize its operations, making them profitable and efficient.

[215] Interview with the research director of Pskov Model Forest, March 2008, St. Petersburg.

For a corporation in this situation, the Pskov Model Forest became an opportunity not only to adjust business to Russia by the simultaneous development and introduction of innovations, but also to advance the development of a new normative base in Russia and to make an effort to change Russian conditions on behalf of businesses. This would change the local context in the space of place to a greater conformity to fit an actor from transnational spaces and allow for the smooth operation of a multinational corporation.

6.3.3 Facilitating role of the foundation: SIDA involvement in the project

The financing of projects promoting sustainability, as well as advancing Swedish businesses in Russia, was one priority of SIDA activity: at the time the project was launched, Sweden and Russia had established timber trade relations. Moreover, the Swedish Parliament had declared a number of initiatives (e.g. rendering help to countries with transition economies, promoting democracy in Central and Eastern Europe, supporting socially stable transitions to market economies, and advancing ideas of sustainable development, etc.) which encouraged SIDA to finance the project.[216] It was a part of the SIDA initiative in poverty reduction in developing countries, based on the idea that if the profitability of the Russian timber industry could increase through the intensification of forest management, this would contribute to the well being of local communities in rural Russia.[217] Another goal was to promote the Swedish forest sector's cooperation with Russia at all levels, making Russian forestry transparent and legal. SIDA viewed such cooperation as important because Russia contains 20% of the world's forest resources and Sweden is a small country, dependent on forestry, making cooperation important for the long term future.[218] Despite the fact that SIDA did not participate directly in the implementation of the Pskov Model Forest, the agency contributed to the project not only financially, but by involving Swedish specialists; scientists and experts visited the model forest for consultation and research. This promoted the use of Scandinavian technologies in the Pskov Model Forest and fostered the intensification of forest management. Since 2005, the PMF has become a part of the larger Russian-Swedish cooperation program in forestry managed by the Swedish Forest Agency. A representative of the Swedish Forest Agency served on the steering committee of the PMF and pushed the project toward development of tools to be used in sustainable forestry, such as satellite imagery to assess stocks, computerized models for calculating economic outcomes, etc'. They were interested in involving Swedish experts in the advising and evaluating of the tools developed. The Swedish Forest Agency also pressed for democratic decision making, emphasizing the need of ordinary citizen's involvement.[219]

[216] Interview with the SIDA representative, March 2008, Stockholm.

[217] Interview with the Manager of the Russian-Swedish cooperation program, Swedish Forest agency, March 2008, Stockholm.

[218] Interview with the Swedish Forest Agency representative, March, 2008, Stockholm.

[219] Interview with the Swedish Forest Agency representative, March, 2008, Stockholm.

Table 6.1. Stakeholders in the space of place and the transnational spaces.

Spaces	Stakeholder	Type	Interest
In the space of place	Rosleskhoz	GO	National forest management authority. Generally supports Model Forests and FSC certification. However, as a rigid and bureaucratic institution, allows innovations only on territories of special status, does not make an effort to introduce innovations in regulations to the new Forest Code of 2007, representative was on the project advisory board.
	Lesoustroistvo Forest inventory unit	GO, since 2008 privatized	Creates plans for harvesting in the region. Regulates a forestry cycle of 10-12 years. Partners of the PMF, helped to include all innovations into the long term forest management plan
	Leskhoz (since 2008 lesnichestvo)	GO	Manages the forest on local level[1]. Leases land to the logging companies, monitors logging operations. Interested in responsible forest users and in their compliance with legislation. Some Leskhoze members are interested in innovations while others are not.
	Pskov regional governments	GO	In different times different reactions, mostly positive. The Department of Education supports the project. Some representatives are on the project's advisory board.
	District and town administration	GO	Supported the project and its developments as well as STF-Strug as a tax payer
	Russian timber companies	Commercial company	The small logging companies that leased the land on the neighboring territories were neither expressing interest in the developments of the PMF nor in forest certification.
	St. Petersburg Research Institute of Forestry	State	Interested in research and the implementation of outcomes, project partner, hosted the MPF office in St. Petersburg.
	Greenforest	NGO	Interested in delivering services to commercial companies in environmental planning and intensive forest management.
	Forest Club	Stakeholder dialogue space	All local and regional stakeholders interested in the project receives information on the PMF's progress, local community activists share information on their activities.
	Eco-tour	Civil society informal organization	Study natural history of the district; organize tours for local activists and all interested parties. Build ecological trails, sustain knowledge on social value forests.
	Local libraries, house of culture, museum, schools	State institutions	Village intelligentsia, small grant recipients, interested in implementing their own ideas and initiatives, interested in funding to support their work, generally support the project.
	Community	Local villagers	Mostly against any logging, however support the WWF and the PMF when they promote nature conservation and community initiatives.

Spaces	Stakeholder	Type	Interest
	Friends of the WWF	Network of high school students and teachers	Interested in nature conservation and ecology, interested to be part of the WWF and in participating in the network activities.
Transnational spaces	Stora Enso	Commercial company	Economic interest in Russian forests. Interested in changing Russian legislation in a way that implementation of Scandinavian technologies with intensive forest management would be possible. Interested is consistency and standardization of its business operations, interested in maintaining good relationships with Russian and international NGOs. Interested in low risk wood sources that would satisfy their buyers, final consumers and shareholders, interested in maintaining a good image and brand.
	SIDA	Swedish GO	Long term cooperation between Russian and Swedish forest sectors, facilitates operations of Swedish industry in Russia, increase profitability, legality and transparency of the Russian forest sector.
	Swedish Forestry Agency	Swedish GO	Cooperation with the Russian forest sector, development of the tools that would intensify sustainable forest management in Russia, standardization of approaches to forestry and legislation in Russia and Sweden.
	Publishers	Commercial company	Intangible value in the project, confidence that the paper is from legal sources and well managed forests, reduced risk of buying from 'unconventional wood markets'.
	Buyers of wood products	Commercial company	Confidence that the supply chain is transparent, legal and wood is coming from a sustainable forest operation, reduction of risk in buying fiber from Russian forests (from 'unconventional wood basket').
	Global Model Forest network	NGO	More Model Forests created around the globe, information sharing, compliance with general global principles.
	WWF network	NGO	Promote sustainable forestry in the form of Model Forests and FSC certification, partnerships with companies committed to sustainability.
	Rainforest Alliance Smartwood Program, Nepkon	NGO	Interested in clients and certifying all Stora Enso operations, good quality certificates.
	FSC International	NGO	To certify as many forests as possible, good quality certificates, interested that companies choose the FSC scheme.

[1] Only until 2008, after 2008 all reforestation, thinning and other forest management operations became the responsibility of the companies that lease the land according to the new forest code.

6.3.4 Buyers of Stora Enso products: drivers of company transparency and accountability

Buyers of Russian wood and wood products played a significant role in fostering the transparency, accountability and sustainability of operations of the companies involved in the supply chain. In the early 2000's, when the forest management of Stora Enso subsidiaries in Russia was not yet certified by the FSC, buyers of Stora Enso products were interested in their own risk reduction coming from the supply chain, involving a country with poor ratings in corruption and legality.[220] Some media and catalogue industries, especially in the USA, were aware of NGO campaigns against Staples, Home Depot, Victoria's Secret, and Sears in the late 1990's and early 2000's and developed very proactive sustainability policies, such as disclosure of the fiber supply chain and complete transparency along the whole chain of custody. Time Inc., for example, which was targeted by Greenpeace in 1999 for using chlorinated pulp paper, took an active role in improving its environmental profile and made a commitment to have 80% certified fiber in their products by the year 2006. They actively encouraged their suppliers, especially those using 'non-conventional' wood sources, such as Russia, to certify their forest management as well as the chain of custody operations.[221]

For all the media companies interviewed, not only legality and corruption, but also social responsibility in the space of place, such as worker's health and safety, accident rates and the business ethics of their suppliers were important issues. Axel Springer's representative called this 'intangible value of the product', that includes both the social and environmental pillars of sustainability. They also emphasized the importance of the company's ability to communicate with local stakeholders to build a 'consensual value' into the product. By requesting a democratic way of decision making, they, to some extent, fostered democratic development in rural areas of Russia. As a media company they considered it important to ask such questions as 'Where did my newsprint grow?' 'Where did my magazine's paper grow?[222]' They stated that their high profile on the environmental and social aspects of production can help to attract the best journalists to the company.

Random House UK, in the same way as Axel Springer and Time Inc., stated the importance of health and safety issues as well as strong environmental standard's implementation during harvesting operations in Russia. They described how important their own environmentally strong purchasing policy is for their young employees, who understand the differences between certification systems. Random House UK, which was the first publisher in the world to certify chain of custody using the FSC scheme, was paying a premium of 30 UK pounds a ton for FSC certified paper and, therefore, was encouraging Stora Enso to increase the volume of FSC certified fiber coming to the UK, as they strongly believed that FSC is a better standard.[223] Their own customers, especially Marks and Spenser and several supermarkets, were continuously asking

[220] Interview with the Tetra Pack representative, November, 2008, Cape Town.

[221] Interview with the director on sustainability of Time Inc, June 2008, New York.

[222] Interview with an Axel Springer representative, December 2008, Dusseldorf.

[223] Interview with the representative of Random House UK, December 2008, London.

for FSC products.[224] Although Stora Enso's buyers and customers were never directly involved in the PMF project,[225] they, with their interests and stakes influenced Stora Enso operations both in Russia and worldwide.

6.4 Local context

The Strugy-Krasnenski region has a population of 18,500 people, about half of which live in the district's center, Strugy-Krasnie. This settlement lies 68 km from the city of Pskov. The economic situation in the Strugo-Krasnenski region in late 1990s was very weak. Before Perestroika, much of the economic activity in the region consisted of branches of St. Petersburg, Moscow, or Riga enterprises that were specialized parts of the Soviet military-industrial complex. Since the late 1980's, however, many of these operations have ceased. The economy has gone down and unemployment has increased. The average salary in business enterprises was only US$ 25 per month, while pensions reached only US$ 16 and were delayed for months. At that time, citizens could use small plots of forest for their own needs for very little money, and resell the plots to small forest producers (Chernova, 2010: 141).

The economy of the region was also heavily influenced by the proximity of Russia's European borders. Logging enterprises in the region were heavily dependent on international markets in Western Europe. The closeness of borders with the Baltic States heavily influences the type of businesses that come into the region. Many companies were export-oriented, and made use of good railway transportation to Latvia and Estonia. Accordingly, the Strugy-Krasnie district was an important raw material provider for the international timber industry of Europe. The PMF project was also oriented towards the trading of round wood with Europe.

The PMF was developed on a part of the Strugo-Krasnenski forest territory, the managed total area being 18,400 hectares leased by the Stora Enso subsidiary STF-Strug. The choice of the Strugo-Krasnenski district was stipulated by three reasons. First, it is a district with a well-developed infrastructure. Second, the local administration and the population of this area were more interested in the development of the project than the people in the neighboring districts. Third, and decisively, the subsidiary of Stora Enso was located in this district.

6.5 Implementing the PMF: stages, chronology and introduction of new practices

The main objective of the PMF project was to develop and implement a model of intensive forestry which would provide for inexhaustible forest management, combining intensive felling technologies with reforestation. As with any model of sustainable forest management, the PPM was focusing on the three main pillars of sustainability: economic, environmental and social. Concerning economic issues, the project envisaged a system of planning that would predict the economic outcomes of the applied forest operations and would justify the economic efficiency of

[224] Interview with the representative of Random House UK, December 2008, London.

[225] They were directly involved in other Stora Enso projects in Russia, that were focusing on the transparency of the supply chain.

investments into forestry.[226] Social issues covered the local population's involvement in forestry-related decision making. In its environmental component, the project contributed to developing techniques of landscape ecological planning that would sustain biodiversity and preserve valuable ecosystems.

6.5.1 Major stages

The first stage of the project (2000-2004) was focused on three aspects: FSC certification, establishing relationships with the local population, and developing methodologies of intensive forest management and organizing test plots for experimenting with different techniques.

The second stage of the project (2004-2008) focused on landscape-environmental planning, educating companies and stakeholders, and dissemination of technologies for intensive sustainable forest management in the Russian northwest among large timber holdings. Companies were interested in training in nature conservation planning, as many of them had started the process of FSC certification. In 2009 the project was finalized, outcomes published and disseminated.

6.5.2 Project activities: brief chronology

In 1998-1999, the company Stora started negotiations with SIDA on possible financial support for the project. Stora was studying Russian woodlands, road networks and the ways of possible log transportation abroad. The company had found the person who had created the Baltstrug enterprise in the Strugo-Krasnenski district, which had leased forest land for 49 years. Baltstrug and the company Stora had formed a joint venture named 'STF-Strug'. In 2002, when the project was in full swing, the management body of the corporation undertook modifications which affected its affiliated branches. With these new conditions, the director of 'STF-Strug' sold his share. However, he proceeded with his work in the project: first, as an expert, later on as an expert on model forests and finally as a consultant on the model experimental plots, including their creation, maintenance, development, and monitoring. He was responsible for the monitoring compliance of STF-Strug with the logging techniques that were developed during the project.

In 2000, the project was officially launched, with all interested parties invited; among them the Chamber of Commerce and Industry of Pskov, the Pskov Forest Producer's Association, representatives of the Department for Education of the Pskov region and others. A Forest Club was founded within the project for establishing a friendly relationship with the local population. At the same time, the project began actively working with journalists. In the framework of the project, a small grants program for supporting local citizen initiatives was launched, with special emphasis on ecological education. In addition to these PR-actions, project participants initiated meetings devoted to landscape-ecological planning.

In 2001, in cooperation with Pskov regional TV, the project contributed to the creation of the ecological TV program 'The World Around Us'. Within the framework of the project, Russian civil-service specialists in forestry attended relevant training in sustainable forestry in Estonia, Latvia, Sweden and the PMF. Foreign specialists were involved in the project as experts: a

[226] Interview with the research director of the Pskov Model Forest, March 2008, St. Petersburg.

representative of the Nordfor Swedish consulting company helped to plan forestry-related actions on the demonstration plots. Contacts with the IKEA Company were established and possibilities for cooperation were discussed.

In 2002, 'STF-Strug' passed the FSC pre-assessment. The certification body Smartwood (a program of the Rainforest Alliance) was involved in conducting the audit. One of the aspects of sustainable forestry are multiple users. The WWF initiated a small project on the development of ecological tourism in the Pskov region, which envisaged special eco-trails for bird-watching, for both ornithologists and the general public. Travel agencies from Sweden and the UK, a Swedish consulting company, the Swedish Public Forest Council, and Russian ornithologists participated in the project. In addition, a documentary movie 'Malachite Walls of the Pskov Forest' was made. In April, public hearings were held, devoted to discussion on possible scenarios for forest management on the territory of a model forest. The same year the project attracted the attention of regional committees of natural resources throughout Russia, and 30 leaders and specialists visited the PMF for studying the model forest experience.

Since 2003, the model forest has gradually become an educational center. A series of workshops were held within the project. In February, 2003, the most prominent workshop, 'Timber Monitoring in the Russian Northwest' was organized in Pskov. It was intended for timber enterprises, trading houses, and logging companies operating in Russia. Other seminars were also conducted on the basis of the Pskov Model Forest; e.g. a workshop within the WWF-IKEA project was organized for representatives of certification centers, experts and auditors. In September, 2003, 'STF-Strug' received a forest management certificate. Financial support of the first stage of the project was terminated near the end of 2003, and the activities for the second stage of the project were negotiated with its donors and partners.

In 2004, there was a 'transition' from the first phase of the project to the second. On the initiative of WWF International, specialists of the World Bank and independent experts from the UK examined the project and formulated recommendations for the next stage of implementation. Despite low funding during the period of transition, the model forest continued to host workshops. For example, the Stora Enso company initiated a workshop on sustainable forest exploitation for experts from the McDonald's company. A four year long project on the creation of a textbook for high school students on sustainable forest management was completed. The text book was tested at schools in Pskov and the Pskov region at grades 8-10 and was prepared for publication.

In April 2005, the financing of the 'Pskov Model Forest' project was renewed, paving the way for implementation of its second stage, which was completed in 2008. One of key phases of this stage was the development of the normative basis for the assessment of the quality of wood on the stem, reforestation, and the conservation of biodiversity, commercial thinning and final felling. These standards take into consideration ecological landscape planning. The new normative basis was consistent with the FSC's standards; however many of the new developments had contradictions with the existing Russian forest code. They were tested in the Pskov region, but it was assumed that they could be used in the Vologda, Novgorod and Leningrad regions as well. In July-October 2005, an expedition to the north of the Pskov region was organized within the project, with local specialists participating. During the expedition the participants gathered material for nature conservation planning and preservation of biodiversity, noted the taxation characteristics of forest stands and collected data about the quality of wood on the stem. Closer contacts with the Svetwood company (a branch of IKEA) were established in 2005. The small grants program, which was suspended during the transition period, was resumed in June 2005. This program has gone beyond the limits of the model forest and was intended for the population of the whole Pskov region. In 2005, the project was integrated into the Russian-Swedish program of forest sector cooperation. According to this program, the PMF has become responsible for the dissemination of its experience in sustainable forest management in the Russian Northwest.

The development of methods for determining the taxation characteristics of forest stands using satellite imagery started in 2006. Such techniques also allowed for the economic assessment of forests. In 2006, two brochures were published by the project: 'Criteria For Assessment of Vertebrates Biodiversity' and 'Ecological Trail in the Pskov Model Forest'. In 2006, the discussion on standards for thinning in the framework of an intensive forest management model began on the federal level.

In June 2006, the PMF joined a Russian model forests initiative network and became a member of its working group. In the same period, the PMF implementers started to conduct seminars for logging enterprises aiming at preparing them for FSC certification. Since 2006, four logging enterprises working in northwestern Russia have been trained in depth at the Model Forest Training Center. These companies included three subdivisions of Stora Enso, 'Russkii les' (Russian Forest), 'STF-Gdov', and the 'STF Kingisepp' (Timber proceeding complex), as well as 'Svedwood', a daughter enterprise of IKEA. Other companies have also participated in training seminars with 'Russkii les', such companies as 'Terminal' and the affiliated company of UPM Kummene participated in the workshop on nature conservation planning during preparations for FSC certification. In the summer of 2006, field work was assigned in southern and central parts of the Pskov region on creating taxation standards for this area. In December 2006, the textbook for high school students in grades 8-10, titled 'Principles of Sustainable Forest Management' was published. The training seminar on forest certification for future auditors and advisers was held in the 'WWF-IKEA' partnership project. On behalf of the Russian model forest initiative network, the employees of the Pskov Model Forest project started developing the concepts and guidelines for model forests in Russia.

During 2007, cooperation with timber enterprises continued and new standards for thinning were tested on the territory leased by 'Russkii Les'. With the lack of action in creating a new Forest Code, in January 2007, the project attempted to participate in the development of statutory acts and the normative base to include the innovations developed in the Pskov Model Forest. The educational part of the project provided for four standardized two-day seminars devoted to the application of new standards in forestry, preparation for FSC forest certification, carrying out commercial thinning, and advancing GIS-technologies. The main impetus of working with local people (forest clubs, small grants, etc.) also continued in 2007.

In 2008, the funding of the PMF ended, and at the end of 2008 Stora Enso sold all its subsidiaries in the St. Petersburg and Pskov districts, largely because of the high export tariffs introduced by the Russian government. In 2009, the PMF implementing team continued working on producing and disseminating the project outcomes.

6.5.3 *Introduction of global practices into localities*

In the PMF project, the introduction of global practices into individual localities took place in several crossing and complementary streams (Figure 6.2). On one hand, it occurred through the FSC forest certifications being promoted by the WWF all over Russia, and on the other hand, through the application of the corporate practices of 'Stora Enso' in its subsidiary 'STF-Strug'.

The forest experts within the PMF were the most important – the implementing team that was adjusting the Scandinavian model of intensive forest management to conditions in Russia. In order to develop a living laboratory of sustainable forestry on Russian soil, the PMF involved many experts, both international and local, in economics, natural sciences and sociology. The expert community served as an additional stream for adjusting global discourses to the local context.

The scientific director of the project, from the St. Petersburg Research Institute of Forestry, treated certification as part of the wider measures needed for constructing a model of intensive forest management. 'STF-Strug', the enterprise which was prepared to receive the forest management certificate, was obliged to adhere to all of the organizational principles of a model forest and implement the innovations developed.

Stora Enso was in charge of the preparations by STF-Strug to comply with the FSC requirements. Stora Enso adheres to common corporate policies implemented worldwide, which include environmental, economic and social components. Along with the others, corporate policies include labor standards and maintenance of health and safety standards for employees consistent with the ILO convention. Thus, the realization of the FSC standards at STF-Strug in training workers, supplying them with uniforms, fulfilling requirements for accident monitoring and prevention, and complying with workers' rights have been ensured by Stora Enso. At early stages, fellers and operators of forwarding machines were trained in Finland and specialists from Sweden visited 'STF-Strug' for conducting training on the spot. During the certification process, the enterprise did not have any problems connected with conformity to requirements for fair wages, which where above the local average and where paid in time.

Stora-Enso managers have contributed to teaching employees to treat spilled petroleum products and to use adsorbents, and how to organize waste disposal. In addition, widespread explanatory work was carried out. It is illustrative that in interviews, workers used ecological terminology, which at some certified enterprises employees may be not accustomed to. 'Formerly we cut everything without fail ... And now we leave seed trees, biotopes, and old trees to preserve biodiversity'.[227]

In the framework of the project, the PMF implementing team increased the expert community available for serving companies in preparation towards FSC certification and for intensification of their forest management.

---

[227] Interview with the operator of a skidding machine, August 2006, Strugi Krasnie.

6.6 Interaction with stakeholders in the process of accommodating the global rules to the locality

From its start, the PMF implementing team mediated different issues and tried to solve conflicts between STF-Strug and the other stakeholders: the local forest management unit (leskhoz) and its workers, government officials from federal, regional and local levels, the regional administration, and the local public (Table 6.1). The PMF implementers made an effort to involve each stakeholder in the project. The goal was to bring all of them into the decision-making processes for the management of the region's forests. The intention was to involve stakeholders in a network that would endure beyond the timeframe of the PMF.

6.6.1 Relationships with national and regional forest agencies as a component of promoting new forest practices

Both the WWF and Stora Enso have shown a strong interest in establishing good relations with state authorities aimed towards creating opportunities to apply the standards developed in the PMF into other localities. The introduction of these standards at legislative levels is necessary. The PMF project treated legislation as one of the possible ways to cooperate with national state agencies, along with the direct lobbying for their own interests, stakeholders' participating in conferences, etc.

In the early 2000's there was resistance from government officials at many levels to FSC certification. For example, one official from the Ministry of Natural Resources believed that it was a political and economic mistake to promote FSC certification in Russia. He thought that it would be costly, need too much investment and therefore cause limitations to Russia's forest industry and hurt the economy.[228] The fact that certification would actually provide increased opportunities in the European timber markets has been widely recognized by state agencies only since 2005.

For the institutionalization of new practices, much attention was paid to building policy networks and searching for partners at national, regional and local levels. The aim of these networks was to prepare Russian territory for the implementing of new practices, some of which contradicted existing Russian legislation. For example, the scientific director of the project worked at the St. Petersburg Research Institute of Forestry, which formerly was a structural unit of the Ministry of Nature Resources, and his affiliation and contacts facilitated his access to the federal agencies.

The Northwest Forest Inventory unit was at that time a state agency, and became a partner of the project as long term planning was a key for sustainable forestry. Their aim was not only to include innovations from the PMF into forest inventory and planning, but also to correct existing pitfalls in the state inventory system.

The first step in gaining permission to create the PMF outside of the normative base was to convince various levels of the forest agencies that sustainable forestry is necessary and beneficial. The WWF used its own channels for cooperation with state authorities. Representatives of the WWF Forest Program are also members of Public Ecological Council at Rosleskhoz, where they gradually advanced the ideas of sustainable forest exploitation. In 1999, prior to launching the

[228] Interview with the representative of the Ministry for Natural Resources, January 2002, Moscow.

project, the WWF and Rosleskhoz signed an agreement on cooperation in the field of forest governance and preservation of biodiversity. According to the agreement, the WWF assumed the conceptualization of model forests. Interaction with Rosleskhoz has contributed in part to obtaining a special status for the model forest. This status favored experimentation with felling and the development of new standards for forest exploitation (Chernova, 2010: 20). To avoid problems with the local authorities, relevant agreements with regional representatives of Rosleskhoz related to forest management were signed as well. These agreements envisaged involvement of the state structures with the advisory board and working groups of the project. According to these agreements, the WWF was allowed to undertake experimentation on the development of regulatory approaches to forestry.

The PMF implementers used various methods of educating and befriending the government in order to gain its support. Their efforts were aimed at members of the leskhoz (state forest management unit) who directly supervise the logging operations, the regional government officials who enforce norms and legislation, and the officials on the federal level who establish forest management policy for the whole of the Russian territory. They held seminars and workshops, including trips to Sweden to study sustainable logging sites, to which government officials at different levels were invited.[229] It was, however, not always convincing to the invited governmental officials. Some of the differences between the Scandinavian and Russian forestry cultures caused misunderstandings between the Model Forest and the government officials from the Russian forestry sector. One of my informants from the Pskov branch of the Ministry of Natural Recourses said:

> *Last year I was in Sweden. I saw that they don't have forests like we have. They have parks with planted trees, not wild forests. All the conditions are different. That is why adopting their system is very complicated. They have trees standing in rows and columns like in a garden, and their machines are used for such plantings ... We have real wild, natural forests, that is why this system is very hard to transfer here, and there is really no need.*[230]

In conjunction with this sentiment, officials wanted to promote Russian science and scientists. While wishing to focus on Russian forest science, the government officials admitted that the sector never had the funds for practical experimentation. The members of the Ministry of Natural Resources thus acknowledged that the PMF was very necessary in putting theory into practice, and so they supported it by helping it to work around the normative base. Cooperation with federal structures ensured relative freedom in performing experiments within the project. After making considerable efforts, the Forest Agency at the Department for Natural Resources of the Pskov region and administration of the Pskov region have become partners in the project at regional and local levels, and constructive relationships with local forestry management units have been established.

By negotiating with different levels of the government, the PMF received permission to log in certain ways so that the local forest management units did not penalize them for their

[229] Interview with the vice-chair of the Department of Forest Use, Pskov oblast, August 2002, Strugi Krasnie.

[230] Interview with the chair of MNR, Pskov oblast, August 2002, Pskov.

innovations. According to our informant, 'we need to coordinate with all the agencies on all levels for every step, which is not according to Russian norms'.[231] An example of the conflicting practices between the STF-Strugy and Russian norms was the leaving of aspen trees on the plots. Aspens (*Populus spp.*), a traditional problem tree for Russian foresters, decay very early and so are not valuable commercially. In Sweden, aspen is valued and preserved for biodiversity. While the Russian foresters consider aspens a nuisance, 'a Swedish specialist who came and saw how many aspens we have in the forest was just crying'.[232] As one of our informants said with irony: 'In Sweden, managers are preserving aspen because they think that it is something so beautiful. They run around aspens and build fences and on each tree they put a ribbon'. The PMF used an ecological argument in this issue. Because aspens rot from the inside, they can provide good homes for animals, especially in the winter. Girdling the trees and leaving them on the plot does not diminish the important ecological role they play.

Companies in Russia were required to remove aspens from the plots. The STF-Strugy did not wish to pay for this extra work. Before the project started, they were constantly fined by the local forestry unit. This penalty was less costly than actually removing the trees, so the STF-Strugy left them in the forest and paid the fines.[233] Later, the PMF coordinated this activity with the state forestry unit and the Ministry of Natural Resources, and so permission to leave aspens was granted. The STF-Strug workers simply had to 'girdle' the trees by cutting a ring around and all the way through the bark. This killed the tree with minimal effort, and then it was left to stand until it naturally fell over.

Since long-term planning and implementation of innovations in Russia is feasible only through forest regulations, the project made efforts to cooperate with the Ministry of Natural Resources and Rosleskhoz (the Russian Forest Management Committee), as one of its tasks was to implement its findings at the federal level and have them included in the new forest code regulations of 2007. In 2006, when the forest code was in the process of development, the WWF organized regional press-conferences where representatives of the Pskov Model Forest familiarized participants with their innovative approaches. In addition, the WWF made many efforts to communicate with key government officials by providing them materials about the PMF and sustainable forestry.

The project's research director took an active role in promoting the standards the PMF developed at federal level. In 2008, there were still no tangible results in introducing these standards at the federal level; however, a certain amount of progress had been achieved. For example, standards for commercial felling in the Pskov region were approved by the scientific council of Rosleskhoz in 2006, and were admitted to testing in the territories of the Leningrad, Novgorod and Pskov regions. Enterprises operating in these territories were permitted to do experimental felling using the new standards. On the other hand, some important standards for nature conservation planning, including preservation of key biotopes in the forest plots, as well as measures for the preservation of biodiversity, were not approved, although the description of methods was sent to Moscow for authorization. To comply with the requirement of this certification, the companies are compelled

[231] Interview with the head of Pskov WWF office, September 2002, Strugi Krasnie.

[232] Interview with the Pskov WWF office manager, September 2002, Strugi Krasnie.

[233] Interview with the Pskov WWF office manager, August 2002, Strugi Krasnie.

to find alternative terminology and designate key biotopes as 'noncommercial forest land', which is allowed by Russian legislation.

The WWF and Stora Enso, in many aspects, have achieved a constructive and favored relationship with state authorities. However, the level of engagement of both the national and regional forestry agencies in the PMF was very limited; state representatives were acting mostly as observers of the process. The problem of introducing the new standards developed within the PMF project into forest legislation had not been solved by the end of the project in 2008. However, work in this direction has been continued by the Research Forest Institute and GreenForest.

6.6.2 Relationships with local forest management units (leskhozes)

There were four state forestry units in the territory of the Strugokrasnenskii region. Although the model forest occupied only part of the land leased by 'STF-Strug', located in Strugokrasnenskii leskhoz, this project has become of interest for all neighboring local forest management units. At different stages of the project, relations with Strugokrasnenskii leskhoz were developing differently. At the very beginning, they refused to recognize the project as a partner, and treated it as something alien that was trying to break the existing order with their innovations and practices. As a forest management staff person said, 'At first it was a little hard to understand the goals of the project'.[234]

At that time, reforestation was a prerogative of the forest management unit, and if case leasers were doing the reforestation, they were supposed to be reimbursed. 'STF-Strug' was contracted by the forestry management unit to do reforestation, and the Strugokrasnenskii leskhoz had to pay them for the service. The firm's efforts at reforestation were hindered by the financial limitations of the forestry unit, and reimbursement was rarely paid.[235] Every year the forestry unit found violations in the firm's operations and fined them. According to an informant, the fines were the exact amounts that the forest management unit was supposed to compensate the company for tree planting.[236] The forest management unit tried to solve their own financial problems at the expense of the logging company. In addition, they felt resentment against a wealthy foreign firm, and so were especially strict in their penalizing.

Attempts to introduce some elements of sustainable forest management practices into neighboring territories (also leased by 'STF-Strug' but without special status), which contradicted existing Russian legislation, were actively opposed by the Strugokrasnenskii leskhoz. Its employees treated such innovations as resulting from the 'STF-Strug' workers' poor qualifications, rather than as attempts to disseminate practices of biodiversity maintenance: 'their managers are not specialists, they all have come from Soviet Collective farms, and their vision of forestry is therefore, absolutely different'.[237]

However, during the research expedition in 2006, changes could be noticed in attitudes towards the 'STF-Strug' practices. Some informants remarked that the 'STF-Strug' was the oldest

[234] Interview with the director of the Strugy-Krasnie leskhoz, August 2002, Strugi Krasnie.

[235] Chernova, E.B. 'Regional Assessment of the Territory of the Model Forest, Strugy-Krasne', expert working-paper (unpublished manuscript), St. Petersburg, 1999.

[236] Interview with the WWF Pskov office manager, August 2002, Strugi Krasnie.

[237] Interview with a forester, 2006, Uzlinski rayon.

and one of most responsible leaseholders. 'They keep the forest roads in order; they are engaged in silviculture, they leave seed trees and undergrowth on forest plots'. [238] Another informant remarked that this company is even engaged in economically unfavorable activities, but necessary for sustainable forestry, such as thinning in young forests. He does not consider the foreign origin of the company as evidence of its being 'alien' to Russian realities, but on the contrary treats it as indicator of its responsibility: 'They are influenced by the West, indeed, and are more concerned with forestry issues, and even reforestation and non-commercial thinning are of equal interest for them'.[239] Relationships between the Project and the Strugokrasnenskii forestry unit were clouded only by slight divergences, such as concerning the removal of branches from the plots after logging operations; as before, the forestry unit did not recognize Scandinavian technologies in this sphere: 'It's terrible, but they practically do not clean-up after logging'.[240]

The attitude of the second, interdepartmental forest management unit[241] to the project was mostly neutral. Both of the informants from this forestry unit saw no difference in the efficiency of practices promoted by the model forest and those used in their own territory. The director and employees of the third unit, Mogutovskii leskhoz, on the contrary, admired the model forest outcomes. This can be explained by the fact that in past Soviet times, this forestry unit hosted the experimental station of the Leningrad Research Institute of Forestry. According to the informant, the current Institute's professional network has promoted a positive attitude towards the project.[242] The fact that the director of this forestry unit worked for some time as an expert in the project also contributed to the project's positive image. Later, this experiment station was transformed from the Research Institute of Forestry into the forestry unit Mogutovskii leskhoz, yet it still maintained an experimental status. It is aimed at innovations and has tried to propagate model forest practices. For example, its employees maintained biodiversity at clear-cut areas, tried to achieve natural reforestation, and undertook graduated felling in recreational zones. The director of Mogutovskii leskhoz's former experience and professional interests helped find alternative legal ways for introducing innovations, and helped in having a free hand in decision-making: 'I am entirely for the model forest. It imparts modern tendencies to forest management ...'[243]

The head of the fourth unit, a Military Forestry unit, was also enthusiastic about the concept of the model forest. This person comes from a family of forestry specialists. He has especially welcomed the idea of preserving biodiversity, such as leaving remnants on plots, and the idea of natural reforestation. He treats the Forest Club, organized by the project, as a place where professionals can exchange opinions and listen to the public's opinion on issues.

As is evident from the above, a paradoxical situation has developed around the project. The project cannot introduce innovative practices in the territories where they have planned to be involved (i.e. territories leased by the 'STF-Strug') that were not included in the Model Forest.

[238] Interview with forester from the Uzlinskii region, 2006.

[239] Interview with director from the Strugokrasnenskii forest management unit, July 2006, Strugi Krasnie.

[240] Interview with director from the Strugokrasnenskii forest management unit, July 2006, Strugi Krasnie.

[241] Formerly it was called the agricultural forestry enterprise.

[242] Interview with the director of Mogutovskii experimental forest management unit, July 2006, Strugi Krasnie.

[243] Interview with the director of Mogutovskii experimental forest management unit, July 2006, Strugi Krasnie.

Hence, the enterprise has had to resort to various tricks in order to introduce innovations there.[244] But on the other hand, its practices were being disseminated rather successfully in other territories by personnel in the forestry units that were inspired by the project. Thus, the institutional rules prevalent in certain places are not isomorphic and more likely depend upon individuals who represent relevant organizations. Accordingly, the company cannot resort to organizational isomorphism, as it is compelled to act differently in each special case, depending on the existing situation.

6.6.3 Cooperation with the local administration

Efficient cooperation with both regional and local administrations has been a necessary condition for the PMF's successful embedding in an existing context in the space of place. Being guided by this idea, the project's implementing team searched for ways to establish partnerships with regional and local authorities.

Relationships with the Pskov regional administration developed gradually, as after each election cycle it was necessary to reestablish relations and to re-introduce the project once again. At the last stage of the project, close cooperation with the Department of Education was established; this entailed a textbook on sustainable forest management developed and published within the project for high school students.

Relationships with the administration of the town of Strugi Krasnie developed favorably right from the beginning of the project.[245] These relations strengthened in the late stages, partly because the first director of 'STF-Strug' (who later became the adviser for the Project) simultaneously became the people's deputy of the town and formed a positive attitude towards the Project in his administration.The administration of the Strugokrasnenskii region has supported the 'STF-Strug' long before the Project was launched, as at that time this enterprise was the main taxpayer in the region. It even made attempts to remove the threat of penalties imposed by the Strugokrasnenskii leskhoz, as the size of the administration's regional budget is directly related to the size of the company's profits. This means that penalties charged by the leskhoz against the company would subtract from the budget. According to the director of the STF-Strug, the administration would try to protect them from these charges. For example, the company is required by Russian norms to harvest a certain amount of wood each month, 'but it could be raining the whole month, and we have deadlines when to get the wood out of the forest. The leskhoz is preparing to penalize us, but the administration says they will help get the wood out of the forest for their own heating'[246]. In addition, under agreements with the heads of administration, the 'STF-Strug', along with other enterprises in the region, rendered assistance in the organization of community holiday celebrations. The administration has not often approached the enterprises for such help, trying not

[244] Interview with the research manager of the Project, March 2008, Siktifkar.

[245] Only interviews collected in 2006 were taken into consideration. In 2002 there was unified regional administration, the city administration was formed later.

[246] Interview with the director of STF-Strug, July 2002, Strugi Krasnie.

to abuse administrative powers by begging for financial help from its significant taxpayers.[247] The informant explained that the administration understands that: 'It is a commercial enterprise that aims to earn money, not to be engaged so much in social issues, and does not aim to substitute for the State'.[248]

The district administration perceived the PMF implementing team and the STF-Strug as separate actors. The WWF and the PMF were regarded as nature protection and research organizations which additionally supported local schools and libraries with grants. Thus, representatives of the administration could not understand the essence of sustainable forest management as a combination of components equal in importance: economic, social and environmental. They saw economic benefits for the region stemming from the timber enterprises' activities, with the social and environmental innovations coming from the PMF implementing team and the WWF. Therefore, the reputation of the STF-Strug was based only on its performance as a logging company, employer and taxpayer, and the credit for bringing sustainable forest management to the region was given to the WWF.

6.6.4 Building relationships with local people

At the first stage of the project, when the office of the PMF was opened in Strugi Krasnie, the task was to secure the sympathy and trust of the local population. People looked suspiciously at the project's implementers. It was necessary for the project to overcome the local population's prejudice against foreign companies, which the local people considered as 'coming to cut down the forests, get profits and abandon the place, having left a desert behind'.[249]

The PMF was actively engaged in informing the population about the Project and its purposes through the media. They published informational press releases about the project in regional and federal publications, created a website, and aired television and radio programs.[250] The PMF used media not only to focus popular attention on the project itself but also on ecology and environmental issues in general. In using this strategy, the PMF was introducing the topic of the environment into all levels of Russian media, which indirectly attracted attention to the activities of the PMF at the local level and its emphasis on the environmental aspects of sustainable forestry. In this way, with popular understanding of ecological concepts, the groundwork was laid for an understanding of and a support for sustainable forestry. In addition to publishing about themselves, the project implementers encouraged journalists to write about the PMF and the environment. They organized press tours through the forest, to which Russian, Swedish and Finnish media were invited. The biggest press-tour took place in 2005 with 33 journalists participating (Chernova, 2010: 136). This was subsequently proven to be a very successful way of information dissemination about the project, in that immediately following these tours, in October 2001, publications about

[247] Interview with the deputy head of the social department at the district administration, August 2006, Strugi Krasnie.

[248] Interview with a representative of the forestry department at the district administration, November 2007, Strugi Krasnie.

[249] Informal conversation with a woman at the market 2002, Strugi Krasnie.

[250] Report on Public Relations of Project Pskov Model Forest from June 2000 to December 2001.

the Model Forest reached a 'peak of informational density' with 48 articles published in Russia and Finland.[251] Many of them were on the first pages of periodicals. The PMF also invited Russian journalists to participate in a competition concerning ecology in journalism. First prize for the best article on ecology was a trip to Sweden.[252] This technique of rewarding activities that promote the Model Forest project played a significant role in introducing the concepts of sustainable forest management in the space of place.

The PMF made additional efforts to develop non-wood resource businesses in order to increase the wellbeing of local communities. They hired a specialist from Sweden to suggest local products that can be certified and sold abroad. He suggested a chocolate product made with cranberries from the local forest. The project made an effort to find business partners and establish links between them and those in the Strugi-Krasnie community who collect cranberries. This Swedish expert also suggested an ecotourism business for bird watchers. Both initiatives finally failed, as it was too hard to develop markets for those products and services.

As the project's implementers were interested in the natural embedding of the PMF's ideas in the local community, they actively participated in various actions related to local cultural traditions. Within the project, the PMF supported construction of an orthodox chapel, organized a Pancake festival and a school graduation party.[253] Environmental actions, such as forest festivals, ecological and sport events, have become another strategy of engaging the population. The PMF improved the quality of community celebrations by sponsoring food service and encouraging all the stakeholders in the forest to attend. They turned the traditional holiday involving the saying of 'goodbye' to winter into an 'environmental goodbye'.[254] Such activities gave the project a convenient opportunity to raise support as the festivities are large gatherings of the local people. Another tactic that helped relate the WWF to a broad section of the population was the sponsoring of a local football team. This tactic gave the WWF the double opportunity of gaining support from the public and advertising their project. The team was called Panda, and the uniforms displayed the WWF panda logo as well as the label of the PMF. The PMF reached much of the local population by bringing the famous football team Zenit from St. Petersburg to play against the Panda team. Many people expressed excitement about this game, which also had a theme and symbol for nature preservation.

The WWF's small grant program was a tool used to link the local population with the PMF. Under this program, citizens with ideas for environmental education or research were invited to apply for a grant. By providing funds, the WWF established a link with the grant's recipient. This translated into ties with the rest of the stakeholders – the local population if the recipient was a local activist. For example, a local teacher received a grant to create an ecological trail through the forest. She said, 'This was my old idea but it was very hard to do without money. I received the grant and realized my dream'.[255] The same teacher later became a president of the Forest Club where stakeholder interaction took place (Figure 6.3: Forest Club). One of the strategies of the

[251] Report on Public Relations of Project Pskov model Forest from June 2000 to December 2001.

[252] Interview with the director of the PMF, October 2002, Strugi Krasnie.

[253] Interview with the director of the Project, October 2002, Strugi Krasnie.

[254] Interview with the Pskov WWF office manager, October 2002, Strugi Krasnie.

[255] Interview with the president of the Forest Club, October 2002, Strugi Krasnie.

grant program was to take activities that already existed and enhance their quality while steering them towards support of the model forest. A summer camp run by the Ecological-Biological Center received financial assistance. Before the grant, there was no money for children's lunches or teacher's salaries.[256] Using the grant, the director improved the camp's environmental education program by inviting expert forest scientists, and the camp's attendance by providing food and lodging for children from outside the area. In another instance, the local museum curator, who received a small grant to create a brochure about the Strugy-Krasnie's cultural and natural history, later became an all-around supporter of the PMF's efforts. Her museum displays the major activities of the PMF. In an interview she said, 'We will be sad when they leave. We have a lot of sympathy for them'.[257]

The popular support raised by the small grant program is limited by the fact that it targets people with ideas. Thus, the resulting links are formed only with local leaders. These are the people in the community who can grasp the PMF's beneficial objectives right away and who perhaps would have supported their efforts even without grant money. An expert-consultant on the social aspects of forest certification said, 'From my experience, the first to participate are always the local intelligentsia. Teachers and librarians support the project first'. The same expert called such citizens a 'golden fund' which 'help to form public opinion'.[258] With the Small grant support, teachers created programs as such recycling, education, nature calendars, and computer education. In the framework of the project, a network of clubs called 'Friends of the WWF' has been created, their logo appearing in practically every school and library of the region. Owing to the Fund's participation, the ecological educational part of the project became central at a certain stage of project's implementation. The WWF contributed to the Russian adaptation of a Swedish textbook on forestry, 'Principles of Sustainable Forest Management'. In 2007, after its approbation in secondary schools, the textbook was published and disseminated throughout northwestern Russia. In 2008, Stora Enso printed more copies and disseminated them in schools situated close to their leases in the different regions. Over the Project's lifetime, 32 small grants were financed. They were an effective tool of involving the local population and a means of disseminating information on the project and on the Fund's activities.

One of the objectives of the Model Forest project was the inclusion of the public in the forest management decision-making. It was a necessary measure, on the one hand, because this was one of the requirements of certification; and on the other, all experts and visitors coming to the model forest from abroad were primarily interested in questions of public participation in forestry-related decision-making: 'In the West it is a favorite subject. They come and immediately inquire whether our public is involved in the decision-making process'.[259] Involving the public, however, met many barriers. For instance, the PMF tried to create a genuine, wide spread interest in managing the forests, however, often people only got involved after their interests were harmed. Public participation, as defined under the international certification norms, must be preemptive of conflict. It was hard to achieve this. For example, the PMF tried to consider hunter's interests in areas where the STF-Strug logged. They made an effort to involve hunters in the development of

[256] Interview with the director of the Ecological-Biological Center, October 2002, Strugi Krasnie.

[257] Interview with the museum curator, Strugi Krasnie November 2007, Strugi Krasnie.

[258] Interview with the expert on social aspects of forest certification,October 2002, Strugi Krasnie.

[259] Interview with a participant of the Project, October 2002, Strugi Krasnie.

the logging plans, however, they received little input. Hunters only raised a voice after the logging plan was published and their hunting areas were threatened.[260] The project implementers were themselves suspicious of the issue of public participation: 'Maybe it is important to involve the public in Western countries, but here we have a different mentality'.[261] Some felt that, despite the FSC's promotion, public involvement was unnecessary for sustainable forestry. One informant asked, 'what can non-specialists do, anyway?'[262] Another informant criticized the local public for 'crying about why we don't remove the branches from the forest'.[263]

In order to involve the public in discussions about forest issues, the PMF organized a forest club, a idea borrowed from the Model Forest Priluzie (see Chapter 7 in this book),[264] another WWF project. The task of the forest club was to inform the population and the stakeholders about the implementation of the Project and answer relevant questions. One of the PMF's implementers said: 'When visitors ask the question, 'how is the public involved?' We know the drill, we have a Forest Club, and so when such people come we refer them to the Forest Club and they are happy because we have involved all the possible stakeholders'.[265]

Club sessions were attended by invited representatives from the administration, leskhoz, local schools, libraries, and the grantees and eco-activists. The meetings often progressed into heated discussions. The population was hostile towards cuttings, and they had lots of questions and advanced their claims, many of which were addressed not so much to the project as to the village administration. Those present wanted answers to their questions: who cut the forests, why they were cut, where was the wood hauled, and how the forest's exploitation is organized in the region. Their questions also concerned firewood, sawn-wood, removal of waste, etc.: 'At the beginning, there was much talking because everything was unclear and it gave rise to a lot of questions concerning forest management in our region'.[266] The Forest Club became a new venue for building the stakeholders' relationships. According to our respondent, the past arguments between the forestry unit and the local citizens had consisted only of back-and-forth newspaper articles. A citizen would publish an article criticizing the forestry unit's activity; in response a forestry unit employee would publish another article refuting the citizen's claim. A few more citizens would then join the discussion by publishing their own articles. Our respondent criticized this type of written discussion, saying, 'at the Forest Club they can just communicate orally'.[267] According to another informant, the Forest Club is an interesting experiment in that, without it, ordinary community members would never meet with the industry people and scientists.[268] The average meeting involved 30-45 people, and it consisted mostly of those who were involved in the small grant projects, forestry units, experts, scientists, regional and local governments and

[260] Interview with a social expert in forest certification, October 2002, Strugi Krasnie.

[261] Interview with one of the PMF staff, October 2002, Strugi Krasnie.

[262] Interview with the PMF staff, October 2002, Strugi Krasnie.

[263] Interview with the Sosnovetsky leskhoz director, October 2002, Strugi Krasnie.

[264] Interview with a PMF staff member, October 2002, Strugi Krasnie.

[265] Interview with a PMF implementer, October 2002, Strugi Krasnie.

[266] Interview with an activist of the forest club, November 2007, Strugi Krasnie.

[267] Interview with the president of the Forest Club, October 2002, Strugi Krasnie.

[268] Interview with the Pskov WWF office manager, October 2002, Strugi Krasnie.

those who were directly involved in the PMF project. There were also some retired people and members of the local intelligentsia.[269] The club was led by the project implementing team. The STF-Strug was also involved, but they never played an active role. The director of the STF-Strug attended the meetings but had no enthusiasm for interaction with other stakeholders. He said, 'I don't like to talk with them, but I sit there to know what is going on'. In the second stage of the project the club held biannual meetings which became quieter over the course of time. From the director's interview: 'At present, there are fewer questions than before, because all is clear now – what are the functions of the model forest, and what are those of the STF-Strug and Stora-Enso. People come here for information they are interested in'.[270] The Forest Club became a space for information exchange between the PMF implementers and the stakeholders. The last meeting of the Forest Club took place in December, 2010, when the book 'Intensive Forest Management for Russia: Innovations of the Pskov Model Forest' had been published and presented to all interested stakeholders.

The attempt to involve the public in decision making was made only once over the lifetime of the project. Public hearings were held to discuss the forest management plan during the state forest inventory process in 2002. The PMF used a scenario method developed in a pilot project of the World Bank on the Karelian Isthmus.[271] The discussion during the hearings resulted in a plan which represented a compromise between the economic component, on the one hand, and the environmental and social on the other. The more environmentally-oriented scenario (of the two most reasonable options) was accepted, providing for the preservation of wood grouse's mating areas.[272] Such a hearing model is hardly applicable to the other regions, since the existing rigid federal regulations in forest management have limited the range of possible scenarios.[273]

6.7 Outcomes of the WWF-Stora Enso strategic partnership

The PMF was directed both inward to the space of place and outward to transnational spaces (sensitive western markets). When building the strategy for adaptation of its business to Russia, Stora Enso had to cope with pressure and requirements put forward by western environmentally and socially sensitive markets. It also responded to the impact of the turbulent institutional field of Russia which has developed under the influence of both the new and old socialist norms. Therefore, Stora Enso was exposed to double pressures: from the new FSC global standards and from the Russian state with its old norms and regulations. Both interested parties have surveillance bodies for supervising compliance subsequent of national and global rules, which often contradict each other. To satisfy each party, Stora Enso, as well as the other corporations that preceded in obtaining FSC certification, has had to research different individual methods of approaching the representatives of these parties.

[269] Interview with a local school teacher and Forest Club activist, October 2002, Strugi Krasnie.

[270] Interview with an activist of the forest club, November 2007, Strugi Krasnie.

[271] Romanyuk, B.D. 'Pskov Model Forest: Public Participation in Forest Planning', published by the WWF 2001.

[272] Interview with the research director of the Project, March 2008, St.Petersburg.

[273] Interview with the research director of the Project, March 2008, St.Petersburg.

When implementing the PMF project, the WWF contributed to the legitimization of the company within the community of international stakeholders. The results of the project were most tangible on foreign markets, as the project completely satisfied the stakeholders in transnational spaces.

Generally, the SIDA representative was very satisfied by the development of innovations in intensive forest management and the documenting tools for sustainable forestry. However, he was not completely satisfied with the dissemination of project outcomes, as only a small community of Russian experts and a small number of companies were taught to use the tools mentioned above.[274] He also stated that project innovations are useful not only for Russian, but for the Swedish forest sector. The project enhanced collaboration between Russian and Swedish foresters.[275] The 19th working group of the Mintriol Process on boreal forests visited the PMF in November 2008, and accessed highly its outcomes for applying in practice principles and criteria of sustainable forest management (Chernova, 2010).

In the space of place, the WWF helped in building relationships with national, regional and local civil society actors, scholars, NGOs, and state agencies. Thus, it has contributed to a positive image of the company and enriched the record of corporate social responsibility in the place of operation.

As was mentioned above, transnational corporations functioning in the different countries propagate their global strategies and approaches in specific localities. In our case, Stora Enso was a 'transmitter' of Scandinavian forestry technologies, and the PMF was a key actor in the adapting of these transnational practices to locality as well as developing innovations that suit local contexts. The role of the WWF was to establish relationships with stakeholders at different levels for introducing these transnational practices.

Among the positive results of the PMF project were the intensive forestry technologies for a total cycle (100 years) that were developed within the project. This would allow logging most effectively, combining economic benefits with environmental expediency.[276] Therefore, the intensive forest management model developed by the PMF is very important for reforming the forest sector in 2011-2020.[277] The PMF developed a concept on how to proceed with FSC certification in Russia with all necessary documentation; it developed an extensive package of documents that would advance the model of forest management and was a 'learning opportunity' for the partners.[278]

Stora Enso obtained a certificate for its subsidiary 'STF-Strug' in 2003 and at the same time obtained experience in certification in Russian conditions. After the certificate was received, the Stora Enso undertook certification of all the other Russian enterprises and by 2006 had certified all their forest management and chain of custody operations. The company standardized the approaches to certification for its divisions, in part by using its experience in the PMF project.

[274] Interview with the representative of the Swedish Forest Agency, Stockholm, March 2008, St.Petersburg.

[275] Interview with Per Hazel, published in the book 'Intensive Forest Management for Russia: Innovations of the Pskov Model Forest', 2010: 175.

[276] Interview with the research director of the Project, St. Petersburg, March 2008.

[277] Interview with Elena Kulikova, WWF forest program director, published in the book: 'Intensive Forest Management for Russia: Innovations of the Pskov Model Forest', 2010: 176.

[278] Interview with the manager for certification at Stora Enso, March 2008, St.Petersburg.

During the project, an expert community was formed and the NGO Greenforest created. Greenforest became a registered consultant at the FSC office in Russia and started to reproduce the developments of the PMF in other forest management operations for companies that were preparing for FSC certification. In particular, they prepared and implemented nature conservation planning, including the designation of High Conservation Value Forests (HCVF), in preparation for FSC certification on territories that included those leased by Stora Enso.[279]

The research has shown that the PMF was Stora Enso's strategy of adaptation of Russian realities to its own business, rather than *vice versa*. The WWF, on its part, was also interested in the transformation of the Russian forestry institutions, aiming to achieve their greater conformity with the idea of sustainable forest management. There were attempts to adjust Scandinavian forest management practices to the Russian context, adapting them to Russian conditions and creating new standards. The further introduction of these standards at legislative levels was a goal. This would be favorable for both foreign and domestic companies working in international markets. However, eventually these practices could be institutionalized only on the territory of the model forest. Up to the end of 2008, the problem of applying new standards in other subdivisions of Stora Enso had not been solved. Moreover, in any region there are different systems of forest management, different attitudes of forestry unit's staff towards nature conservation, different stakeholders and different levels of development of civil society. That is why the company has had to allocate a significant amount of resources in order to understand what was happening in the regions and what would be good strategies of operation. The absence of a unified forestry system impedes the exchange of experience between the regions, making it impossible to have a consistent and standardized business model and therefore, increases the transaction costs. However, a small number of the lessons learned were possible to reproduce. Since 2006, Stora Enso has replicated the methodology of nature conservation planning in its subsidiaries. Although this methodology was not approved at the federal level it was still possible to find legal ways for its implementation through the allocation of special protection zones. In each region of Stora Enso's operation, its subsidiaries had to solve the problems they encountered in different ways, often through ordering special forest management plans from the forest inventory agency. The allocation of key biotopes has broached the question of the state forestry unit's representatives understanding. When allocating logging plots together with the forestry unit staff, the company had to term the areas with key biotopes as 'noncommercial forest zones', which they are allowed to leave on the cutting plots. In such a way, biodiversity could be preserved.

The replication of the lessons learned working with the local population within the PMF project to other territories was also complicated, although for other reasons. As was mentioned above, the scenario approach used in the public hearings during the project is generally not applicable in ordinary conditions. To reproduce the Forest Club's operations, many resources are needed for creating an interested stakeholder community.

One of the results of the project was an official state decision to create a Russian model forest network to be bankrolled by the state. This implies a victory for the WWF, which considers the development of model forests as an efficient strategy for promoting sustainable forest management.

[279] Participant observation during the training seminar for Stora Enso personnel in charge of certification, November 2007, Kingisepp.

This also means the creation of an additional channel for applying global practices in the Russian locality. However, with changes in the jurisdiction of the Russian state agencies and the economic crisis of 2008-2009, there were no developments in the direction of building new model forests.

6.8 Discussion and conclusion

6.8.1 FSC certification through TNC-NGO partnership with outside funding

The translation of global designs, concepts, rules and discourses from transnational spaces to localities goes through several channels and governance generating networks (GGNs). The case study allowed understanding of how FSC certification happens if facilitated by a NGO implementer with outside donor funding (Figure 6.5).[280] Certainly, the two major channels are the TNC, interested in implementing sustainable practices that would benefit its business, and the NGO – the implementer of the project. The TNC is influencing practices in the space of place predominantly through its local subsidiary, which has to comply with the global sustainability policies of the company. The TNCs directly control compliance with their policies, and by doing so change practices in the space of place. To ensure sustainability on the ground, the TNC adopts a FSC certification scheme and automatically becomes a facilitator of the FSC-GGN, as the TNC's logging subsidiaries implement the FSC standards, and by doing so expand the FSC-GGN sites of implementation. The FSC-GGN provides another channel through which globally developed

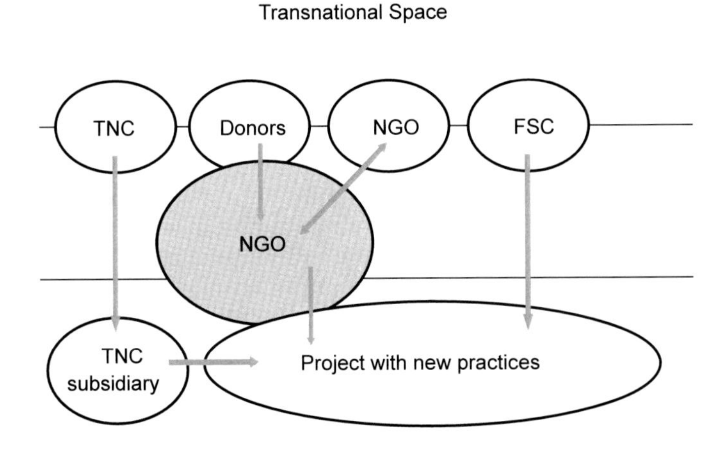

Figure 6.5. Channels of translation of FSC rules to local practices: TNC-NGO partnership with donor funding.

[280] Figure 6.2 represents real actors and their relationships in PMF development, while Figure 6.5 models FSC certification of TNC subsidiaries, using donor funding NGO implementer.

rules are transformed into local practices. The implementation and compliance with rules are verified by the accredited certification body that interprets the standards and directs the company toward better compliance.

The NGO-implementer is operating between transnational spaces and the spaces of places. They need to absorb the interests of the TNC, the demands of the donors, concepts and discourses of NGOs that are involved in the network and process all these global designs into the space of place in a way that fits local context. They assist the TNC and its subsidiary in the space of place in their interaction with local and regional stakeholders, and involve local communities in the decision making process.

6.8.2 Analysis of the PMF project using the GGN concept

The PMF is a special case where the impact of global rules on national institutions was simultaneously realized through several channels – through the FSC-GGN, the foundation GGN, TNC-GGN and NGO-GGN. Therefore, representatives of several GGNs formed a top-down network for co-designing, redesigning and negotiating new practices to be implemented at the PMF. All of the above mentioned GGNs participated in the design of the PMF (Figure 6.6: nodes of design, TNC-F-NGO-NGO) The physical territory of the PMF became a site of implementation for several GGNs engaged in facilitating relationships in their joint effort of fostering sustainable forest management.

The leading actor of this top-down network was the TNC Stora-Enso with its goal to adapt its business to both transnational and local pressures. As was described, they negotiated with SIDA for a large grant for testing and introducing innovations in the space of place. Practically the whole PMF project promoted Swedish technologies of forest utilization that would enhance the Stora Enso business. Thus, in this project, the TNC was the major agent of institutional change, while the other participants took facilitating roles. The TNC fostered institutionalization of new norms and concrete practices related to the processes of production, employee's safety and new logging technologies. The implementing team, built under the aegis of the WWF, fostered implementation of their own mission related to achieving sustainability through conservation planning and intensification of forest management. The priority of the foundation was the strengthening of public participation, but because of the nature of their participation (they were the main sponsor of the project, but did not take part in its actualization) they were less directly involved in changes.

The forums of negotiations took place both in transnational spaces and the spaces of places. The character of the forums of negotiation was dispersed. Transnational actors used the forums for periodically developing strategies for the project and monitoring it (Figure 6.6: forums of negotiation, NGO-TNC-F). On the federal level, in Russia, negotiations were sporadically organized with federal authorities for the purpose of lobbying the models of intensive sustainable forest utilization and preservation of biodiversity to be introduced into Russian legislation (forums of negotiation, NGO-FG). On the regional level the forums of negotiations were organized with the regional state bodies and stakeholders (forums of negotiation, NGO-RG-SH). On the local level, forums were designed to be a form of information exchange between the project's implementer and the local stakeholders (forums of negotiation, NGO-SH). In this case, the forums did not generally play a decisive role at any of the levels.

Initially it was intended that the innovations developed would be implemented in the whole of Northwestern Russia, however, the territories where the TNC had its subsidiaries were a priority. However, as was described in Section 6.6.6, the best practices of the PMF were impossible to reproduce because of their inconsistencies with Russian legislation. Only tools developed for FSC certification were reproduced at Stora Enso subsidiaries (Figure 6.6: sites of implementation, FSC-FM). Therefore, the mimetic form of organizational isomorphism mostly failed, despite the efforts of the actors to reproduce lessons learned.

The strategic partnership of Stora-Enso with the WWF allowed for certifying forest operations using the FSC scheme and establishing meaningful relationships with both transnational and local actors. However, the coalition of private actors confronted the forest sector state authorities during the process of transformation and reform and failed in achieving its major goal of changing Russian legislation towards accommodating innovative forest management practices. The PMF was a demonstration project, which became an arena for renegotiating both for power relationships and forest management practices between global, national, regional and local actors. Only the innovations related to the FSC were accepted by all the stakeholders. Therefore, the process of introducing global norms into national institutions was far from being smooth in Russia, as the state's norms (Forest Code 2007, regulations in the sphere of forest management, etc.) 'resisted' the changes introduced by the transnational NGOs and corporations. This 'resistance' is neither consequent nor intentional; it is more likely connected with the institutional turbulence in transition processes in the Russian economy over recent years.

Through the project of a model forest, Stora Enso attempted to create an optimum algorithm for successful business operations in a particular country. This algorithm was intended to be applied in all the subsidiaries working there. Thus, mimetic isomorphism would be exploited. However, instability and turbulence of organizational fields in regions have become obstacles in carrying out this plan in its original form. Only measures such as worker's safety and nature conservation planning were possible to standardize and reproduce.

The situation was aggravated by the fact that Russia, as a country with a transition economy, often encounters measures which can be neither predicted nor tested and which can sharply raise transactional costs. For example, the imposition of high customs tariffs on export of round timber, which is one of the main items of business activity of the STF-Strug in the Pskov region, made the company unprofitable and therefore, it was closed in October, 2008, as along with the other Stora Enso subsidiaries, STF-Gdov and Kingisepp.

For its part, the WWF also tried to create a model of sustainable forest management which could be further disseminated throughout the country. In this case, the aforementioned solution of Rosleskhoz on the creation of a model forest's network, if implemented, would open up opportunities to reproduce lessons learned and by doing so foster a mimetic form of organizational isomorphism.

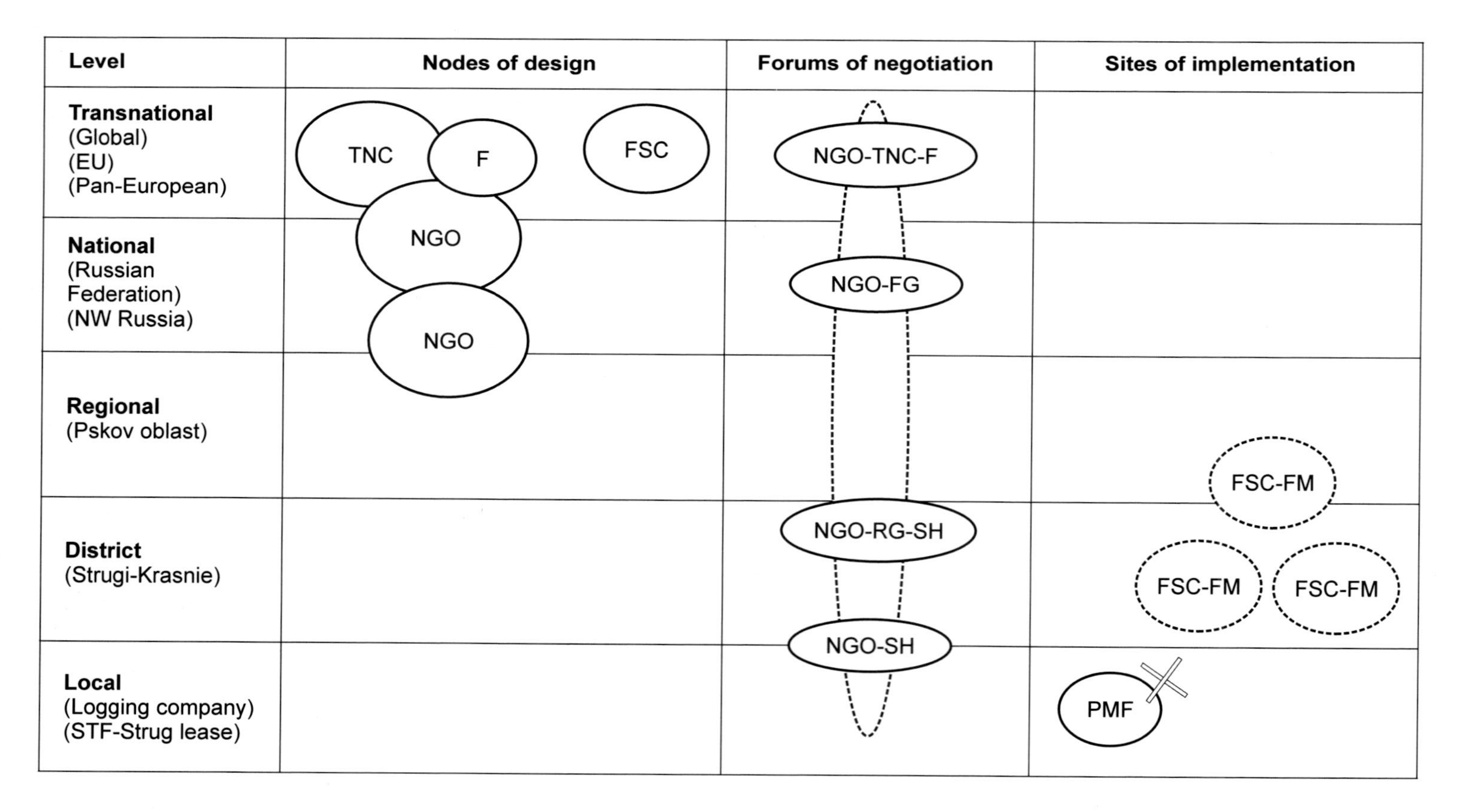

Figure 6.6. Analysis of the PMF network using the GGN concept. In the corner PMF is crossed to express that the project does not exist any longer, more explanation in the text.

Chapter 7. NGO implementing sustainable forest management in the space of place: building the Model Forest Priluzie

7.1 Introduction

This case study explores how the NGO Silver Taiga Foundation (a WWF branch until 2002), in partnership with regional and local state agencies, and with funding from the Swiss Agency for Development and Cooperation (SDC), built a model of sustainable forest management, the Model Forest Priluzie (MFP) on the ground in the Republic of Komi. The actors of the MFP network initiated an intervention in the Komi Republic in order to establish environmental, social and economic innovations in forest management. Although the MFP network engaged transnational, regional and local actors, the initiative and resources came from transnational stakeholders. Therefore, the MFP builders fall into the category of a top-down network (see Chapter 2, ideal types of networks), linking transnational spaces with the spaces of places.

There were many historical and economic barriers that an NGO needed to overcome in the process of implementing innovations in the Komi Republic forestry sector over the life of the 10 year project, which lasted from 1996 until 2006. Komi's distance from Western Europe has been a historical barrier to the region's forest industry development. In the early 19th century, enterprises in the republic began exporting logs to countries such as Norway, England, Germany, Belgium, and Holland; however, this was not always profitable. Wood from Russia, in the form of unprocessed logs, sold for comparatively low prices. Often the market value did not compensate for the cost of transporting the product from Komi to Western Europe, thus making profits low.[281] Processing infrastructure was not well developed, because the strategy was to raft logs on the rivers from Komi to Archangelisk forest enterprises. This was not always profitable and negatively affected both the river ecosystems and the forests.

When the Soviet state crashed, resulting in regional separatism, the forest sector of Komi suffered further. Between 1990 and 1994, Komi lost many of its traditional forest markets in Central Southern Russia, Moldova, and the Ukraine, and production decreased catastrophically.[282] Therefore, when the MFP project started, the forest industry in the Komi Republic had been in a state of crisis. People living in forest settlements suffered high unemployment rates and out migration from forest dependent villages to the cities increased. In 2000, the general economic situation in the Komi forest sector had improved, and the context for the MFP changed. A TNC, Mondi Business Paper, purchased the old Soviet Siktivkar Pulp company and became a key player in the Komi Republic. The TNC greatly stimulated the development of the wood-processing industry and simultaneously introduced sustainability requirements to their suppliers. The forest

[281] Karakchiev, A.A. 'First Vice-chair of Head of Republic of Komi, Syktyfkar' In: Proceedings of the Conference 'Forest Certification and Sustainable Forest Management in the North of European Russia', Syktyfkar, February 2-4, 2000, pp. 64-74.

[282] *Ibid.*

industry's interest in gaining access to European markets increased, and subsequently increased interest in introducing sustainable approaches to forest management.

The MFP was simultaneously a specific territory in which the new practices were tested and implemented and a forum where the innovations were negotiated between multiple stakeholders with the aim of disseminating them throughout the Komi Republic.

This chapter will focus on how the Silver Taiga foundation served as an implementing agent of the top-down network, stretching from transnational spaces to the space of place for introducing innovations such as landscape ecological planning for forest operations, preservation of biodiversity (in particular old growth forests), soil erosion prevention, citizen involvement in decisions related to forestry, economic assessment of the forests, and FSC certification of forest management.

The FSC certification involved the whole state forest management unit (leskhoz). Preparing the leskhoz to meet FSC standards was a difficult task, as the requirements of FSC certification differ from the usual tasks and responsibilities faced by leskhozes. Additionally, the FSC standards were stricter than existing Russian legislation and some requirements contradicted Russian legislation. FSC certification involved not only changing logging practices but also establishing a democratic decision making process and multi-stakeholder engagement. The Silver Taiga foundation tried to achieve this by initiating partnerships between all the regional and local stakeholders – government, the public and the logging companies. The chapter will describe the challenges that Silva Taiga was overcoming in the process of fostering and disseminating new practices of forest management in Komi Republic.

The chapter is based on field research conducted between 2002 and 2009. Data were collected during 6 research expeditions to the research area, in November-December 2002, March 2005, February-March and April 2006, and October 2008 (see Appendix A). Participant observation was done at community events, public hearings and community council. Both semi-structured and open ended interviews were conducted (total #470). Interviews involved NGOs (#39), representatives of state agencies at different levels (#54), representatives of local administrations (#26), local state forest management unit representatives (#6), companies leasing woodlands on the territory of Model Forest Priluzie (#77), community activists and ordinary citizens (#148), representatives of local museums (#14), teachers (#27), representatives of houses of culture (#33), and librarians (#36). Transnational actors, including Mondi Business Paper representatives were interviewed in Vienna, London, South Africa, and Bonn, in their offices and during international events (see Appendix A, B, C)

The chapter starts with a brief description of regional and local contexts (Section 7.2) and the project's implementation processes, with its actors, stages and major events (Section 7.3). The major focus of the chapter is on the stakeholder engagement in the MFP (Section 7.4), in particular the involvement of state agencies at different levels (Section 7.5.), and the logging companies (Section 7.6). Special attention is given to the engagement of the MFP with local communities (Section 7.7), how efforts were made for building trust, how key local agents of institutional change have been identified at schools and libraries (Section 7.8), and how different kinds of civic initiatives were set up and/or facilitated by the project in order to build new democratic ways of decision making processes (Section 7.9). Next, the focus of the chapter switches to the interaction of the MFP implementers with transnational stakeholders (Section 7.10). The chapter concludes with theorizing the empirical findings through the lenses of the GGN concept, in particular the role

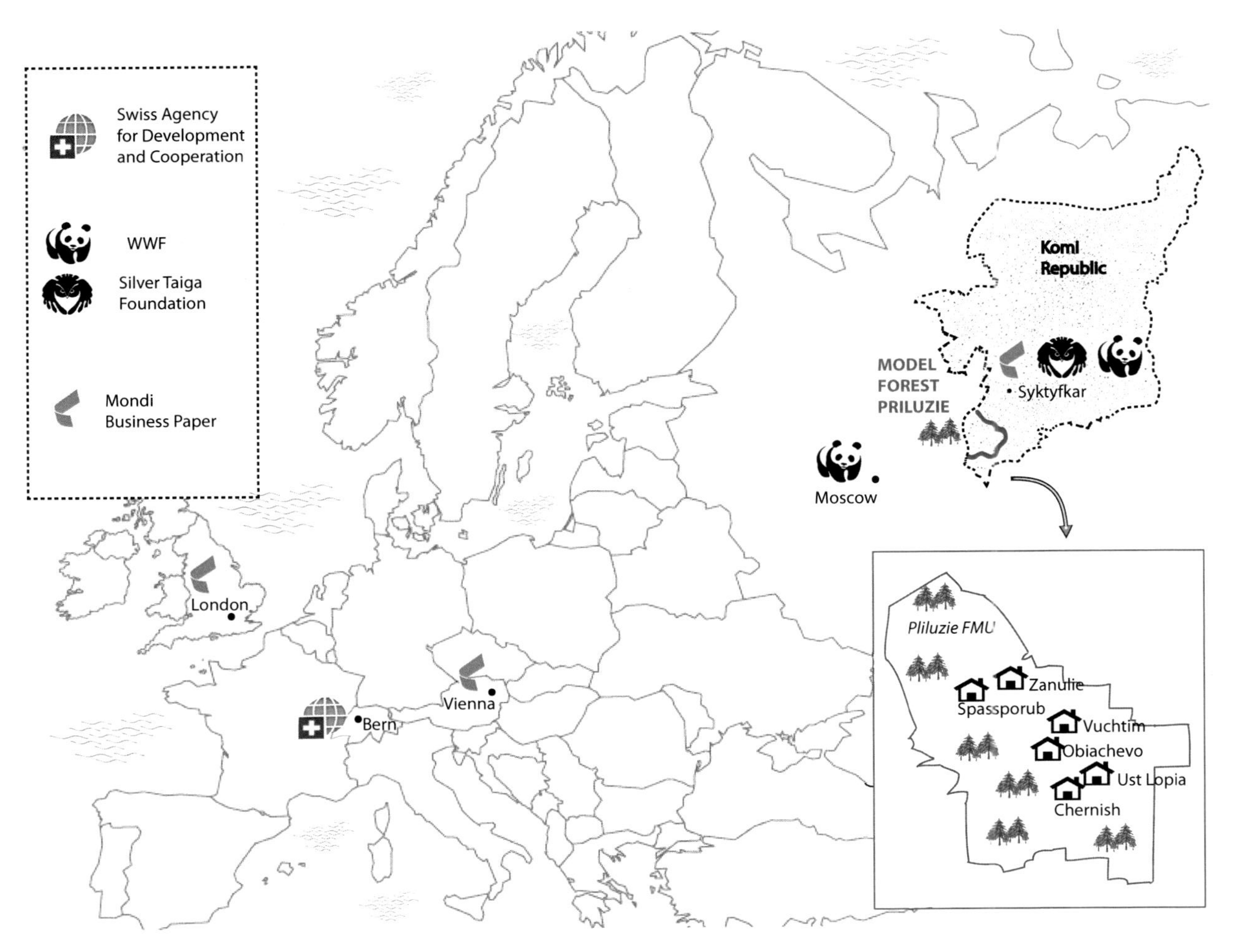

Figure 7.1. Model Forest Priluzie: case study map.

of nodes, and forums of negotiation in setting up new rules in the site of implementation and a discussion on institutional isomorphism of the project's outcomes. The MFP is compared with the Pskov Model Forest, described in Chapter 6 (Section 7.11).

7.2 Regional and local contexts: the Komi Republic and Priluzie

The Komi Republic consists of 416,800 square kilometers just west of the northern Ural Mountains. The forests are a vital aspect of the republic's economy, although oil and coal industries are also well developed. Since 1917, the forest sector has provided the primary source of income, employing one-third of the republic's working population.[283]

In the 1990's, 200,000 hectares of Komi's forests were clear-cut, and trees were planted on 20-23,000 hectares – roughly 10% of the deforested land. The survival rate for planted trees was 30%. Due to a lack of financial, labor, and technical resources, the republic has tracts of unforested land as well as tracts of low-productivity leaf-trees which naturally succeeded after logging.[284]

People throughout the republic rely on gathering berries and mushrooms, as well as hunting in the forests. Komi's unique virgin, old growth forests have been a source of conflict among differing interests. The Republic of Komi contains a large portion of the remaining virgin forests in northern Europe. Some of these areas are officially protected, such as within the National Park Yugyd-Va and Pechora Ilichsky Zapovednik, however, there are also many virgin tracts on land situated in the territories leased by logging companies.

The Republic of Komi is recognized in publications as the first region in Russia to pay attention to virgin forests in policies.[285] However, old growth forests in the republic are still periodically in danger, depending on changing policies to build more pulp and paper mills in the region.

The Model Forest Priluzie consisted of 800,000 hectares managed by the Preluzye Leskhoz (until the new forest code came into force and the role of leskhozes changed). Within this territory are permanent settlements, small industries, remnants of collective farms and logging companies.

There are old traditional villages and forest settlements, established in the 1950's and 1960's, to accommodate workers involved in logging. In the socialist period these logging enterprises supported all the villages' infrastructure, including heating systems, roads, kindergartens, houses of culture, hospitals and other services. With the collapse of the planned economy and development of the market economy, it became very hard for the companies to sustain the village infrastructures. After privatization, the companies were supposed to receive compensation from the state for maintaining the infrastructure, but this money was insufficient and much less than the real costs. Some companies managed to peacefully transfer the responsibility for the infrastructure to the state; others have done this through a series of bankruptcies, most of which took place in the 1990's. During the project's period there were 12 to 17 logging companies working in the territory of the MFP.

283 *Ibid.*

284 *Ibid.*

285 'Virgin Forests' in WWF Bulletin, No. 1 March 2001: 2.

7.3 Project implementation

7.3.1 An overview of the network structure

The donor, the SDC, allocated the resources for building the model of sustainable forest management in the Komi Republic (Figure 7.2: step 2). Their understanding of the model of sustainability was based on the predominant world-wide discourses involving the preservation of biodiversity, maintaining sustainable forest yields, and the economic profitability of forest enterprises which take into account the interests of all the stakeholders, including local populations and indigenous peoples (Figure 7.2: step 1). Concepts designed on the transnational level were translated into places using different kinds of mechanisms, one of which was forest certification. Another enabling mechanism was donor funding. That allowed the transfer of designs into places where there was no capacity to implement such concepts on their own due to existing governmental policies, local contexts and economic situations. The major implementer was the NGO Silver Taiga (Figure 7.2: step 3), which engaged stakeholders, including the Komi governmental agencies (Figure 7.2: step 5), scientists (Figure 7.2: step 6), local community organizers (Figure 7.2: step 7) and businesses (Figure 7.2: step 8), into the Working Group (Figure 7.2: step 4), which developed strategies for the project's implementation. The Working Group, together with Silver Taiga, designed the intervention into the space of place that would bring new practices of forest management into the site of implementation of the MFP (Figure 7.2: step 9).

The coordinating council of Silver Taiga employees implemented the project. At the beginning of the project, the major goal was building trust in the local communities (Figure 7.2: step 10). In parallel, Silver Taiga identified the institutes and actors that can play a role as transmitters of the project's ideas to the population at large (Figure 7.2: step 11). Technologies for mobilizing public participation were used for community engagement (Figure 7.2: step 12).

The Forest Council was created in the territory of the MFP, and included the most active people, who along with the other stakeholders participated in the future governance of the project (Figure 7.2: step 13). The model forest served as an educational playground, a laboratory for social change and a transmitter of innovative ideas to other places in Komi and North Western Russia (Tysiachniouk, 2006 b: 123).

The major uniqueness of this case is that it started with a NGO-state partnership and in the later stages became driven by businesses and a NGO. When the project ended in 2006, the transnational corporation Mondi Business Paper became Silva Taiga's sponsor in its efforts to foster the FSC in Komi, therefore the NGO-state partnership, with its outside funding, shifted to a continually operating TNC-NGO partnership (Tysiachniouk, 2008c; Tysiachniouk & Meidinger, 2007), similar to that described in Chapter 6.

The same shift, from a state agency to business actors taking responsibility for the FSC's implementation, took place on the local level. The peculiarity of this case was that the FSC's certificate holder during the project's implementation stage was a local state forest management unit (Leskhoz) and not a logging company, which is the case in the other regions of Russia. With the new forest code of 2007, the responsibilities of the state forest management units changed and they lost their responsibilities in managing forests and became controlling agents. This made it impossible for them to hold the FSC certificates, so finally a group of companies took responsibility for the FSC-FM certificates.

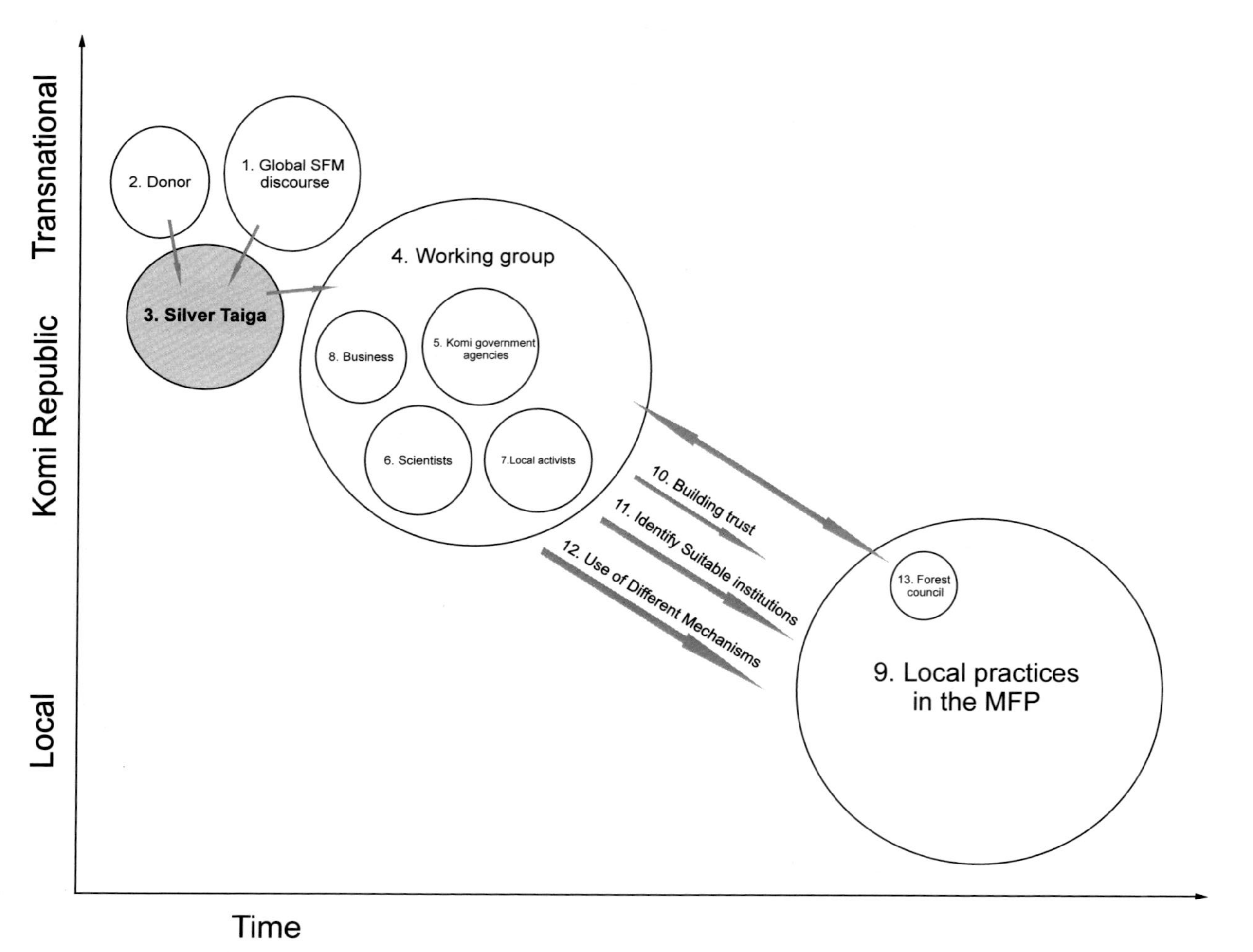

Figure 7.2. Implementation of the Model Forest Preluzie.

7.3.2 Set of events of MFP development

The implementation of the Model Forest Priluzie project had four major phases: 1996-1999, 1999-2002, 2002-2005, and 2005-2006. The WWF's branch in Syktyfkar was originally formed in 1996 as part of a project to protect the Pechora Ilichsky Zapovednik (biosphere reserve) in Komi. Forest scientists from the Finnish Institute and the Swedish Agricultural Institute contacted WWF International in Switzerland after a research expedition to study Komi's virgin forests. Together with the scientific institutions the Syktyfkar Institute of Biology, the Syktyfkar State University, and the administration of the biosphere reserve, these scientists decided that Komi's virgin forests are valuable and should be protected. Mayevsky Pshemislav, one of the scientists from the Swedish Institute (currently the executive director of Silver Taiga), wrote to WWF International with a project proposal for improving forest management in Komi. WWF International, based in Switzerland, proposed a project to Switzerland's Agency for Development and Cooperation, which promotes sustainable development around the world. Three years later, the Swiss government accepted the project. Protection of this preserve was incorporated into the WWF's worldwide program for protecting eco-regions. In this framework, the area of the Ural Mountains is one of 233 biodiversity hot spots, and so Pechora Ilichsky became a part of the Ural Mountains eco-region.

In the 1990s, WWF Moscow was just in its first few years of operation, so the project and WWF Komi were organized through the WWF's international office in Switzerland, but later managed through WWF-Moscow. During the first phase of its operations it worked on nature protection training in Pechora Ilichsky Zapovednik. The goal was to increase local awareness of the value of virgin forests by organizing courses on forest ecology.

During these three years, the focus of WWF Komi gradually shifted towards sustainable forestry in the region, and away from working with a nature preserve.

In 1997, in parallel with the project's original focus, the WWF started the creation of the model forest in the Priluzie district. At the Petrozavodsk conference in in 1998, where FSC certification in Russia was discussed, the WWF-Komi made an agreement with the American company SmartWood (a program of the Rainforest Alliance), which works as an auditor for FSC certification, to conduct a test audit in the territory of the MFP.

In November 1998, the WWF took the initiative in creating the working group on developing regional FSC standards for the Komi republic, to be tested later in the MFP. The Working Group on Forest Certification became an independent NGO and in March 1999 received funding from MacArthur Foundation for standards development. Later, SmartWood took these adjusted standardsinto account for audits in Komi.

With the grant from the MacArthur Foundation, SmartWood conducted a test certification of the MFP between August and November, 1999. They formulated a list of 10 preconditions that must be met before the main certification auditing would take place in March, 2002. These preconditions restructured the agenda of the project's implementers and oriented them toward obtaining the FSC-FM certificate.

During the 1999-2002 phase, the WWF's small grant program supported multiple initiatives on environmental education at schools, libraries, and clubs. Grants were also distributed to students to conduct their thesis research in Priluzie.

In 2001, the WWF-Komi developed recommendations on forest policy for the Komi government and a set of initiatives on how to govern and protect the old growth forests during the period of 2001-2008.

In 2000, the international conference on FSC certification in the European North of Russia was held in Siktivkar and special attention was paid to the results of the FSC's test certification in Priluzie. The WWF worked on the issue of preservation of old growth forests, and their efforts resulted in the Komi governmental decree, issued in 2001, to do evaluations of virgin forests in Komi.

In 2002, on the basis of the WWF-Komi office, the Silver Taiga Foundation was formed and became independent from the WWF. The Swiss International Agency of Development and Cooperation was the founder, and Silver Taiga started managing the project directly with money from its founder and donor. The emphasis of the project shifted from nature conservation to sustainability issues, from youth education towards forester's education. The grant program was closed, but most project partners continued working with Silver Taiga on a contract basis.

The 2002-2005 phases were much less oriented towards supporting community initiatives, although some initiatives were still funded. The project in Priluzie started developing, testing and implementing different kind of environmental and social innovations related to sustainable forest management. On that basis, certain recommendations were approved by the government of Karelia and reproduced at the level of the Komi Republic.

In March 2003, the Priluzie Leskhoz received the FSC certificate. This created a positive environment for logging companies in Priluzie and several of them received chain of custody certificates in 2003-2004.

In 2004, Silver Taiga started to facilitate other FSC certification projects, such as IKEA's daughter enterprise 'Kay'. In 2003-2005, Silver Taiga started to collaborate with the Mondi Business Paper Siktivkar Pulp and Paper Mill and with two leskhozes, (Kaigorodsky and Sisolsky), helping to prepare them for certification. Both received FSC certificates in May, 2006. In 2008, lestnichestvo[286] Kashimskoe received the FSC-FM certificate with Silver Taiga's assistance.

Simultaneous with the promotion of forest certification, Silver Taiga continued working with the Forest Agency, the Komi Ministry of Industry and several scientific institutions in developing the Conceptof Sustainable Forest Governance for the Republic of Komi through 2015. In 2005, the Forest Agency of the Komi Republic and Silver Taiga signed an agreement in organizing educational courses for the forestry staffs of governmental agencies in the Model Forest Priluzie. Silver Taiga continued extensive educational programs for forestry companies and other interested parties (Tysiachniouk, 2006b).

In 2006, the emphasis of Silver Taiga was on introducing forest management innovations through the Forest Inventory Agency (10-12 forest management plans). At the end of 2006, the Swiss International Agency of Development and Cooperation stopped financing the project. Since then, Silver Taiga has been seeking financial support from other foundations for its prioritized activities. Activities related to the MFP slowed down; however some innovative approaches to sustainable forest management continued to be tested and implemented in the Priluzie territory. For example, social aspects of sustainable forest management were implemented in the framework

[286] With the new Forest Code, Leskhoz was reorganized and become Lesnichestvo (in 2008).

of the Forest Village project (2008-2011), sponsored by the Ford Foundation, in which the economic stabilization of Komi villages is linked to the sustainable use of natural resources. Silver Taiga continues to promote the preservation of biodiversity in Komi, and in 2009 set up new demonstration plots on the MFP territory.

Silver Taiga increased its activity in both the International Model Forest Network and in the Russian Model Forest network. They coordinated the Russian Model Forest network until May, 2007 and with the support of the Federal Forest Agency, participated in preparation of the conception of a Model Forest in Russia. The Federal Forest Agency declared support for the development of the Russian Model Forest Network, however they never allocated resources for it.

7.4 Stakeholders and their engagement in MFP

A variety of transnational, regional and local stakeholders were involved in the MFP. Silver Taiga was, on the one hand, a transmitter and adapter of global discourses of sustainable forest management into a Russian region, and on the other hand, a creative designer coordinating regional and local stakeholders in order to both strategically construct and implement the Model. Figure 7.3 visualizes its interaction with the other NGOs in the process of building the MFP. Silver Taiga/ WWF Komi had close relationships with WWF International and WWF-Moscow, collaborated with the Rainforest Alliance Smartwood program on FSC certification, using maps of old growth forests developed by the World Resource Institute (WRI) and Greenpeace with the participation of the Biodiversity Conservation Center (BCC). On the local level they involved experts from the Save Pechora foundation and local activists from the Udora Region in their work with local communities.

More radical NGOs were on the side, playing the role of providing a threat in the case of forest management proving to be unsustainable.

Silver Taiga linked all the possible stakeholders into a working group that was developing a strategy for the model (see the range of stakeholders in Table 7.1). In this section, however, I will focus only on Silver Taiga's interaction with four sets of the most important stakeholders: (a) government, (b) business, (c) local communities, and (d) transnational stakeholders.

7.5 MFP engagement with state agencies

Silver Taiga partnered with governmental agencies at different levels in order to implement the project and reproduce innovative approaches to sustainable forest management in the Komi Republic.

On the highest level the project was supported by the Government of Komi. The Committee on Forests of the Komi Republic was the major partner. The local forest management unit (Leskhoz Priluzie), which is subordinate to the Committee on Forests of the Komi Republic, was the major partner at the local level. The project was supported by the local administration of Priluzie, although the interaction between them was not always smooth going and many misunderstandings took place with different representatives of the administration at different times of implementation.

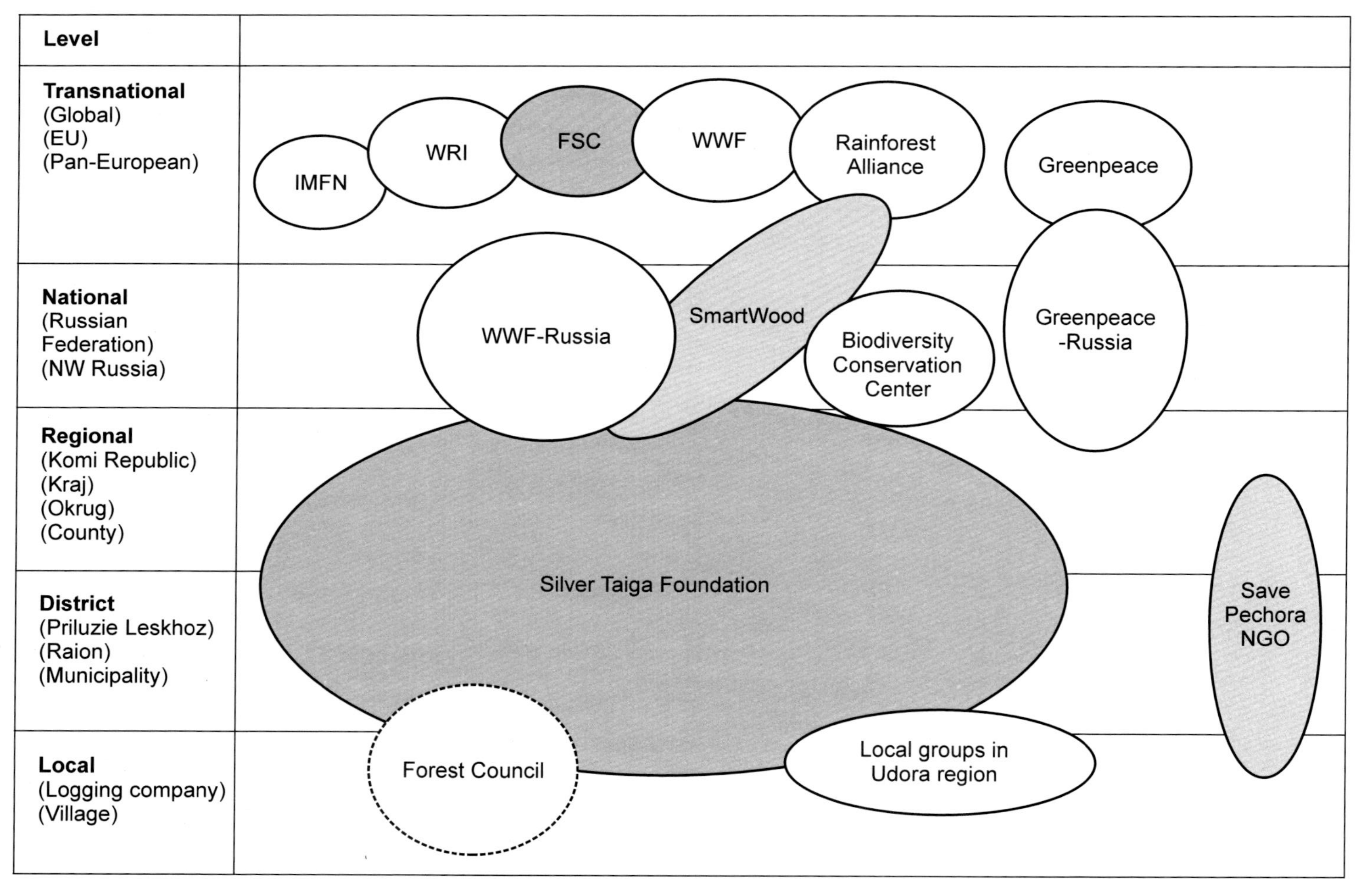

Figure 7.3. NGO involvement in Model Forest Priluzie.

Table 7.1. Stakeholders involved in Model Forest Priluzie.

Spaces	Stakeholder	Type of organization	Interests
Transnational	Swiss Agency for International Development	GO	To help transitioning countries to introduce sustainable Forest Management with an emphasis in social aspects.
Transnational	MacArthur Foundation	Non-profit foundation	To foster FSC certification in Russia.
Space of place/ transnational	WWF/Silver Taiga	NGO	To implement the model of sustainable forest management, to reproduce this model in Komi and other regions.
Transnational	WWF-International	NGO	To preserve as much forest land as possible and on the rest to implement sustainable forest management around the World 'Forest for Life Program', promote the FSC around the World.
Place/ transnational	WWF-Moscow	NGO	Implement Forest for Life in Russia, promote the FSC in Russia, build FSC institutions in Russia.
Transnational	Global Model Forest Network	NGO	Build Model Forests around the world, maintain the network of model forests, reproduce the model.
Transnational	Canadian Development Agency	GO	To support sustainable forest management in Canada and other countries.
Transnational	Barentz Region Network	GO, NGO, scientists	Build up sustainable development projects in the Barentz region.
Transnational	Rainforest Alliance	NGO	Expand non-profit FSC audits worldwide, promote the FSC, gain money for other projects from Smartwood.
Transnational	Smartwood	Non-profit	Certification Body – expand and take a lead in FSC audits in Russia, promote the FSC.
Place/ transnational	Nepcon	Non-profit	Become leading experts in FSC audits in Eastern Europe and Russia.
Place/ transnational	Greenpeace	NGO	Preserve old growth forests.
Place	Save Pechora	NGO	Get the public in the villages to participate in decision making. Get support from Silver Taiga.
Place	Local group in Udora Region	Civic initiative	Save forests, involve people in decision making, protect rights in Komi.
Transnational	Mondi Business paper (UK, Austria office)	TNC	Decrease risks of operating in a country with a transitional economy. Consistency in business operations. Demand all subsidiaries and suppliers around the World to be FSC certified.

Spaces	Stakeholder	Type of organization	Interests
Place/ transnational	Mondi Business paper Siktivkar Pulp and Paper Mill	TNC	Comply with the headquarters' policies, certify forest management and chain of custody, keep good relationships with NGOs.
Place/ transnational	Ilim Group/Kotlas PPM	TNC	Certify forest management and chain of custody, buy pulp wood with lowest prices.
Place/ transnational	Luza Les	business	Profits, international markets, expansion.
Mostly place	Small forest enterprices	business	To survive, to spend less resources on certification.
Place	Private entrepreneurs	Business	To survive, to be part of community.
Mostly place	Komi Ministry of Natural Resources	GO	Economic growth in Komi, for logging companies to be successful in international trade, support certification, support MFP, accept preservation of old growth forests on the MFP territory.
Mostly place	Komi Ministry of Economics	GO	Economic growth of the Komi forest sector, investors for processing industry, to build new pulp and paper mills, and/or expand Mondi Business Paper, does not share the value of old growth forests, support the MFP.
Place	Forest Inventory Agency (Lesoustroistvo)	GO/business, when new Forest Code came into force	Create 10-12 years plans, interested in additional paid projects, open to innovative planning.
Place	Leskhoz (lestnichestvo with new forest Code of 2007)	GO	Manage forests on the local level, monitor and control logging operations, since 2007control without managing.
Place	Local administration	GO	Economic growth, support to the area.
Place	Local schools	Civil society	To receive support from Silver Taiga (financial and educational), to develop courses and summer school programs, to create new opportunities.
Place	Local libraries	Civil society	To have activities in the libraries, to attract more readers and attention, to contribute to the community, to be supported by Silver Taiga and other agencies, both governmental and non-governmental.
Place	Local House of Culture, art centre	Civil society	Support for the art centre activities, to be part of supported projects.

7.5.1 Partnerships on the level of the Komi Republic

The role of the Komi governmental agencies was to provide political, legislative and administrative support and to help resolve the contradictions between Russian legislation and the FSC's requirements.

Komi's Republican government looked favorably at the Model Forest Priluzie's objectives for building the model of sustainability (Figure 7.4). The model forest's working group recruited members of the Forest Committee that would be supportive and cooperative. One of the Silver Taiga employees says this about the five Forest Committee officials participating in the working group,: 'I knew all of them, their levels, attitudes, and personal qualities. The vice-chair of the Forest Committee was recruited to help the group in dealing with a complicated forest code because he was the most well prepared specialist in forest legislation'.[287] Such knowledge benefited the network-building activities of the model forest.

Networking with the government was not difficult and quickly brought to the project support in the form of leniency with forestry norms and participation in the MFP strategy development and planning group. The MFP obtained special permission from the head of the Komi Republic to experiment with forest harvesting outside of the republic's norms and legislation at a site of the

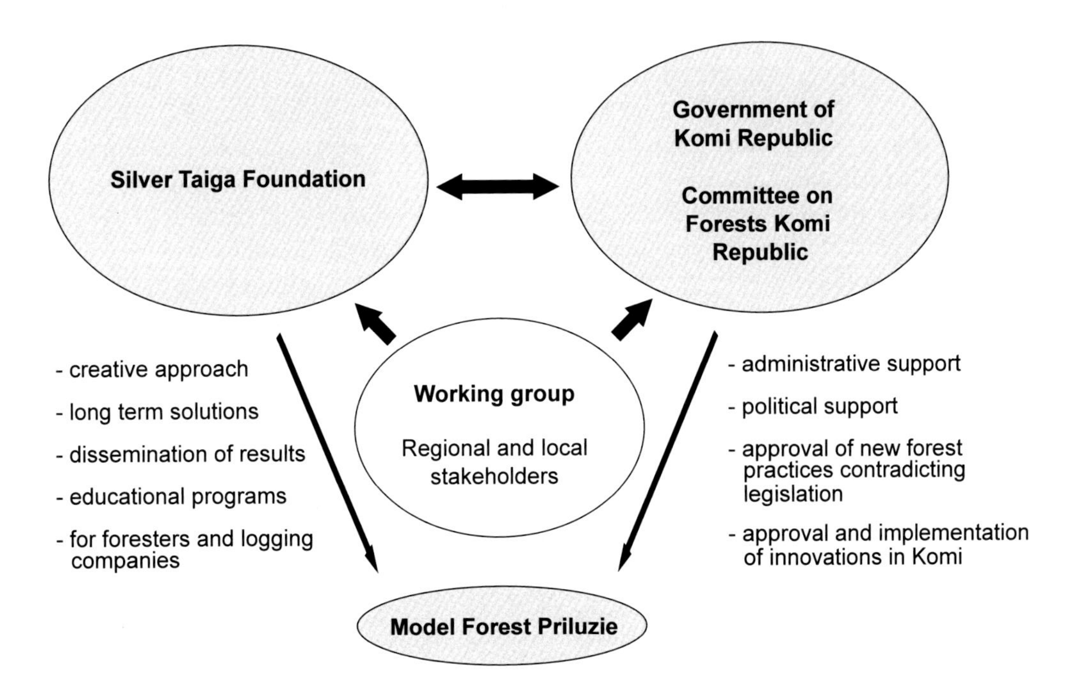

Figure 7.4. Silver Taiga-Governmental interaction (developed by Silver Taiga).

[287] Interview with a Silver Taiga coordinator, December 2002, Obyachevo.

project's implementation.[288] The project cooperated with many departments of the republican government, including the Ministry of Economics, the Forest Committee of the Ministry of Natural Resources, and the Ministry of Transport and Connections. A governmental coordinating and consulting council of 8 people, who met together once a year to discuss the progress of the project and innovations produced by the MFP, [289] was created at the level of the Komi Republic. They were also actively involved in the MFP working group meetings.

The sense of ownership that members of the government show towards the MFP is illustrative of their involvement. The head of Forest Committee said, 'we view the project like our child'[290] while other officials on the Republic level claim that the MFP is a government initiative. The head of the Forest Committee, under the Ministry of Natural Resources, told that government officials from the ministry saw the need to improve forest technologies and practices, so they traveled to Sweden, studied the international experience, linked with forest scientists, and only then brought the WWF into the project. Also, in his words, government officials chose the project's site in Preluzye and initiated the working group. Another official said, 'In this project, everything started with the power structure, with the government'.[291] These officials placed much importance on the success of the project. Such support from the side of the Komi governmental agencies facilitated not only the implementation of the project on the ground but allowed Silver Taiga's contribution to regional forest policy.

Already in 1997, the Concept of the MFP was taken into account by the government as a basis for development of such policies. Preservation of old growth forests was one of the most difficult tasks that Silver Taiga faced in working with the state agencies. The FSC requirements involve the establishment of a special governance regime in such forests and consider them to be high conservation value forests (HCVF-2). In this example it was very hard to achieve common understanding with stakeholders, especially with governmental officials and the old generation of forest scientists, whose Soviet style of education did not include issues of biodiversity conservation. Despite the difficulties, the republic's governor approved a plan for the years 2001-2008 that included inventory and mapping of virgin forests, and assessment of biodiversity levels.[292] Through the process of constant negotiation in 2005, it became possible to make an agreement with the Komi government on a special protection status for old growth forests in Priluzie and to exclude them from logging operations.[293] Since 2001, Silver Taiga has become the key agent in integrating stakeholders for developing a proposal for the 'Concept of Sustainable Forest Management and Use' in the Komi Republic that highlights the main directions of forest policy until the year 2015. This effort was coordinated by the Forest Department of the Ministry of Industry. Attention was paid to creating a business environment that encourages investments in the processing industry and facilitates logging companies to go through FSC certification.[294] Their recommendations on

[288] Interview with a forest policy coordinator, February 2006, Siktifkar.

[289] Interview with the Silver Taiga forest policy coordinator, February 2006, Siktifkar.

[290] Interview with a member of the Forest Committee, 2002, Siktifkar.

[291] Interview with the chair of the Forest Service, 2002, Siktifkar.

[292] 'Virgin Forests: State Approach' in WWF Bulletin No. 4. August, September 2001: 1.

[293] Interview with a Silver Taiga forest policy coordinator, February 2006, Siktifkar.

[294] Interview with the staff of the Forest Department of the Ministry of Industry, March 2006, Siktifkar.

special felling regimes, with ecological planning and biodiversity conservation in the old growth forests, were finally approved by the Head of the State Forestry Agency for the Komi Republic. He made a considerable effort in institutionalizing the FSC's principles in Komi and together with the regional authorities in 2004, in Siktivkar, held a conference called 'Regional Forest Policy Based on the Principles of Sustainable Forest Management'.

During the project period, Silver Taiga suggested to the governmental agencies certain innovative approaches to forest management to be tested in the MFP and afterwards, to be implemented in the whole Komi Republic. In certain cases this effort helped to institutionalize a new practice as a compulsory norm, in other cases to issue recommendations that would suggest the practice.[295] For example, the procedure of public hearings when companies lease the forest land from a leskhoz[296] was approved by the Komi government as a norm, while the measures to prevent soil degradation and suggestions on using water resources were offered as recommendations.

All decisions related to forest management were made by the network of experts that Silver Taiga involved and different governmental agencies. The Northern State Forest Enterprise (a Forest Inventory Agency based in Vologda) developed a long term forest management plan in 2006, using the requirements developed in the USSR in the years 1986-1990, which did not involve the participatory approach. However, the plan was discussed by a MFP working group and presented at several state agencies' meetings. The planning process involved not only the experts in forestry and state authorities, but other stakeholders, including local citizens, who helped to identify forests with social values. In addition, upon the MFP's request, the forest inventory agency conducted a special type of nature conservation planning session, taking into account preservation of biodiversity and the requirements of the FSC. They also took into account the economic measures suggested by the MFP.

All this intensely collaborative work took place only at the level of the Komi Republic and the MFP. Silver Taiga had very few successes with federal level government. Furthermore, both national legislation and the federal governmental policy were barriers for implementation of many of the Komi Republic's initiatives. Some regulations that had been developed in 1930 for logging with tractors are still in place. They are incompatible with logging using modern Scandinavian technologies, but have not changed,[297] despite several new Forest Codes being enacted, since the regulations accompanying them were not developed.

7.5.2 Partnership with the local forest management unit (leskhoz)

Partnership with the Forest Management Unit Priluzie (Priluzie Leskhoz)[298] was essential as they were the receivers of the FSC's certificate. This means that the leskhoz needed to fulfill all certification requirements and to take on all responsibility for forest management and forest protection, enforcing the logging companies that lease the forest lands to comply with the FSC

[295] Interview with a Silver Taiga forest policy coordinator, February 2006, Siktifkar.

[296] This was true only until 2007, when the new forest code required leasing woodlands through auctions.

[297] Interview with the Silver Taiga forest policy coordinator, February 2006, Siktifkar.

[298] Priluzie Leskhoz is a regional representative of the Forest Agency (former Forestry committee of the Ministry of Natural Resources of the Komi Republic and is a part of the Russian Federation Forest Servise.

standards of forest management. Silver Taiga was organized and lead the efforts to make the necessary changes and facilitated the close collaboration between the two actors.[299]

In the early years, the network between the MFP implementers and the Preluzye Leskhoz was hindered by differing perceptions on how to run the project. In 2002, the leskhoz director felt that the project did not focus enough on real, practical changes in the forest territory. He said, 'Preluzye is one of the absolutely ordinary leskhozes in Komi. The project lives its life, supporting community initiatives and school education and the leskhoz lives its own life managing forests and dealing with leasers. The approach to forest management and use is the same as elsewhere'.[300] The leader of the innovation group acknowledged the slow pace of the project; however, he felt that Silver Taiga must proceed carefully. He said, 'Our forestry is under many old rules and traditions and it's all hard to change ... Now there is so much inertia in forestry, we need to act slowly and carefully without rushing'.[301]

Since 2002, many forest management innovations, such as inventory and assessment of virgin forests, the plan for biodiversity conservation, and recommendations for soil preservation were tested and implemented in Priluzie through the partnership with the leskhoz. In 2006-2007, the innovations developed in the Priluzie Leskhoz were introduced into the long term forest inventory plan for the next 10-12 years. The educational center was established under leskhoz supervision and from the start up to 2006, 1346 forest specialists, ecologists and representatives of the logging companies were trained.[302]

There was a considerable amount of collaborative work related to the preparation of the leskhoz for FSC certification, and after receiving the certificate, in maintaining it. Silver Taiga was a consultant and a tutor, while the FSC auditors played the role of creative examiners. Both Silver Taiga and the FSC auditors jointly moved the leskhoz towards sustainable forest management.

A negative result of this close networking with the leskhoz was that when restructuring hurt the leskhoz it also affected the MFP. With the new forest code in effect since January 1, 2008, the role of leskhozes has changed immensely and is reduced to leasing forest land to private companies, organizing forest auctions and controlling the way private companies do logging. Leskhozes were transferred into Lestnichestvo,[303] which then became responsible for monitoring larger areas and were not allowed to be involved in forest management activities, such as thinning, reforestation, etc. In such situations, it became impossible for the Lestnichestvo to hold certificates. So, groups of interested logging companies became the FSC-FM certificate holders and Lestnichestvo became a manager of the group. Areas leased by companies not interested in certification were excluded.

There were good and bad effects of such a transition. Leskhoz certification created favorable conditions for small logging companies that could not afford certification on their own.[304] It

[299] Interview with the head of innovation group, Preluzye Model Forest, 2002, Obyachevo.

[300] Interview with the leskhoz director in 2002, Obyachevo.

[301] Interview with the innovation group coordinator, 2002, Obyachevo.

[302] Http://komimodelforest.ru/pages/proj_map_mlp.php.

[303] With the Forest Code of 1997, lestnichestvo was a smaller unit subordinated to leskhoz. Leskhoz was responsible for forest management, except commercial logging operations, which were done by leasers.

[304] The presentation of a Silver Taiga representative at the meeting of certification bodies with civil society groups, April 2007, Moscow.

became much easier for them to receive a chain of custody certificate when the forest management was already certified using external funding and expertise. On the other hand, the leskhoz does not have economic incentives for holding the burdens and troubles related to certification. One of our informants argued 'FSC is like a drug injection, once you are treated, you get involved and during audits it is stricter and stricter, more and more corrective action requests, and you think does it really make any sense to comply? In my opinion, certification of the leskhoz would make sense in case the leskhoz would be allowed to do the logging and selling of the wood, otherwise its better that logging companies would deal with certification'.[305]

In addition to the overwhelming amount of work related to certification, leskhozes have to rely on logging companies for financing the process. The first test certification was sponsored by the MacArthur Foundation grant; the next, Silver Taiga contributed, and in 2005-2006 the Priluzie Leskhoz charged the companies the certification fee. All the companies were obliged to pay the fee; however, benefits from certification were received only by those companies that had a chain of custody certificate. Others were simply forced to comply with the FSC standards. Since 2008, only interested companies have joined the FSC group certificate.

7.5.3 *Interaction with local administration*

The interaction of Silver Taiga with the local administration in Priluzie was twofold. On one hand, the administration supported the MFP, on the other, especially in early stages of the project, it argued against the preservation of old growth forests on the leased territory. In respect to the future of Komi's forest sector, the head of Preluzye's administration said, 'We really need the model forest to get the certificate' but also 'we need to cut all forests that are ready'.[306] For example, in a conflict over protecting virgin forests, the administration supported a leasing company over the public's protest and the advice of the PMF.

From his viewpoint, the head of administration in 2002 saw the project as a way of increasing production levels and thus helping the region's economy. In the course of the project, the head of administration changed, as well as the degree of support for the project. In 2006, the head of the administration supported the project. Several committees of local administration had differing relationships with Silver Taiga and the MFP; some were more supportive, such as the Committee on industry, some less, such as the Committee on the Environment. Despite some tensions, however, relationships with the administration facilitated implementation of the project.

7.6 MFP involvement with logging companies

During the period of the PMF's implementation, the number of companies leasing the woodlands varied from 12 to 17. They represented a diverse business community with different kinds of relationships with leskhoz, the local administration, TNCs, Silver Taiga and local communities. Figure 7.5 visualizes their location and size. One of the companies, Luzales, is a large enterprise, while the others are fairly small. Most forest producers had their own saw mills, and could process

[305] Interview with the director of the Priluzie leskhoz, February 2006, Obyachevo.

[306] Interview with the head of Priluzie administration, 2002, Obyachevo.

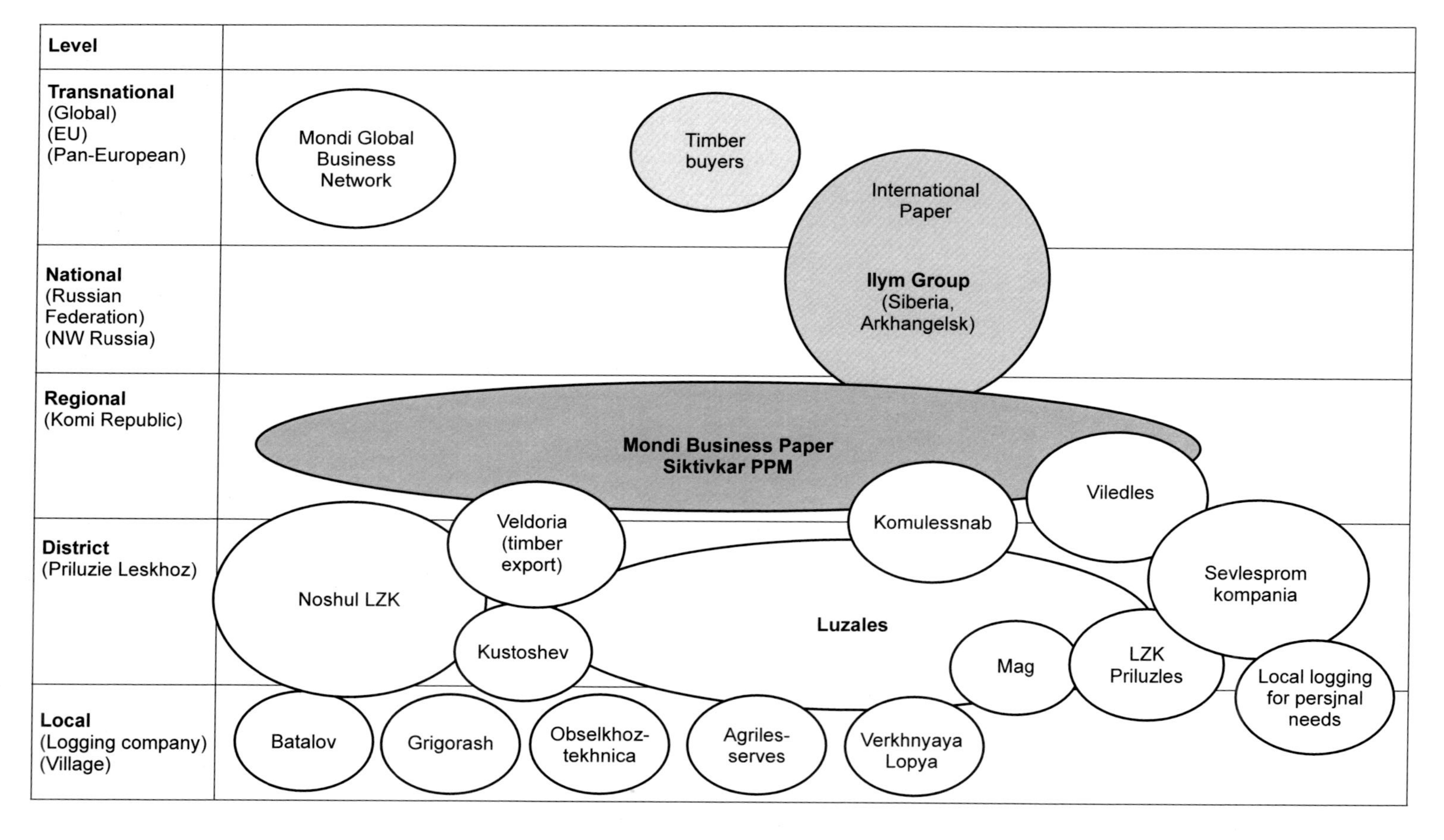

Figure 7.5. Companies involved in Model Forest Priluzie.

wood and sell to both Russian and foreign markets. Some of the companies were owned by individual entrepreneurs,[307] others originated from privatized Soviet state companies, lespromhozes,[308] and a few evolved from Soviet collective farms.[309] All the companies in Priluzie supply pulp wood to the Mondi Business Paper Syktyfkar LPK, which is the only pulp and paper mill in Komi. When Mondi was over supplied with pulp wood, companies had to transport their pulp wood to the Kotlass Pulp and Paper Mill,[310] in the Arhangelsk oblast, which was far away and costly, yet that company was less demanding in terms of pulp wood quality and imposed fewer requirements on its suppliers. Some companies faced falling profits and became subsidiaries of holding companies. For example, the Noshulsky LPK became a subsidiary of the Mondi Business Paper, and Viledles a subsidiary of the Kotlass Pulp and Paper Mill.

During the period of 1990-2006, 5[311] out of the 12 companies in Priluzie passed through bankruptcies, and one[312] was under a bankruptcy threat. Forest companies used the bankruptcy procedure, on one hand, to get rid of their responsibilities for village social infrastructure, and on the other hand, to get rid of all their debts for taxes, electricity and loans. Usually they started the process by organizing a new enterprise with a different name and transferring all their new equipment and machinery to this enterprise. After that, the 'old' enterprise announced its bankruptcy, its remaining old equipment was sold and the company was closed. The 'new' company continued to work under another name with the same director and employees, but without any debts or outdated equipment. This process of bankruptcy, as well as the overall operations of small logging companies, involved both formal and informal relationships between village administrations, other businesses and the company. No one was interested in the complete closure of the business supporting a village; therefore the actors, through multiple informal agreements, achieved compromises throughout the process.

Informal relationships were also very important when the logging companies needed to lease forest territory. Companies would make 'voluntary' financial contributions for the social needs and community development with local administrations, before the leasing competitions began, in order to ensure support. Leasing authorities gave priority to those companies that provided more jobs and contributed to the social infrastructure.

At the same time, local administrations supported some of the weak companies in the event they were the only ones who could support a certain village. In 2006, there was such a situation with the company OOO Verchnaya Lopya. Our informant from the local administration said:

[307] Kustoshev, Batalov, Grigorash (Figure 7.5).

[308] Noshul LZK, Veldoria, Viledles, Komilesnab, Severnaya Lesopromishlennaya Kompania, LZK Priluzles, Mag, Verhniaya Lopia (Figure 7.5).

[309] Obselkhoztechnika, Agrilesservis (Figure 7.5).

[310] Company belonging to the Ilim Group, with 50% of the shares owned by TNC International Paper (Figure 7.5).

[311] OOO Priluzles, OOO Noshulski LDK, OOO Severnaya Lesopromishlennaya Kompania, OOO Veldoria.

[312] OOO Verhnaya Lopia was close to bankruptcy, but not yet in the process, there was a small possibility for avoiding it (when follow on research was done in 2008, company was still alive).

> *The village is alive only while the company is alive. They could not pay the bills, their electricity was switched off, the saw mill stopped. I made the phone calls, we found the way, electricity is on again and everything works, and even the debts for electricity are annulled.*[313]

7.6.1 Interaction between the certificate holder leskhoz and the logging companies in Priluzie

For the logging companies, the leskhoz is a major controlling body on which they depend for their lease. In early 2000's, small companies were not interested in FSC certification. With leskhoz FSC-FM certification, logging companies faced increasing requirements and costs. For the leskhoz, it was a challenge to administer small businesses in such a way that they would comply with the FSC requirements. The leskhol director told us:

> *We are pulling them, all the time pulling, they comply, but without any interest, especially when something costs money. We are trying to renew the lease only for those companies who can come up with payments.*[314]

One of the hardest issues was related to enforcing worker's safety standards. Logging companies were forced to buy safety equipment complying with the ILO convention, but the most difficult task was to make workers wear the new equipment, especially in hot weather. The second hard issue was related to enforcing that the workers take away garbage from logging plots and to take precautionary measures against gasoline leaks into the environment. The leskhoz developed a step-by-step process where the leasers completed each of the recommendations of the leskhoz in order to minimize violations.

Bigger companies, such as Luzales and Noshul LZK, demonstrated greater interest in certification and did not have tensions related to certification with the leskhoz. Luzales was trading in the European market, and acknowledged interest in certification within the European buyers' community and felt that certification is a 'payment for the image'.[315] Noshul LZK is a subsidiary of Mondi Business Paper SLPK, and Mondi was administrating certification of their subsidiaries. Mondi Business Paper SLPK had announced that after 2009 it would stop buying non-certified wood,[316] so the eight companies in Priluzie that already had a chain of custody certificate had an advantage compared to the others. The other companies began to envision certification as a necessity.[317]

[313] Interview with the chair of the committee on industry in Priluzie administration, February 2006, Obyachevo.

[314] Interview with the director of Leskhoz, March 2006, Obyachevo.

[315] Interview with Luzales vice-director, February 2006, Obyachevo.

[316] The requirement for all suppliers to be certified in 2009 turned out to be unrealistic and has been lifted, however, it played a significant role in stimulating companies to get certified.

[317] In 2009, the company realized that its requirement to be certified was unrealistic for its suppliers, so it was postponed.

The situation where the leskhoz is an FSC certificate holder and external funding is involved in certification has its peculiarities. It creates favorable conditions for business development, as it allows companies to get certified even if they have significant debts, as did OOO Verhniaya Lopya, or are bankrupt, as was OOO Priluzles. This does not fit the whole idea of certification as a market driven instrument, in which businesses should theoretically take the lead.

7.6.2 Interaction between Silver Taiga and logging companies

The interaction between the Silver Taiga and small logging companies working in the Preluzye region remained undeveloped during the project's implementation. Logging companies received opportunities to get training in the Model Forest and almost all of them took courses on certification. However, when Silver Taiga organized events involving the stakeholders in decision making, small businesses were passive and almost always skipped the events.

The relationship between Silver Taiga and the bigger company Luzales differed during the course of the project. During the early stages, there was a conflict related to the virgin forests. Silver Taiga was able to convince Luzales to give up logging the old growth. This was an important step in building a relationship with the industry. Luzales' representative said:

> *It would be easier if the Model Forest didn't exist ... They took a whole piece from our territory which was already rented. We had to comply with that. They would not make us do this using power, but I kind of understood that I have to leave something for future generations. They don't have the right to take rented land but I still gave it.*[318]

The situation was resolved in such a way that all parties were satisfied. Virgin forests were preserved, but Luzales received equally profitable plots for logging in another location.

A few years later, the situation in the markets changed in favor of the FSC. It became common knowledge for the companies that virgin forests have to be preserved if the company wants to trade in the international markets. Luzales became more cooperative with Silver Taiga, and began acknowledging the benefits of certification.

When Mondi Business Paper bought Siktivkar LPK, it started to impose on their suppliers the requirements, policies, and codes of conduct that were demanded from their headquarters. They immediately took the path to certification even though the Siktivkar plant was trading only in the Russian internal market. They fostered the demand for certified wood and terrified all the suppliers in the region that they would stop buying non certified wood after 2009. Mondi Business Paper SLPK promoted the certification of three other leskhozes and certified all of their subsidiaries. They used Silver Taiga as consultants in the process. When the support from the SDC ended in 2006, Mondi started to support Silver Taiga's projects. To some extent, Mondi Business Paper SLPK brought new life into the MFP and renewed interest in sustainable forest management of small business enterprises.

---

[318] Representative of Luzales, participant observation, Priluzie Leskhoz meeting, 2002, Obyachevo.

7.7 Interaction of the PMF implementers with the local community

7.7.1 Using the media for building trust

In working with the public, the Silver Taiga's strategy included promoting a positive attitude toward the MFP and sustainable forest management, educating and spreading information, and encouraging members of the population to become active stakeholders in forest relations. From the beginning, the MFP encountered obstacles in linking with the local population in Priluzie. The Model Forest was perceived by the local people as a new forest enterprise that would export wood from their area.[319] It was quite a challenge to explain that the MFP was a new governance arrangement with a new set of rules in which the public has a voice. There was a widespread attitude among the public that when many trucks are seen carrying logs out of the region, the forests are being depleted.[320] The MFP staff explained the difference between sustainable and unsustainable forestry, and that the appearance of many cut logs does not necessarily mean that the forests are in poor health. In addition, they had to clarify the social aspects of sustainable forest management and show the public that they can benefit from implementing this concept. In early stages, the Silver Taiga employees were called 'spies', a common reaction to foreign environmental organizations working in regions of Russia.[321] The common opinion was that if an NGO is funded by Western sources it would work on behalf of those Western interests, and fostering the introduction of stricter forestry rules and the additional expenses would thus make Russia less competitive in the international market place.[322]

Media on various levels were important tools for reaching broad sectors of the public. Silver Taiga sent press releases to regional publications, set up interviews with journalists, and organized MFP tours for film crews.[323] By the beginning of 2002, a total of 18 television programs about the project had been aired. However, according to one informant, most of these aired on a channel that does not broadcast in the Priluzie region.[324]

Silver Taiga used media not only to focus popular attention on the Model Forest, but also to promote an interest in ecology and environmental issues in general. This technique also benefited the MFP by raising support among the public for nature protection, which could then be translated into an interest in the ecological aspects of sustainable forest management. For instance, Silver Taiga funded a one-man photo exhibition that portrayed the unique nature of Pechora-Ilich Zapovednik.[325] While this zapovednik is not closely connected with the Model Forest in Priluzie,

[319] Interview with a local journalist, 2006, Obyachevo.

[320] Interview with the coordinator of a public outreach group, Preluzye Model Forest, 2002, Obyachevo.

[321] Interview with the coordinator of a public relations campaign, Model Forest Preluzye, 2002, Obyachevo.

[322] Interview with the director of the museum of natural history, 2005; interview with a local journalist, 2006, Obyachevo.

[323] 'News of the Working Group of Model Forest – Local Population and Ecological Education' in WWF Bulletin No. 6. January-February 2002: 5.

[324] Interview with an expert on social aspects of certification, 2002, Obyachevo.

[325] 'News of the Working Group of Model Forest – Local Population and Ecological Education' in WWF Bulletin No. 6. January-February 2002: 5.

appreciation for nature in general was valuable for the project. Another example of this technique was Silver Taiga's promoting the television journal Nature.[326] They provided funding and equipment for the journal's TV ecology programs, one of which ended up winning a Golden Nika award for the country-wide All-Russian Competition of Journalists – Ecology of Russia, 1998.

Regular information about the project was delivered by an supplement to a local newspaper, Banner of Labor (*Znamia Truda*). In the early stages of the project, Silver Taiga sponsored one page of the newspaper. Later on, approximately until 2004, Model Forest Priluzie had its own newspaper. After 2004, it again became an addition to the Banner of Labor, which was published once every 2-3 months. By 2005-2006 this had become the only regular media effort to inform the population of Priluzie about the progress of the MFP.

With a popular understanding of ecological concepts, the groundwork was laid for an understanding of and support for sustainable forest management. Despite all the efforts made, it would be wrong to state that hostility towards the project ceased. Many of the villagers remained sceptical about building this model of sustainability on their land.

7.7.2 Revitalizing the old traditions of self-governance

Another technique used to motivate and involve the local population to engage in the sustainable forest management discourse is a revitalization of the region's traditions of self-governance. This effort, like most efforts to empower the public, encountered hindrances left over from the Soviet legacy. Self-governance is an old concept in Komi, however, it was largely lost during the 20th century. One technique used to change this attitude was an appeal to the history and traditions of the Komi people. Silver Taiga linked with the leader of self-governance in the village of Bolshaya Pisa in the Udorsky region of Komi, where the old system was still alive. He was then brought to the Priluzie region to help villagers establish their own self-governance structures.[327] He educated the people of Priluzie about Russia's traditions of self-governance (*schodi, zemstva*). He also wrote an article about Russia's and Komi's past experiences with self-governance, its current state, and a discussion of current efforts towards its revitalization.[328]

7.7.3 Supporting community non-wood resource businesses

Another activity that Silver Taiga used to gain the support of local citizens is the promotion of businesses based on non-wood resources. Here, the strategy was to place an advertisement in the newspapers inviting entrepreneurs already working with non-wood resources to apply for micro credit loans. Silver Taiga channeled money through the Tichenov Foundation and helped it choose one small chain, the Spassporub Consumer Council, and funded and enhanced its operations through micro credit. This council was created in the 1920's as a commercial organization for satisfying the needs of Soviet cooperatives, and during socialism operated very effectively. In 2001,

[326] 'News of the Working Group of Model Forest – Local Population and Ecological Education' in WWF Bulletin No. 6. January-February 2002: 8.

[327] Interview with a local leader, Udorsky region, 2002.

[328] Interview with a local leader, Udorsky region, 2002.

Silver Taiga gave the council the first credit in its history (290,000 rubles interest-free, equivalent to US$ 10,000), which helped to greatly increase production. Our informant claimed that before this credit, the council's operations were based mainly on a barter system, where citizens could, for example, trade gathered berries for food. In 2001, however, 'we could pay people money right away. For blackberries and cranberries, the population got real money this year. People were very happy because it was really essential for their family budgets'.[329] In 2004-2005, Spassporub spread out with small branches to several villages across Priluzie and became a good resource for local communities. The strategy of the micro-credit system was evaluated by a Silver Taiga working group as effective, and works in the same direction continued in 2006.

Silver Taiga, in conjunction with the Institute of Economics in Siktivkar, in 2005-2006 did a survey in four Komi regions[330] in order to evaluate to what extent non-wood resources provide subsistence and seasonal employment for people in rural areas. The Institute of Economics designed the study based on quantitative sociological methods and developed the questionnaire.[331] The survey was conducted by community organizers, who worked with the MFP from its start.[332] Therefore, community members themselves participated as researchers in order to evaluate the future capacity of non-wood resource business so they could include this information in Sustainable Forest Management regional strategy.[333]

One of the principle objectives in forest certification for the MFP was giving the local population some role in the decision-making of forest management. For this, the public must have a stake in the forests and at least something of a voice. A public stake in the forests had always existed, but Silver Taiga tried to enhance it. Businesses using non-wood resources were one way, and promoting environmental awareness was another. These efforts were not directly related to decision-making, though. The next step was to activate the local public into participating in discussions and decision-making.

7.8 Finding local key transmitters: relying on institutional infrastructure created during socialism

7.8.1 Relying on libraries

As the MFP was trying to promote a long-term planning strategy that will benefit future generations, education on all levels can be considered as a key for the project's success. The primary Silver Taiga tactic in involving the public in forest governance was linking with institutions and organizations already established during socialism in the region and, through the use of small grants, slightly tweaking their message to include information from the MFP.

Silver Taiga thus strongly networked with the libraries as local agents for gaining the support of much of the population. Since 1979, the library system in the Soviet Union was centralized, and

[329] Interview with the chair of Spassporub Consumer Council, 2002, Spassporub.

[330] Pechora, Izhemski, Priluzie, and Kotliarovsky.

[331] Information from the working group meeting, 2006, Siktifkar.

[332] Interview with the representative of the Institute of Economics, 2006, Siktifkar.

[333] Interview with the representative of Institute of Economics, 2006, Siktifkar.

library governance, funding and book supply were organized in a hierarchical system. For example, 20 central regional libraries, one of which is located in Obyechevo, belong to the National Library of the Republic of Komi. The 24 libraries situated in the district's villages belong to the central library in Obyachevo. The system was designed in such a way that it is very easy to distribute information from the top to the bottom – to every small forest settlement.[334]

Libraries in Russia, especially in rural areas, are well integrated into the communities and oriented towards spreading education. The Soviet hierarchical infrastructure was used by the MFP as a quick channel for information dissemination. The libraries also gained from such collaboration in many ways: they were supported financially by small grants, the attendance statistics were improved, and their plans and reports were enriched by the new activities that they were conducting to promote the MFP.

Silver Taiga started with a number of educational courses for librarians which would enhance their understanding of the concept of sustainable forest management.[335] Priluzie's regional library, in the settlement of Obyachevo, was used by Silver Taiga as a meeting place and educational centre and served as a base for distributing information throughout the region. The library's staff was employed on small grants from Silver Taiga. Silver Taiga supplied the library with all the publications and information concerning the MFP, as well as new technology, such as televisions and VCR's, along with educational model forest videos.[336] In addition, the libraries were involved in conducting seminars and roundtables on sustainable forestry for all interested citizens. One effective technique was having all the branches of Preluzie's central library system participate in a competition to produce the best informational production on the MFP's sustainable forestry initiatives. Silver Taiga supplied all the branches of the Priluzie libraries with posters that displayed information about the MFP. The collaboration with the library's branch in Loyma went even further. With Silver Taiga's support, they held an annual environmental summer camp for the village's children of different ages and involved them in gardening, river clean ups, village garbage management and doing work for the nature reserve.[337]

Librarians in Priluzie also developed hand written books with illustrations (albums), such as 'By doing good – live in harmony with nature' (*Tvoria dobro zivi v soglasii s prirodoy*) by collecting local material about people who are enthusiastic and knowledgeable in gardening, bee farming, herbal medicinal plants, mushroom and berry collection, fishing and hunting, and who are doing all these activities in an environmentally friendly way. The libraries also collected biographies, poems, stories and field notes from the local people. Each of the village libraries kept this information for their own records, and provided it to the central library. The central library in Obyachevo collected all this information and made colourful albums. In 2004-2005, the staff of the central library went to the villages with a series of concerts and thematic ecological programs, and displayed materials mixed with songs and poems. The sustainable forest management theme was even linked with the celebration of May 9th (World War II victory day), a very respected holiday in all the village

[334] Interview with the head of the library in Obyechevo, 2005; interview with the representative of National Library of Republic of Komi, 2006, Syktifkar.

[335] Interview with a Silver Taiga innovation group former staff member, 2005, Siktifkar.

[336] 'Work with Local Population' in WWF Bulletin No. 4. August-September 2001.

[337] Interview with the librarian 2006, Loyma.

communities. Many of the events were prepared in collaboration with the House of Culture. People in the villages were deeply touched by the attention paid to their respect for nature.

However, when the small grants to the libraries were finished, in most cases the library posters about the model forest were no longer updated, and in some cases they were removed. The MFP events disappeared from the library planning and reporting system. Nevertheless, Silver Taiga has continued to supply libraries with information – this information is available upon request, but is not on the front shelves as it was before. In 2006, library informants spoke about the MFP and Silver Taiga programs as if they had taken place in the past.

7.8.2 Networking with schools

The network between the MFP implementers and local schools was used by the project to supplement environmental curricula with knowledge that brought the experiences of the Model Forest into the classrooms. The MFP helped financially to equip several schools. In Obyachevo, the biology-ecology classroom was totally renovated and equipped with computers. In the early 2000's the MFP organized an initiative group, based in the Obyachevo School, which included all the teachers dealing with the environment in the Priluzie region. Silver Taiga held courses for teachers in forest ecology, biodiversity and sustainable forest management. Separate courses were conducted for teachers of high schools and primary schools and were adjusted to the interests of the teachers. The courses included field trips to the educational forest management plots that were developed by Silver Taiga, and at the end, Silver Taiga organized a tour for teachers to Pechoro Ilichski Zapovednik, where they studied biodiversity in old growth forests.[338] The number of events held by schools and related to the environment increased greatly. For example, in November 2001, 14 out of 19 schools in the Priluzie region attended a party for young naturalists called 'Forest as a Temple, Home, and Workshop' organized by a teacher's initiative group. The number of school children that were winners at the Komi republic level and at national level competitions, such as the scientific Olympics, increased, and many students received awards and the schools were rewarded financially from Russian governmental agencies, as well as several teachers who were promoted to the next salary level.[339]

As a result of Silver Taiga's educational efforts, teachers developed new programs of environmental education that combined ecology, sustainable forest management and included information about the MFP.[340] A teacher in the village of Chernish became very advanced in environmental education. She developed several courses, such as Village ecology (for a primary school), Village history and nature (for a middle school), and Forest ecology (for a high school). Courses included field trips and the extensive use of the school's forest museum. The Forest ecology course involved role playing games on stakeholder engagement in forest management decisions, for example, related to old growth forests, and the analyses of how the public can participate in the decision making process. She developed several ecological trails and educated

[338] Interview with a teacher, 2005, Obyachevo.

[339] Interview with a teacher, 2005, Chernish.

[340] Interview with the coordinator of the public outreach group, Preluzye Model Forest, 2002, Obyachevo.

other teachers in Priluzie on how to build such trails. As a result, six ecological trails were built close to different villages.

Another educational activity was the annual environmental summer camp 'Boomerang', which existed for three years, growing into an all Komi Republic camp, with 30-50 high school students participating. Silver Taiga's staffs were guests, educators and tour guides for the children. The first camp was completely sponsored by Silver Taiga, but later camps were co-sponsored by the republic's Committee for Youth Affairs. Since 2003, it has been totally sponsored by the Committee for Youth Affairs. When the sponsorship was taken over by the Russian governmental agency, the environmental education programs were shortened, and the camp involved more sporting and entertainment activities compared to the years when Silver Taiga had the lead.

Silver Taiga's support for schools has significantly diminished since 2002. Despite this, some enthusiastic teachers have continued their educational programs. The Chernish School continued to educate children from the whole region using its ecological trails, held its own camps and even presented its educational programs at the international level.[341] However, in a number of schools, environmental education has slowed down, many of the ecological trails are used only occasionally because some are relatively far from the villages and transportaton costs are high, therefore it is hard to organize bus field trips and maintenance for the trails.[342] In addition, highly qualified educators from the outside are not serving as teachers anymore and the village teachers lost not only financial support, but the energizer in the form of Silver Taiga. Silver Taiga's efforts to educate teachers in fundraising and help them receive other grants failed, mainly because of the teacher's unpreparedness to combine a high work load at school with taking time for grant writing.[343]

Two teachers were part of the MFP working group. They voiced local issues to the MFP working group and transmitted ideas and innovations from the working group back to the local level.

7.9 Building democratic decision-making mechanisms

7.9.1 Forest discussion clubs

One of the major efforts to initiate stakeholder dialogue around forest issues was through a club in the Priluzie district called Shuvgy Parma (translated to 'the sound of wind through the taiga forest'). The idea was to build a foundation for cooperative policy networks and to get stakeholders into the habit of discussing important issues within a democratic structure of dialogue.[344] The meetings of this club included various members of the local public, leskhoz workers, scientists, and the Priluzie administration in discussions about forests and their uses. Topics of the meetings have included the multipurpose use of forest resources and modern forest politics in the Komi Republic.[345] Another topic, the importance of forests in traditional Komi culture, allowed the discussion group to link itself with the local culture. This was a tactic for gaining local support

[341] Interview with a teacher, 2005, Chernish.

[342] Interview with a teacher, 2005, Chernish.

[343] Interview with a vice director of Silver Taiga, 2005, Siktifkar.

[344] Interview with Silver Taiga staff, 2002, Siktifkar.

[345] Interview with a Shuvgy Parma organizer, December 2002, Obyachevo.

because all over Russia people are interested in revitalizing the cultural traditions of the past. While the discussions did not result in concrete solutions, the club was a tool that allowed various stakeholders to understand each other's values and opinions. The club's first meeting was on April 17, 1999. Since then, the club has convened more than 30 times in villages throughout the Priluzie region.

The club's coordinator was the chair of the Department of Culture of the regional administration. She was doing most of the organizing, and Silver Taiga only approved her ideas and provided grant money. This strategy gave Shuvgy Parma the appearance of a local initiative, which may serve to better attract the local communities. Shuvgy Parma's limited attendance, however, demonstrated a barrier to the MFP's efforts for public participation. The large majority of those participating in the discussions were either people already connected to the Model Forest, the elderly, or members of the local intelligentsia such as 'teachers, librarians, journalists, members of people's councils, scientists, and members of the working group', there were very few workers coming to the club.[346] One of our informants complained that some club attendees were sceptical towards the MFP and were resistant to education and innovation. She said: 'they don't believe people will ever do things differently'.[347] Several of my informants believe that this is linked to growing up under the Soviet system of forest management. One said, 'We have logic from the old times that the boss is always right. Years need to go by before the situation is changed'.[348]

There were many marginalized groups in the community that were not reached through links such as the teachers or libraries. However, these groups are nevertheless stakeholders. One of our informants asked, 'How to involve those who steal from the forest, who litter and who burn it? This is a real problem not solved in Komi'.[349] The failure to include such people in any of the events organized by Silver Taiga was a hindrance to the achievement of the inclusive participatory process.

Working with youth was also a part of the MFP's efforts toward public participation. Familiarizing the youths with discussing forest management issues was important for improving the efficiency and effectiveness of cooperative policy networks in the future.[350] One such effort was the discussion club named Beta Valna ('beta waves', as in what the human brain produces while thinking and discussing). This club met in regional libraries and villages in Priluzie.

This effort to educate and inform youth met with a barrier similar to the one faced by the club Shuvgy Parma. According to our informant, media announcements were not effective in recruiting participants, so she went through personal contacts to find interested youth.[351] This usually means 'young specialists, engineers, doctors, and teachers'. This group is the community's future intelligentsia and so the Model Forest again encountered the barrier of limited 'ordinary' people's participation.

The coordinator of the discussion club tried to work around this limitation by purposely targeting certain members of the public, such as working youths and young families. A discussion

[346] Interview with a teacher, December 2002, Chernish.

[347] Interview with a club participant, December 2002, Obyachevo.

[348] Interview with the coordinator of the discussion club, December 2002, Obyachevo.

[349] Interview with a Silver Taiga expert, December 2002, Obyachevo.

[350] Interview with Silver Taiga staff, 2002, Obyachevo.

[351] Interview with the coordinator of a youth discussion club, 2002, Obyachevo.

club worked with this sector of the population, but it focused on attitudes towards nature rather than forest management issues in order to maintain attendance. According to the club's convener, 'We were worried about discussing forest issues with them'.[352] This illustrates the difficulty of turning the love of nature, which most people have, into active citizenship roles in the political, economic, and social realms.

When the coordinator of Shuvge Parma left Obyachevo for Syktyvkar, the remnants of both Shuvgy Parma and the Youth club were formed into a single discussion club which worked with elderly people, adults and youths.[353] Since funding for community events has diminished and gasoline prices have risen, the events have become rarer. However, in the spring of 2005, the club still operated, but at a slower pace compared to 2001-2002. Later, Shuvge Parma revitalized its activities in the framework of the Forest Village project, initiated by Silver Taiga in 2009.

7.9.2 Negotiating special regimes of logging in the territories where local people hunt and collect mushrooms and berries

One way that Silver Taiga tried to involve the public in a decision making process is by making maps of the Preluzye forest management unit that show the areas where villagers hunt or pick berries and mushrooms. The goal was to use these maps as a planning tool for decision-making on forest management. Preluzye Leskhoz supported this effort from its start as there were many conflicts arising around the issues of non-logging buffer zones surrounding the villages. The first draft maps were made by leskhoz employees, as they are very familiar with the forest habitats in the area. These maps were distributed to people active in the villages and displayed in libraries and village councils.[354] Members of the public were invited to draw dots on the forest management map showing where they often collect non-wood products, and where many dots appeared in the same spot, the area was circled and marked accordingly.[355]

For each valuable species of mushrooms and berries, special logging regimes were developed. Some Forest quadrates around the villages were designated for local people's use only and companies could no longer lease them. In 2004, the Forest Agency of the Republic of Komi approved special regimes of logging in the territories of mass collection of berries and mushrooms for the whole republic of Komi. In such a way, innovative decisions developed and tested on the MFP's territory were transmitted to the level of the whole republic.[356]

With the advice of Silver Taiga, the leskhoz developed a procedure for conflict resolution with the population, which is required by the FSC. Many villagers continued to be concerned with issues of fire wood and building wood supply around the villages. Along with the MFP's development, the leskhoz was continuously becoming more open to conversations with public.[357]

---

352 Interview with a coordinator of youth discussion clubs, 2002, Obyachevo.

353 Interview with the discussion club coordinator, 2005, Obyachevo.

354 Interview with a community organizer, 2006, Obyachevo.

355 Interview with a coordinator of public outreach group, Model Forest Preluzye, 2002, Siktifkar.

356 Reference to the Silver Taiga brochure 2004.

357 Interview with a community organizer, 2006, Obyachevo.

7.9.3 Public participation in decisions related to virgin forests

A successful example of public participation and activism is the case of the protection of virgin forests in the MFP territory. In 1990's, Russian governmental agencies were against the protection of virgin forests that were situated outside of specially protected areas. According to the governmental approach, these virgin old-growth forests were 'old' and 'ready to be cut' and otherwise would be in increased risk from disease and fires.[358] Usually the companies that lease land have to log all forests in that territory. Therefore, it was a challenge for Silver Taigar to change the attitudes towards old growth forests and create an understanding of them as a source of high biodiversity, subject to preservation. In this case, Silver Taiga was able to mobilize members of the local population to protect the area. In 1998, LuzaLes, the largest company then currently operating in the MFP, leased a virgin tract of forest and had already begun building an access road for harvesting it.[359]

Silver Taiga linked with the influential members of the local population to oppose LuzaLes. The discussion club Shuvgy Parma held a meeting about virgin forests in the closest village to the leased virgin tract, Spasporub.[360] There, scientists and participants of the MFP explained the uniqueness of the tract and influenced many local villagers to write protest letters to the press and to the head of the regional administration, as well as to the LuzaLes company.[361] Activists went from house-to-house collecting signatures and explaining the issue, which was based on education provided by Silver Taiga and common knowledge of the valuable species of herbs in that area.[362] The village's activist also attended meetings with industry and the local administration as a representative of the indigenous organization Komi Voiter.[363] He linked an already existing interest in forests and local Komi culture, and used education to direct his effort towards the virgin forest issue.[364]

The dialogue surrounding this issue involved a lot of conflict and fighting. One of our informants said, 'From the beginning, they [representatives of LuzaLes] were against us when we said that the forest needs to be preserved ... very unfriendly'.[365] A local villager described one of his meetings with industry, saying, 'We were shouting and there were scandals. We were yelling personal insults and attacks at each other'.[366] After eight meetings, a compromise was reached. This example shows that the Russian public can be successfully mobilized if the right tactics are used.

[358] Interview with a coordinator of the innovation group, Preluzye Model Forest, December 2002, Obyachevo.
[359] Participant observation, meeting at the Preluzye Leskhoz, December 2002, Obyachevo.
[360] Interview with a director of Shuvgy Parma, December 2002, Obyachevo.
[361] Interview with a coordinator of the ecology group, Preluzye Model Forest, December 2002, Obyachevo.
[362] Interview with a community organizer, March 2006, Obyachevo.
[363] Interview with a local leader, village Spasporub, December 2002, Spasporub.
[364] Interview with the head of Silver Taiga Foundation, December 2002, Siktifkar.
[365] Interview with the Silver Taiga staff, December 2002, Siktifkar.
[366] Interview with a local leader, village of Spasporub, December 2002, Spasporub.

7.9.4 Public hearings for leasing decisions

One of the obligatory requirements of FSC forest certification is the informing of the public on forest management plans. The Silver Taiga Foundation decided to use the procedure of public hearings for this purpose. Silver Taiga organized a small working group and involved an expert from the Committee for Saving the Pechora River, who had already participated in organizing a public hearing in Pechora around oil issues. This expert developed recommendations for public hearings, trying to avoid the too formal approach that took place in Pechora. The working group also involved a lawyer to ensure that the newly developed procedures corresponded to existing Russian legislation. The working group decided to organize hearings to take place when the company applied for leasing the woodlands. In this case, the interests of the community are considered, as most operations of the company directly affect the community. Silver Taiga invited experts from the St. Petersburg NGO ECOM, who conducted training workshops for the Priluzie community activists. The community activists in turn made an effort to explain to the forest company representatives that a high level of community participation in public hearings can help to avoid future conflicts and can benefit both the company and the community.[367]

One of the first public hearings that took place, in the settlement Vaimos, was the most challenging, where both the Siktivkar plant and a small forest company applied for the same lease. The industrial plant promised to do logging using new technology, which does not create much of an environmental impact, but also does not create local jobs, while the small company offered to work using old technology, but to create jobs. The community was more interested in the jobs. The public was active and expressed a variety of opinions. The special commission which makes decisions on who will receive the leases for the land choose to give some land plots to both applicants, but also took into consideration some of the public's requests. The industrial plant was asked to maintain the road and to participate in several community social programs. Social aspects were included in the leasing agreement.[368]

In general, the local population usually does not discuss issues related to forest management during public hearings, yet raises issues such as the delivery of fire wood to the community, or asks the company to buy a computer for the local school, or musical equipment for the local House of Culture. Only occasionally are the community requests more focussed on land use and the production process. Once a community asked the company to build a new road from the village to the main road and in another case asked that 80% of the people employed would come from the local community.[369] Decisions on the leasing agreements are done by a special commission, which does not always include community requests in the leasing agreements. Another problem is that there is no mechanism to follow compliance and monitor to what extent community demands are being met by the company.[370] However, most governmental representatives and community organizers that I interviewed were generally satisfied with the public hearings. When public hearings became regular events, the public lost interest in them and many hearings took place

[367] Interview with a Saving Pechora Committee expert, March 2006, Siktifkar.

[368] Interview with the representative of the Silver Taiga innovation group, April 2005,Siktifkar.

[369] Interview witha local community organizer, April 2005, Obyachevo.

[370] Interview with a Saving Pechora expert, March 2006, Siktifkar.

without much community input. Therefore, community organizing and mobilizing continues to be critical for successful public participation in decision making.[371]

Based on the the MFP experience, public hearing procedures have been developed for the whole Komi Republic. Whole groups of interested parties, such as the Priluzie People's Deputies, regional administrations, representatives of the Komi Ministry and representatives of local communities have participated in this work. Recommendations were approved by the People's Deputy Council, and public hearings became compulsory for the whole Republic of Komi and became part of the procedure for leasing agreements. A number of public hearings were organized in the summer and fall of 2006, when the forest inventory agency was working on the new Forest management plan.

According to the survey conducted by the 'Toy Opinion', 37% of the citizens of Priluzie participated in at least one of the public hearings, 42% expressed their willingness to participate, and from those who would like to participate, 70% are interested enough to read materials about the subject in advance, while 60% learned about public hearings from newspapers, and the balance from friends.[372]

With the new Forest Code of 2007, the leasing of woodlands became possible only through auctions. Therefore, since the new Code came into force, the general public has stopped its participation in the leasing procedures.

7.9.5 Civic conference on FLEG implementation

One of the innovative forms of development of participatory democracy in Priluzie was the civic conference, where representatives of local communities discussed possible mechanisms for controlling and monitoring illegal logging and illegal trade in wood products. This event was organized as one of the meetings in preparation for the Ministerial Conference of Europe and Asia on Forest Law Enforcement and Governance in St. Petersburg in 2005, known as the regional (stands for) ENA-FLEG process. The purpose of the civic conference in Priluzie was to develop and submit suggestions for the conference of Ministers and therefore, to involve local actors at the village level in the participation of global forest governance. The civic conference was set up and prepared by an expert from the NGO ECOM, based in St. Petersburg, together with Silver Taiga. The initiative was from ECOM, as they received the grant for developing mechanisms of public participation in FLEG and were willing to test them in the MFP. In addition, representatives from both ECOM and Silver Taiga were selected to participate on the NGO side in developing St. Petersburg's Declaration on FLEG during the Ministerial conference.[373]

ECOM, in part, used the methodology of public engagement developed in the US by the Jefferson Center in 1974, which was modified in Denmark and commonly called the 'conference for reaching consensus'.[374] In Priluzie, in the first stage, a telephone survey was conducted with 500 citizens in order to find those interested in participating in the civic conference. Organizers sent documents about FLEG to 200 citizens who during the survey had expressed their interest

[371] Interview with a community activist and participant of many public hearings, March 2006, Siktifkar.

[372] Research of 'Toy Opinion', supported by Silver Taiga, http://ecominfo.spb.ru, 14 April 2006.

[373] Interview with a Silver Taiga expert, Soktifkar.

[374] Http://www.ecoinfo.ru, 20.06.06 documents on the organization of the civic conferences, 2006.

in participating in the process. 24 informed organizers that had read the documents, and seven who were the most prepared, were then invited to attend the conference and participate in the document's development. During the conference, participants discussed issues related to illegal logging and trade and developed the final document. The final document has been sent to local and regional governmental agencies, forest companies and the secretariat of the Ministerial FLEG conference.

Our informants assessed the conference differently. One of them thinks that the civic conference is a very interesting, innovative mechanism of public engagement in the decision making process which can involve completely new people.[375] Another informant feels that this form is less democratic then public hearings, as only selected representatives of the public participate.[376] A representative of Mondi Business Paper who attended the conference considered the conference very useful, as Mondi Business Paper is interested in making their chain of custody totally legal and transparent in order to be successful in European markets. Other participants assessed the civic conference positively as a form of public engagement but were critical of the topic.[377] Some felt that the local people were interested in discussing illegal logging done by their neighbours or small firms, while the conference was dedicated to discussions on how to make the whole chain of custody transparent, which seemed too removed from local issues. One of the informants felt that even on the local level, citizens themselves cannot develop mechanisms for preventing illegal logging, as small violations are part of the everyday life of almost all the villagers. He said:

> *People won't say anything about their neighbour even if he does something illegal, because one day one person does something illegal, and the next day another does something illegal. Everyone does illegal things all the time. No one ever says anything to the inspectors, so it is impossible to stop illegal cutting and impossible to punish anyone'.*[378]

This situation complicates the interests of the local population as stakeholders. Our informant said, 'They have a double morality. Today they want to protect the forest and tomorrow they cut illegally themselves or don't mind if their neighbour does it'.[379] According to the opinion of most civic conference observers, the organizers failed to maintain an interest in participation in global forest policy development. It seems that it was too hard to globalize the local actors from the Russian forest communities.

7.9.6 Open houses

In Russia, leskhozes traditionally are fairly closed structures. Usually the information on illegal logging, statistics, and even the information on leasers is available only for business purposes

[375] Interview with a member of the forest council, March 2006, Siktifkar.

[376] Interview with a leskhoz employee, March 2006, Obyachevo.

[377] Several participants of the civic conference, March 2006, Obyachevo.

[378] Interview with a community member, March 2006, Obyachevo.

[379] Interview, coordinator of citizen involvement, the Model Forest Preluzye, December 2002, Obyachevo.

and is closed to the public. The FSC auditors require transparency, therefore with certification, the situation with respect to openness of the leskhozes changed. However, this does not mean that the problems between leskhozes and the local citizens are resolved in the certified territories. People complain about leskhoz policies, but this happens mostly because of the lack of knowledge by the local public on how a leskhoz operates and what functions are allowed by legislation. Therefore, Silver Taiga decided to develop mechanisms that would promote dialogue and mutual understanding between leskhozes and the public. This goal was achieved by organizing Open Houses for the MFP project. These open houses took place both in Obiatchevo and in other villages in Priluzie, such as Noshul. Several buses delivered people from different villages to Obiatchevo. Several sessions, such as on biodiversity and environmental education were organized. The leskhoz staff answered all kinds of questions coming from communities and there is no doubt that the event fostered a constructive dialogue.[380]

7.9.7 Forest Council

One of the goals of Silver Taiga was to establish some sort of institution that could continue the MFP after financial support from the Swiss Development Agency ended. The special unit 'Forest Education' was created at the leskhoz for organizing different kinds of educational projects and maintaining a forest museum. The unit worked effectively, but was closed with the advent of state reforms, as all state agencies were required to cut their staffs.

In 2004 the Forest Council was created with the purpose of organizing and moderating public participation in forest governance. The Forest Council consists of the most active people, who from its start were supportive of the project, in particular; a journalist from the newspaper 'Banner of Labour', dedicated teachers and librarians, a leskhoz representative and an expert from the Save Pechora NGO. It happened that the Forest Council consisted of leaders from different branches of the project.[381] From its start, the Forest Council worked effectively and was dealing with the issues that were not resolved by the Obiatchevo administration and the leskhoz.[382] The Forest Council developed guidelines for fire wood supply for the local population. These guidelines were approved by both the administration and the leskhoz and were incorporated into practice.[383] With the initiative of the Forest Council, the recreational site called 'The Meadow of Brides' was built by community volunteers and supported by small businesses, and currently is used not only by brides, but also by high school students for discussions related to forestry and as a tourist attraction.[384] Despite the fact that the Forest Council is an interesting experimental institution, our informants feel that it is fragile and non-stable. 'When someone active leaves the community or Silver Taiga stops funding certain issues, the project dies like a flower without water'.[385] However, it serves its role as a local stakeholder's consultative platform. Silver Taiga used the Forest Council in Priluzie

[380] Interview with a participant of the Open House, March 2006, Obyachevo.
[381] Interview with an expert from Save Pechora NGO, April 2006, Obyachevo.
[382] Interview with the librarian, forest Council member, April 2006, Obyachevo.
[383] Interview with a journalist, Forest Council member, April 2006, Obyachevo.
[384] Interview with a teacher, Forest Council member, April 2006, Chernish.
[385] Interview with a local community organizer, October 2005, Chernish.

as a model and created a similar council in the Udora district in order to bring stakeholders to the roundtable to discuss issues related to old growth forests.

7.10 Interaction with transnational stakeholders

The major transnational stakeholder groups engaged in building the MFP were: (a) foundations, in particular the SDC – a key supporter of the project, as well as other foundations, which were less involved; (b) NGOs, in which the WWF network was most significantly involved in implementing the project; (c) the TNC Mondi Business Paper, which became especially important in the later stages of the project; (d) other transnational networks involved in similar activities, such as the International Model Forest network and the High Conservation Value (HCV) Network.

The SDC, the main sponsor of the project for eleven years, was an essential part of the MFP network. As I already mentioned, the first portion of the grant was channeled to the site of implementation through the WWF, another major transnational actor. As the main financial supporter of the project, the SDC was indirectly involved in the project's management. At a time when the project's management was carried out by the WWF, the local project coordinators of the Komi WWF office (Silver Taiga, since 2002) were 'between two fires', as they had to reckon with both the requirements of the WWF and with the desires of the SDC. The WWF considered it to be more promising to partner with large businesses in developing models for sustainable forest management. For example, the WWF suggested moving the office from Komi to the territories where the major holding companies' leases were concentrated, such as the Archangel Region. In addition, the WWF focused its efforts on environmental issues to a much larger extent than economic or social issues. The priorities of the WWF as a GGN were consistent with their global designs, as expressed in the 'Forests for Life' and 'Ecoregions' programs. When discussing sustainable development, the WWF always put the environmental issues first.

For the SDC the environment was not a priority, but simply part of sustainability, along with economic and social issues, while giving preference to social issues. In 2002, the SDC started the Silver Taiga Foundation and became its only source of funding. Silver Taiga continued collaboration with the WWF, but a representative of the SDC, in the Swiss Moscow Embassy, supervised the project. To make the project more democratic, an international board of trustees, comprised of representatives from Sweden (a representative of the GFTN), Switzerland, Canada (a consultant on forest economics and forestry), Finland, and Russia (the Komi Republic), the FSC Moscow, has been formed. The Board of Trustees met 2 times a year and its tasks included supervision and evaluation of the effectiveness of project funds' use, decisions on core activities, and advising on specific issues related to forest management. Operational management of the project was carried out by Silver Taiga.

At the level of the MFP Working group, the project's main direction of development, as formulated by the Board of Trustees, has been taken into account as a general framework, but specific measures have been developed by regional stakeholders represented in the working group. During their regular reporting, Silver Taiga, in turn, provided suggestions for the project's future development. There were constant interactions between the Board, the working group and Silver Taiga.

At the last stage of the project, when it became clear that the SDC would stop its financing, Silver Taiga became more active in seeking collaboration with other donors. They applied to the UN Development Programme (UNDP) to conserve biodiversity and promote sustainable management of the forests in the headwaters of the Pechora River. They started working with the Program of the Barent's Euro-Arctic Region (BEAR). The Komi Republic is in that region, along with the Scandinavian countries. This program attracts investments from international financial institutions for various projects, particularly in the environmental field. In 1999, the BEAR 'International Environmental Forum' program was created. In 2003, in Kuhmo, Finland, the Environmental Forum of BEAR first raised the question of preservation of intact forest ecosystems. In 2005, at the next forum, held in Syktyvkar, Silver Taiga presented information on the conservation and monitoring of old-growth boreal forests and sustainable forest management, involving the preservation of biodiversity outside protected areas.

The MacArthur Foundation was not involved in the MFP, however, they financed two projects indirectly related to the model forest. The first project funded the working group for developing regional FSC standards for the republic of Komi. The second funded the testing of the FSC standards at the MFP. The working group that developed the FSC regional standards was set up according to FSC requirements, involving business, environmental and social chambers. It was not common in Russia in early 2000's to have dialogues between business, social and environmental stakeholders. Therefore, both the working group for the development of the FSC standards and the working group for developing the MFP contributed to the development of a democratic dialogue and new ways of decision making. Thus, transnational actors, both foundations and the FSC, played a decisive role in establishing democratic institutions in the forestry sector of the Komi Republic.

Since 2008, the Ford Foundation has become essential for Silver Taiga's operations in Komi with its support for the project called 'Integration of rural communities around economic stabilization based on sustainable use of natural resources', commonly called the Forest Village project. The Ford Foundation's involvement has allowed for continuing the discussion club Shuvge Parma in the territory of Priluzie, as well as dissemination of the PMF's democratic institutional innovations, such as Forest Councils, to other districts in the Komi Republic.

The TNC Mondi Business Paper Group has become the most important stakeholder since the SDC funding ended. In 2007, Mondi Business Paper Group became independent from the Anglo-American company and started its due-diligence process with the setting up of policies and management systems. Under the Board, the Sustainable Development Committee was formed, focusing on governance, forestry, climate change, product stewardship, corporate social responsibility and safety. The forestry policy was developed and forest standards written along with performance requirements, compulsory for implementation at the mills and logging operations.[386] The standards and policies were channeled to all operations in the Komi Republic, requiring changing practices in the space of place. The practices changed initially at Mondi Business Paper Siktivlar LDK's subsidiaries, such as Noshulski LDK in the MFP's territory; however, they also affected the company's suppliers, which included almost all the forest producers in Komi.

Mondi Business Paper accepted PEFC certified wood from Austria, Slovakia and Poland, however it made a strong commitment to have its own forest operations certified by the FSC.

386 Interview with the Director of the Sustainable Development Committee, July, 2009, Austria, Vienna.

A member of the Sustainable Committee in the UK said: 'We always believed that the FSC is a 'Golden standard' that will always minimize our reputation's risks'.[387] Mondi Business Paper made a public commitment to foster the FSC in Komi and produced the first FSC certified paper, Snegurochka (Snowmaiden), in Russia. Mondi requested that all their suppliers become certified by the FSC, and became a major driver of the internal FSC market in Russia. Silver Taiga became a major consultant in this process. Mondi sponsored certification of three leskhozes using the lessons learned in the MFP, and certified all their own subsidiaries. They introduced the regular socio-economic assessment toolbox review, the same as in their South African operations, as a large consultative platform with stakeholders. Such reviews were conducted in 2006 and 2009. Silver Taiga was involved in interviews with the local communities, brought attention to the ecological sensitivity of certain districts, and collected suggestions from the local population regarding improvements in fire wood supply. Social agreements have been signed between the company and local communities in Komi. Mondi Business Paper was striving for the standardization of its practices around the world and made an effort to regulate the forest operations in South Africa and in Russia in a similar way, despite considerable differences in the natural and organizational environment. It became a fully functioning GGN, whose designs reach the sites of implementation.

In October 2008, Silver Taiga hosted the Forest Dialogue conference on poverty reduction, funded by Mondi Business Paper, in which TNCs, representatives of the World Business Council for Sustainable Development, major transnational and Russian NGOs all participated.

Another initiative of the Silver Taiga was including the MFP in an international network of model forests. The headquarters of the network are in Canada, as the Canadian government initiated development of the Model Forests. Today, the Model Forest Network unites 40 model forests in 19 countries. The international network includes several regional and national networks. Silver Taiga presented the MFP at the World Forum of model forests in Costa Rica in 2005, where they expressed a desire to join the international network. Since then, Silver Taiga has received several grants from the international Model Forest Network to assist the newly emerging Russian model forests.

In April 2009, Silver Taiga also hosted the HCV network, which discussed issues related to preservation of the old growth forests in Russia and Canada. Thus, Silver Taiga used every opportunity to attract the attention of transnational actors to the MFP, and particularly to the issue of the conservation of old-growth forests, and continued seeking funding for the continuation of its efforts in promoting sustainable forest management.

7.11 Discussion and conclusions

The case study represents an example of a top-down network in which a regional NGO played the role of a transmitter and co-designer of a model of sustainable forest management. An NGO, Silver Taiga linked stakeholders from transnational spaces and the spaces of places in their joint efforts in building the MFP. Being a key player in the network, Silver Taiga needed to acknowledge the interests of the donor and later of the TNC, accumulate input from stakeholders and process both global designs and regional/local suggestions into the space of place in a way that fit the

[387] Interview with the staff of the Sustainable Development Committee, March 2008, Addelstone, UK.

local context and involved actors in the institutionalization of new practices on the ground. As it was described, when the donor funding expired, the TNC became a major actor in the network. The FSC rules of the game and its approach to sustainable forest management were essential for the MFP from its start. The FSC's certification institutional infrastructure and the incentives they provide to the business actors ensured sustainability of the institutionalized practices.

A peculiarity of this case for the MFP was that the territory of a whole leskhoz, having many different forest companies located in it, was certified. That led to a situation where small businesses were under pressure, because for some of them trading in internal markets did not have any economic benefits from certification, but bore the large costs of certification anyway. At the same time, companies interested in trading in Europe benefited, because for them it became easier to certify the chain of custody when the forest management was already certified. This was a specificity of this case, as the parties interested in certification, along with those who had no interest, all took part in certification. Thus, the process was voluntary for only several of the logging companies. For the rest, certification was an additional burden imposed by the state authority: the leskhoz. As a result, many small forest companies left their leases in this region.

Afterwards, pressure was applied by the Mondi Business Paper network in the form of a strict requirement for their suppliers to become certified. These requirements were negatively received and provoked conflicts between space of place and transnational space. Support for the local small projects that the MFP was sponsoring eased these conflicts, giving new rights and advantages to local actors. In the case of the MFP, as with the previous case of the Pskov Model Forest (PMF) described in Chapter 6, the implementers of the project used the programs of small grants for the development of democracy and public participation. In the MFP's example, the institutional changes in the field of public participation were more notable than in the PMF case, as more effort and resources were invested in building the capacity of local actors. Initially, a weak civil society got a powerful incentive due to the intervention of the transnational network.

7.11.1 A generalized model of a top-down network: from transnational spaces to the spaces of places: channels of transmission and agents of institutional change

From the empirical finding described in this chapter, it became possible to create a more general model of the network, in which the NGO used external donor funding for implementing sustainable forest management on the ground and the same network succeeded to a more 'natural' stage, in which the TNC became interested in FSC certification and started promoting its institutionalization.

Figure 7.6 visualizes the top-down network, in which a foundation creates a channel for transmission of global 'designs' to the space of place and a NGO is an implementer and co-designer of the network. The NGO partners with the regional state agencies, which operate top-down and create an additional channel for the transmission of global discourses about sustainable forest management into concrete practices on the ground. An NGO mediates relationships with the stakeholders and facilitates implementation of the FSC certification on the ground. In such a network, the FSC-GGN, with its principles and criteria, determines the rules and practices and simultaneously becomes an additional channel for transmitting innovations to the space of place. In such a network, however, innovations in forest management implemented on the ground may go

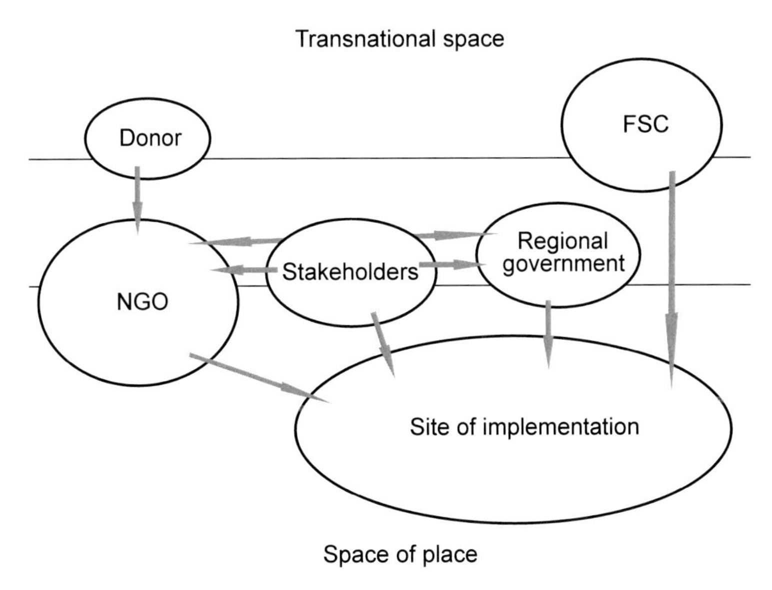

Figure 7.6. Top-down NGO network with donor funding: channels of transmission and agents of institutional change.

beyond the FSC requirements as they accumulate designs from the funding agency and initiatives coming from the stakeholders. Donor funding allows for the implementation of such innovations.

Figure 7.7 visualizes the 'follow on' network, which shows that when the foundation funding ends, the TNC, interested in certifing its subsidiaries and suppliers, takes the lead in the network.

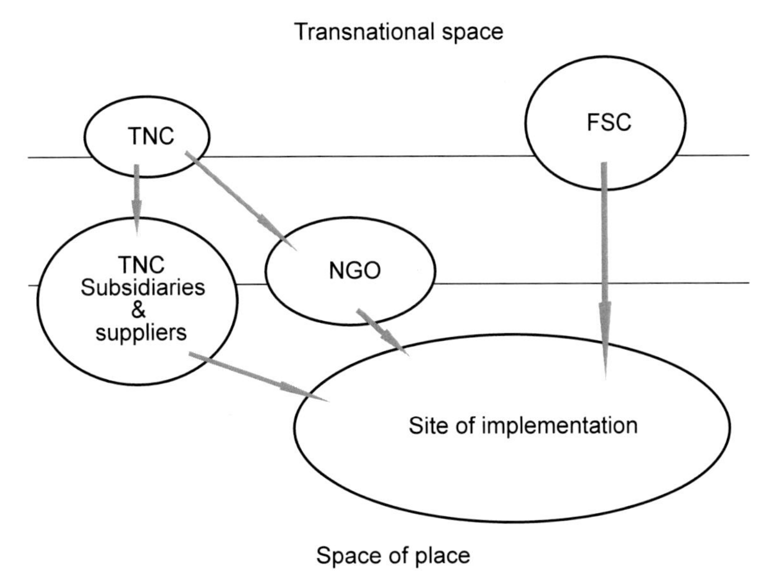

Figure 7.7. Top-down TNC-NGO network without outside funding.

When the donor funding ends, the network structure changes. The TNC influences practices in the space of place predominantly through its local subsidiaries, which have to comply with the global sustainability policies of the company and the FSC rules. The TNC fosters certification by demanding that its suppliers become certified. As the TNC requires its suppliers to become certified, it automatically becomes a facilitator of the FSC-GGN. The FSC-GGN provides a framework and creates the institutional infrastructure through which globally developed rules are transformed into local practices. The implementation and compliance with the rules are verified by the accredited certification body that interprets the standards and directs the company towards better compliance. The NGO in such a network plays the role of a facilitator and consultant in the FSC process.

7.11.2 Analysis of the case study using the GGN concept

The specificity of this case is that several supplementary networks were facilitating and reinforcing each other in the process of implementation in the MFP. In the beginning the network was the SDC foundation, which enabled Silver Taiga as well as the regional and local actors to implement the model of sustainability (Figure 7.8: nodes of design). The second phase was the NGO network, lead by the WWF/Silver Taiga. The FSC rules were essential for keeping the MFP in compliance with the global 'golden standard' of sustainable forest management. The state hierarchy of the Komi Republic helped in implementing innovations and in introducing them to the whole republic of Komi. After 2006, Mondi Business Paper, interested in FSC certification for its subsidiaries and suppliers, took the lead in promoting sustainable forest management in Priluzie and the whole of Komi. This affected the entire business network in Komi.

Speaking in terms of the GGN concept, the MFP became a site of implementation of several GGNs engaged in facilitating relationships (see Chapter 3 on facilitating relationships between GGNs). The top-down network, involved in building the MFP, represents the implementing agency for several GGNs: (a) the SDC, a foundation-GGN, channeled resources through the NGO (regional WWF office until 2002) for designing, building and disseminating the model of sustainable forest management, (b) the WWF, a GGN itself with its own agenda of designing model forests and promoting FSC certification, (c) the TNC, also a GGN, became the major agent for institutionalizing the FSC in the Komi Republic, and (d) the FSC-GGN, which was involved as certification of forest management was part and parcel of the model on the ground.

As was described in the previous chapters, the major components of the GGN; nodes of global governance design, forums of negotiation and sites of implementation, differ in various networks in terms of structure and significance. The composition of the nodes themselves can be different. In some networks, forums of negotiation are more intense and important than in others. Sites of implementation can range in their ability to be reproduced and subjected to organizational isomorphism.

Figure 7.8 visualizes nodes, forums and sites of implementation of the GGNs involved in building the MFP. The foundations GGN-SDC, the FSC, and the WWF contributed to the design of the MFP, operating mostly in transnational spaces, as well as the regional NGO, the WWF/ Silver Taiga, which played the role of a sub-node and co-designer (Figure 7.8: nodes of design).

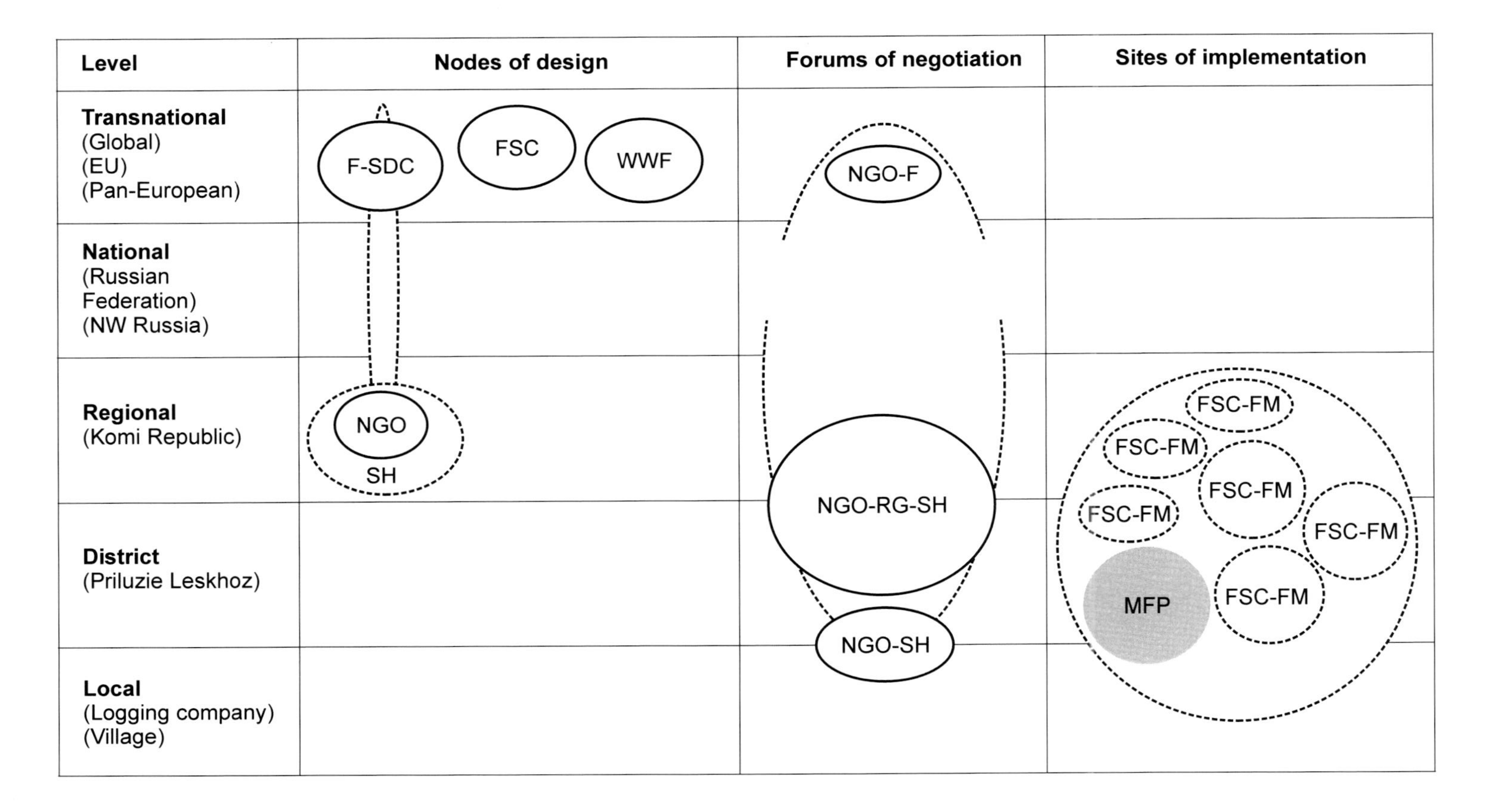

Figure 7.8. Analysis of the case study using the GGN concept.

Forums of negotiation were very important in building the MFP. Forums functioned on the transnational level in the form of steering group meetings for the preliminary design of the intervention. The granting foundation emphasized the importance of stakeholder involvement from the start of the project. Silver Taiga became the agency of stakeholder involvement, partnered with regional and local governmental agencies and organized wide stakeholder dialogues. The working group, where all regional and local stakeholders, including representatives of the site of implementation took part, represented a regional forum, contributing to the MFP's design. Therefore, the forums of negotiation at the level of the Komi Republic became an extremely important component of this particular network (Figure 7.8: forums of negotiation, NGO-Regional Government (RG)-Stakeholders (SH)). It has to be noted that forums at the level of the Russian Federation were missing;[388] that can partly explain the fact that innovations in forest management developed in the MFP did not spread beyond Komi.

In the space of place, forums of negotiation in the form of public hearings on decisions related to the leases of the forest territories were organized between businesses and local stakeholders. At that time, due to existing legislation and the rights brought by certification, local stakeholders had a real opportunity to influence leasing decisions, hence, to present their requirements to future leasers – logging companies. However, those forums turned out to be unstable and did not continue for long because of changes in the legislation. Yet, in the course of those forums, a new civil initiative was created called the Forest Council, which continued the dialogue between stakeholders at the local level. Several Forest dialogues were created in Komi and became a platform for stakeholder forums of negotiation (Figure7.8: forums of negotiation (NGO-SH)).

The lessons learned and the best practices of the MFP were successfully reproduced throughout the Komi Republic (Figure 7.8: sites of implementation). This success was aided by the participation of the state agencies in the project. Research demonstrated that co-operation with state authorities and converting them into an agency of the network itself allows mimetic organizational isomorphism. Three leskhozes were certified using the lessons learned from the MFP (Figure 7.8: sites of implementation, round circles). At later stages, as was mentioned above, the TCN (Mondi Business Paper) came to be the leading actor influencing spaces of places. In the same manner as in the PMF case, the TNC strove first for the standardization of forest management practices, both in its subsidiaries and its suppliers. Being the most important business actor in the Komi Republic, Mondi Business Paper became the key 'player' of promotion for FSC certification. The company stimulated certification for its subsidiaries and required certification from its suppliers. On account of this, the sites of implementation came to be largely determined by the end of the project. Since the appearance of the TNC as the main partner of the NGO, the latter was forced to prioritize TNC business interests and to help certify the leased territories of the company's subsidiaries. This allowed a highly mimetic isomorphism in the FSC-FM sites of implementation (Figure 7.8: sites of implementation, oval circles FSC-FM).

[388] That is why the dotted line on the Figure 7.8. is broken.

Chapter 8. Forest certification and institutional change: the case of the Northern Logging Company

8.1 Introduction

Institutional changes in local places are currently occurring not only under the influence of local, regional and national processes, but also as a result of the imposition of global rules and practices of transnational actors into localities. In this chapter I will analyze how Forest Stewardship Council (FSC) certification of the area leased by logging enterprises belonging to the North Logging Company (NLC), a subsidiary of the Russian forest holding Investlesprom (INP), has changed local practices, and what actors are involved in transferring and adjusting global rules to local contexts during the process of certification.

Prior the economic downturn of 2008, INP holding was a dynamically developing and rapidly growing transnational corporation of Russian origin. The NLC, with its numerous subsidiaries, was engaged in logging on several leased areas in the Republic of Karelia. The subsidiaries of NLC, together with their leased lands, were acquired by Investlesprom over various times, and hence, during the period of research (2006-2012) they were at different stages of certification. Forest management certification in many of these newly acquired territories started as a requirement of INP's interest in trading in environmentally sensitive European markets. INP was pressed by British buyers with company policies to purchase only certified products when dealing with large-scale deliveries of sawn timber to the UK. Thus, it happened that the local Russian logging companies affiliated with a big international holding company were forced to change their practices in order to meet relevant requirements, which derived from a global marketplace and which are localized and materialized in the specifications of an international certification system.

Contrary to other cases described in previous chapters, no outside funding was used in the FSC certification. Both environmental and social NGOs were involved as experts in the process of certification, along with other stakeholders. The Regional Nature Conservancy named SPOK, which originally begun as a student group supporting Greenpeace, was both assisting the NLC in biodiversity conservation and playing a role as a surveillance agent, monitoring the companies' progress and forcing compliance with the FSC. The Centre for Independent Social Research (CISR) also was simultaneously assisting the company in their interactions with stakeholders and ensuring the rights of the local communities within the framework of the FSC. The author of this book had a dual role in this case study, representing the CISR as a practitioner and researcher, and at the same time involved in participant observation.

In Chapter 2 the three ideal network types, e.g. top-down, bottom-up and two-fork NGO networks connecting transnational spaces with the spaces of places are described. In this chapter I analyze the interplay of actors from both transnational and local spaces related to FSC implementation and institutionalization as an example of the two-fork network. This chapter intends to explore the process of FSC certification, as a linkage between local sites of implementation and transnational spaces. As was described in previous chapters, the FSC represents NGO-driven governance generating network (GGN), producing regulations in the form of globally recognized forest management standards (see Chapter 3). Environmental and social NGOs involved in certification provide both expert and FSC quality control functions. Therefore, several NGO networks operating top-down and bottom-up contribute to linking transnational and local actors. This linkage serves both for the introduction of new practices into localities and their correction and alteration throughout the course of this process.

The company, with its chain of custody, represents a business network also constantly connecting transnational spaces where products are sold with the places where wood is harvested. Its involvement in FSC certification provides additional linkages between local and transnational actors. The major drivers for FSC certification are the buyers of sawn wood in Western Europe. The management body of the Russian holding company becomes the major interlink between players in transnational spaces and actors on the ground at the sites of implementation (Figure 8.1). There were two major parallel processes that influenced institutional change in the space of place: the development and modernization of forest operations of the holding company and the process of FSC certification.

The chapter is based on field research conducted between 2006 and 2012. Data were collected during seven research expeditions to the research area, in September-October 2006, September 2007, March 2008, February 2011, July 2011, March 2011 and May 2011. Seven additional trips were done during consulting work, in particular in August 2008, February 2009, July 2009, August 2009, February 2010 and August 2010. Participant observation was done during several FSC audits of companies belonging to the Investlesprom holding company (see Appendix A). Both semi-structured and open ended interviews were conducted during research and practitioner's trips (total #442). Interviews involved NGOs (#27), representatives of certification bodies (#16), representatives of state agencies at different levels (#27), representatives of local administrations (#32), local state forest management unit representatives (#9), representatives of Investlesprom and its subsidiaries (#33), community activists and ordinary citizens (#207), representatives of local museums (#21), teachers (#34), representatives of houses of culture (#8), and librarians (#26). Buyers of Investlesprom products were interviewed in London and Hull (#21), see Appendices A, B and C.

The chapter starts with a background of the formation of the NLC and its owner, INP (Section 8.2), describes their motivation to become FSC certified, and summarizes the chronology of events in the process of certification (Section 8.3). Subsequently, the chapter describes the specific yet powerful role that certification bodies (CBs) played in directing the company towards compliance with the FSC, and by doing so promoting the institutionalization of new rules in locales (Section 8.4). The following section analyzes the role of both transnational and local stakeholders in the process of FSC certification (Section 8.5). It highlights the important role of the buyers of certified wood in the UK in forcing the companies' engagement with FSC certification in Russia (Section 8.5.1). Special attention is given to the role of the NGOs SPOK and CISR and their specifics in assisting the company in implementing social change in the spaces of places (Section 8.5.2), followed by sections on the interaction of the company with state agencies in the process of certification (Section 8.5.3) and on the involvement and attitudes of local stakeholders and local communities (Section 8.5.4). The chapter proceeds with a generalized model for the case, which illustrates the interaction of the actors of transnational spaces and the sites of implementation of

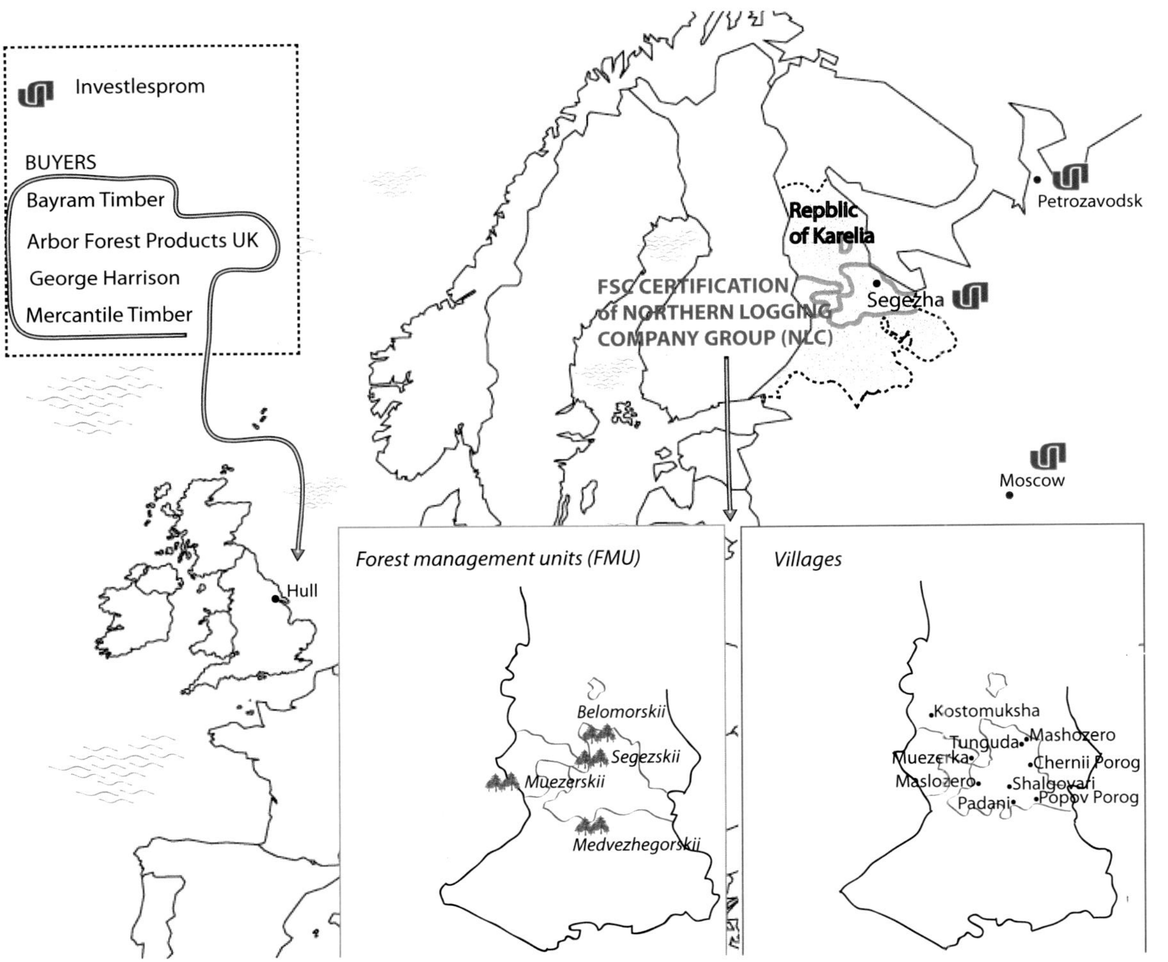

Figure 8.1. FSC certification of Northern Logging Company: map of the case study.

the FSC rules. It concludes with the analysis of the channels that translate global rules into concrete local practices at the sites of FSC implementation as well as the mechanisms of institutionalization of new rules (Section 8.6).

8.2 Context for FSC certification: hardships in the holding company formation

This section describes how the INP was formed through acquisition of multiple pulp and paper mills (PPMs) with their surrounding logging companies and leased territories. I focus on the turbulent process of formation of a vertically integrated and modernized holding, striving for ever more resources for the acquired PPMs. This process resulted in high yield pressures on the spaces of place and on local practices, and it influenced the process of forest certification.

The group declared itself as INP comparatively recently, in 2006. It formed gradually, as the result of purchasing and incorporating of more and more holdings, PPMs, saw mills and logging enterprises under the umbrella of the group. Therefore, INP has a complicated ownership and management structure. Most assets belong to the company's management and Russian private investors; 19.9% are owned by the Bank of Moscow, which is involved in investment projects in all Russian regions. The City of Moscow is involved in the ownership of the Printing and Publishing House in Moscow, as well as pulp and paper production[389] (Figure 8.2).[390]

The holding is composed of four major divisions: Forest Resources division, Wood processing division, Pulp and Paper division, and Paper Sacks division. The Forest Resources division incorporates logging enterprises in the Republic of Karelia, Archangelsk, Kirov, and Vologda regions. The saw mills are situated in the same regions, and logging companies belonging to the Forest Resources division supply them with timber. The Pulp and Paper Mills division includes three pulp and paper mills built in Soviet times, which have been partially modernized by the new owners. The Paper Sacks division incorporates two Segezha Packaging plants in Russia and 11 Segezha packaging[391] plants in Europe.

Since this study examines the certification case of the NLC group harvesting wood in Karelia, I will supply some details of its history. The NLC is closely connected with the Segezha PPM, established as early as 1939 and specializing in the manufacturing of paper bags. At its very beginning, the PPM incorporated a saw mill which was later developed into the Segezha saw mill. The Segezha PPM was privatized in 1992. After Perestroika, the businesses, as well as the state entities, were permanently reorganized, and the Segezha PPM changed its ownership and name several times.[392]

In 1999-2008, the Segezha PPM purchased many logging enterprises in order to ensure a constant resource supply (Figure 8.2). The incorporation of the logging companies together with

[389] Investlesprom Leading Russian Integrating Forestry Group, Power Point presentation of the manager, June 10, 2009.

[390] The graph represents a simplified version of Investlesprom owners, operations in the Karelia Republic and its markets in the UK.

[391] Segezha packaging plants in Europe originate from the Swedish owned Koerstens Packaging group of plants, renamed after purchasing by members of Investlestrom group.

[392] Interview with the top manager of the resource branch at Segezha PPM, October 2006, Segezha.

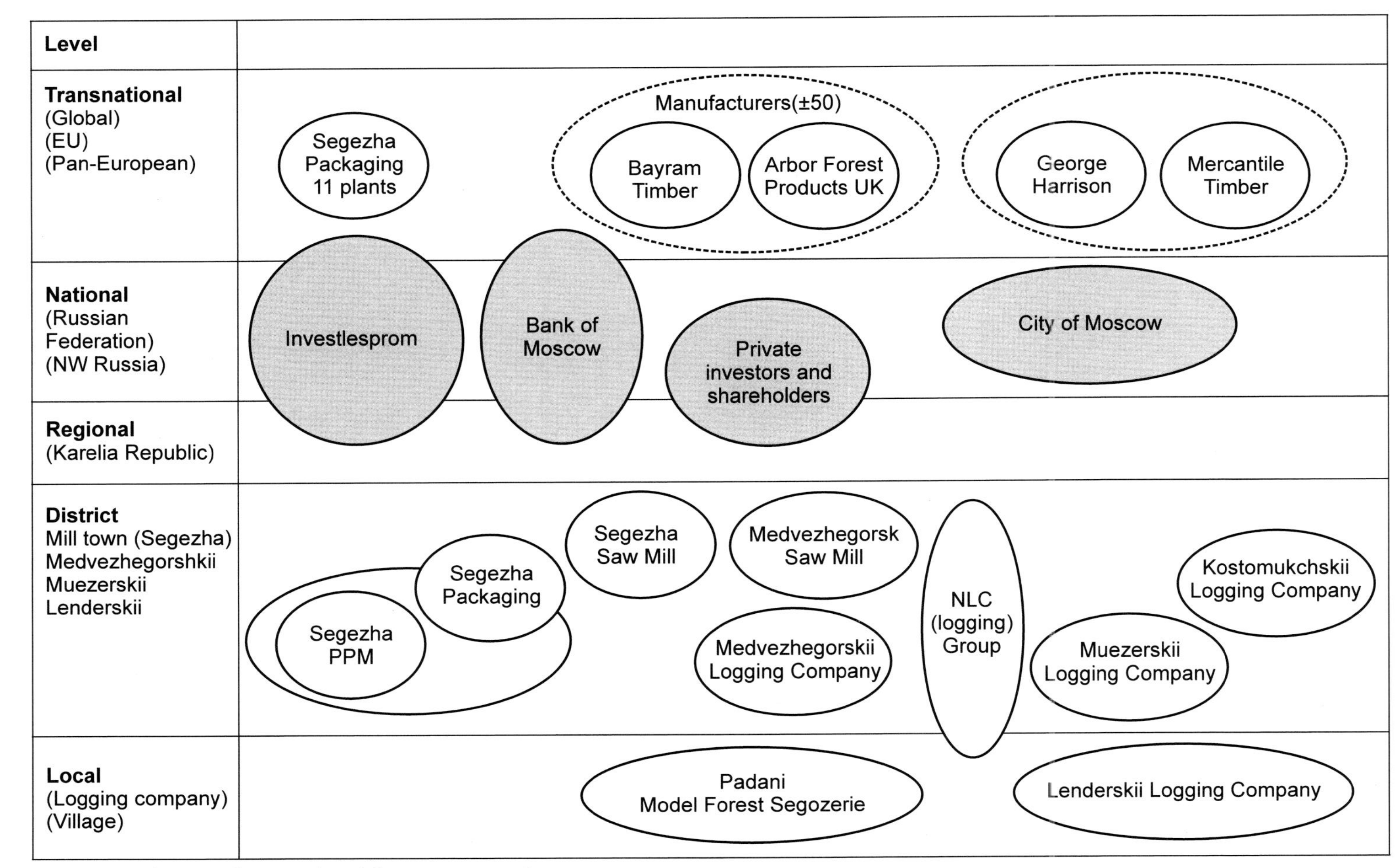

Figure 8.2. The Investlesprom (INP) business network.

their leases into the holding structure was an optimal way to provide a systematic raw material supply. The holding company bought up all available enterprises with short term forest land leases and reorganized those that would work.[393] The newly acquired companies went through bankruptcy procedures and in due course resumed their work under new names under the control of the holding company.[394] Gradually, on the expiration of their short-term lease contracts for logging, the holding company took over the lease contracts.

In 2003, to increase the efficiency of the acquired logging enterprises, the holding company established a managerial body for forest resources. As a result, the Segezha PPM received a 49-year lease contract for 1.8 million ha in its own name. In order to manage the logging process on these territories, a decision was made to detach a forest managerial entity from the PPM and retain it as a separate company. The NLC was founded on the June 1, 2007 on the basis of the bankrupt 'Segezha Les' and formed one part of this new company. It administers all former logging enterprises (Segezha, Padan, and Valdai) located in the territories of the Segezha PPM. At the same time, the Segezha PPM was an official leaseholder of forest lands. By then it had become part of INP group. The head of the forest director's office at the Segezha PPM moved to Moscow and became a top manager of the whole Forest Resource division for the holding. In this way, the management of all forest resources was outsourced from the Segezha PPM into several structures: the INP office in Moscow, the INP office in Petrozavodsk and the NLC.

Subsequently, in addition to smaller timber enterprises, the INP group began to buy up entire holding companies in the Vologda, Arkhangelsk and Kirov regions. In Karelia it continued to purchase logging companies that had their own lease contracts. New logging enterprises in Karelia began to accede to the NLC and formed a group. The Segezha PPM continued to be a forest land-lease holder for 1.8 million ha of logging operations, because it proved difficult in Russia to rearrange the leases. However, all forest management and all logging operations and related financial flows were administered by the NLC. At the time of writing (April, 2011), the NLC has incorporated the Muezerskii, Kostomukshskii, Medvezhegorskii and Lenderskii logging companies together with their leased territories (Figure 8.2). Newly purchased companies maintained their own leases, but all the decision making and management moved to the NLC. Total woodlands area in lease on which NLC group was harvesting forests by 2011 was 5.3 million ha.

With the purpose of creating a vertically structured entity, four divisions, described above, were formed, with the forest resource division being engaged in timber supply for all the PPMs as well as the holding's saw mills. This division, based in Moscow, ran core regional subdivisions, including the Karelian NLC. The strategic aim of the 'forest division' was to ensure a raw material supply, i.e. to develop a strategy that would decrease the need to purchase wood while providing mills with 'in-house' raw material. Within this strategy, the Bank of Moscow has assigned large funding amounts to be invested in the technical reorganization of the operating logging enterprises. As a result, by 2007 almost all of them have switched over to modern forwarders and harvesters.

The program of resource security also included intensification of forest exploitation by envisaging steps to be taken, such as lobbying the Russian governments to lower the cutting age to 80-100 years (instead of the 120 years according to the existing Russian legislation), an active

[393] Interview with the manager at the office of forest resource directors at Segezha PPM, October 2006, Segezha.
[394] Interview with the bankruptcy manager, office of forest resource directors, October 2006, Segezha.

reforestation and target cultivation of certain wood species, and the execution of commercial thinning, etc. To avoid forest exhaustion under conditions of intensive forest exploitation, it was necessary to take additional measures. One such measure has been to pursue sustainable forest management techniques, including forest certification.

Over the years 2006-2011, constant restructuring and reorganization in the INP took place. The INP made considerable efforts to consolidate the company, while building a vertically managed holding. In some of the newly acquired subsidiaries, the INP faced resistance from local actors to the new management. Local subsidiaries resisted FSC certification in particular, as they had no economic incentives and much additional work to do. Most of the old managers were fired and new younger ones were hired. With the modernization and the world economic crisis that hit the company severely in 2007-2009, the holding fired many people in all its subdivisions, including the logging companies. In 2009-2012 the resistance of local subsidiaries to certification lessened as the INP introduced internal economic incentives for logging companies by obligating processing plants to pay a premium for certified wood to the logging companies within the holding.[395] However, the reorganization caused constant institutional turbulence within the holding, and was painful for all the actors and made the FSC certification process, as well as maintenance of the certificate, more difficult.

8.3 Chronology in the process of forest management certification

This section explains why and how the INP approached the process of FSC certification. It highlights the major events related to the FSC certification in the different Karelian leased territories, the drawbacks of the process that led to the suspension of the FSC certificate of the Segezha PPM, and the necessity of dealing with several CBs.

The possible involvement of the Segezha PPM in the forest certification processes had been discussed internally since 2003. The necessity to get forest management certified was stipulated by the situation of environmentally and socially sensitive European markets, where this enterprise distributed its products. Around that time, sawn timber from the Segezha saw mill was being exported to the UK, Denmark, Spain, France and other European countries. In the UK, where the Segezha sawn timber was used basically in manufacturing of garden furniture, do it yourself (DIY) stores paid special attention to the legality and transparency of wood supply as well as to the social and environmental responsibility of the producers. Inquires concerning certification came first from UK buyers. Although the Segezha PPM had entered this market – as one of its managers put it – 'in the period between the waves of consumer boycotts',[396] and did not experience their effects, it became aware of them in the early 1990s and understood their ramifications. It was the main motivation behind the decision to become certified. The Segezha PPM administration associated certification with the process of forming the enterprise's strategy: 'forests should be treated not as a place where harvesting and related work is conducted, i.e. to be felled, debranched and loaded onto trucks, but rather as a resource to be attended to so as to minimize the danger to

[395] Interview with the INP top manager on forest resources, March 2010, Segezha.

[396] Interview with the Segezha PPM director, responsible for purchasing companies and raw material supply, October 2006, Segezha.

nature'.[397] Certification impacts the employees' understanding of the essence of sustainable forest management at the enterprise.

In 2005, the Segezha PPM launched the process of certification. At that time, the forest directors of the Segezha PPM consisted of a young and dynamically developing team, which started this process with enthusiasm. Based on the results of a tender conducted by the Segezha PPM, the 'SGS Qualifor' was chosen as a certification body (CB). This choice was stipulated largely by the cost of service.

In December, 2005 the enterprise underwent a pre-assessment. Preparatory work consisted of developing the enterprise's environmental policy in cooperation with the regional environmental NGOs SPOK and Greenpeace. After the FSC pre-assessment, the Segezha PPM began to prepare for the main assessment. They did not involve consultants in certification, preferring to prepare for the certification using their own specialists.

The enterprise preserved old growth (virgin) forests suggested by NGO SPOK and Greenpeace in order to comply with FSC standards. They signed a voluntarily moratorium on felling with the group of environmental NGOs.[398]

The enterprise had basic difficulties with certification standard's compliance, connected with the timely payment of salaries to employees at subsidiary logging companies. To adjust these enterprises in accordance with the requirements of certification, the administration introduced[399] 'social security' items into their budgets, and employees started to receive salaries (although minimal) regularly.[400] The main assessment was conducted in July, 2006.[401] Afterwards, the Segezha PPM was afforded an opportunity to correct all non-compliances in due time, and a significant number of them were eliminated by October, 2006.[402] Complying with the International Labor Organization (ILO) Convention requirements for accident prevention has been a difficult process for the enterprise. According to this Convention, all employees should be provided with specialized safety equipment and uniforms. High costs for such equipment were only part of the problem, because it was also necessary to persuade the workers to wear this clothing all the time while they were not used to this.[403] With respect to environmental requirements, the most difficult part was to comply with those stipulations referring to fuels and lubricants collection and waste removal to

[397] Interview with the Segezha PPM director, responsible for purchasing companies and raw material supply, October 2006, Segezha.

[398] Interview with one of the forest directors at the Segezha PPM, October 2006, Segezha, a copy of a signed voluntary moratorium.

[399] Interview with the top manager of the Forest Resource division of Investlesprom, April 2008, interview with the Investlesprom representative in Petrozavodsk, March 2008, correspondence between the company, authorities and the FSC (copies provided by the Investlesprom managers).

[400] Interview with the head of the office of forest directors at Segezha PPM, October 2006, Segezha.

[401] The author of this case study participated in the main assessment of forest management of Segezha PPM as an expert auditor, involved by the SGS Qualifor.

[402] The author of this case study in October 2006 was involved in consulting the Segezha PPM on social aspects of the FSC certification (to avoid a conflict of interest she declared that she will not work as an auditor for the Segezha PPM going forward).

[403] Participant observation during visits to the logging operations on the lease of Segezha PPM, during the field expedition in October 2006, Segezha, villages.

designated dumps. In case, they had to 'fight' against both the employees themselves and the lower management, since in Russia environmental safety issues have traditionally been disregarded.[404] In October 2006, a follow-up assessment was held, and in March, 2007 the Segezha PPM obtained a forest management certificate. In July, 2007 the enterprise successfully went through the first surveillance, and the certificate was approved.[405]

Soon after that, Investlesprom was pressured by the federal authorities in connection with inconsistencies in its activities detected during a satellite monitoring of the territories leased by the Segezha PPM. The monitoring revealed waste on the logging plots of land leased by the enterprise; uncompleted thinning in the young growths; and insufficient replanting of trees. Rosleskhoz accused the enterprise of extensive forest exploitation and has sent a letter to the FSC office in Moscow with a request to examine the situation. The Russian FSC office informed the Accreditation Service International (ASI) about the incident. This resulted in a decision to inspect both the Segezha PPM and the CB SGS-Qualifor during the next surveillance audit in May, 2008. The validity check confirmed only part of accusations. However, there were other non-compliances with the FSC criteria, and the certificate was suspended.[406]

In September, 2008 the company allocated significant resources to improving compliance with the FSC requirements. The personnel of the Segezha saw mill, together with NLC personnel, were involved in cleaning up the waste at the logging sites.[407] The INP FSC manager got personally involved in the internal audit of forest operations. In the fall of 2008 an additional audit was conducted by SGS Qualifor that showed compliance with the most of the standards. However, the SGS Qualifor surveillance report was not approved by their own office in South Africa as High Conservation Value Forests (HCVF) were not fully identified and designated.

In January 2009, the company improved its documentation and involved experts from the WWF and the CISR as trainers on HCVF.[408] In February 2009, the SGS Qualifor auditor conducted the next inspection visit and found the company in compliance with the FSC. However, the FSC Forest management certificate wasn't returned to the company until August, 2009.

The INP and NLC learned the 'lesson' from the negative experience of Segezha PPM's forest management certification. From the very start, they allocated significant resources to prepare for certification of the other newly purchased subsidiaries, such as Muezerskii LC, Medvezhegorskii LC and Kostomukshskii LC and involved both environmental and social experts in consulting activities.[409] Lenderskii LC had been purchased with an existing certification, so the goal was only to maintain the certificate.

---

[404] Participant observation, October 2006, Segezha, villages.

[405] SGS surveillance report, interview with the manager responsible for forest certification in Karelia, September 2007.

[406] Participant observation during the surveillance audit, May 2008, Segezha, villages.

[407] Interviews with the FSC Investlesprom and the FSC NLC managers in the fall of 2008, Petrozavodsk, Segezha, Moscow.

[408] The author of this case study served as a trainer in the HCVF for NLC employees.

[409] The author of the case study worked on preparation of the Muezerskii, Kostomukshskii and Medvezhegorskii LC's to meet social standards of FSC certification during summer and fall of 2008.

In 2008 the INP started to work with several certification bodies for differing reasons that will be described below. As I mentioned in the previous chapter, if national standards are not approved by the ASI, the CB adjusts the generic FSC standards to the local context. Although the FSC Principles and Criteria are the same for the whole world, the indicators and verifiers differ and can be developed by CBs in consultation with local experts and stakeholders. Therefore, CBs have much flexibility in adjusting the standards to the local context (Malets & Tysiachniouk, 2009). Even when the national FSC standard is approved and all CBs have to use it, their prioritization during the assessment continues to be different. This results in sharp consequences of practices in the spaces of place.[410]

The Lenderskii LC was purchased with the FSC certificate given by a CB of Russian origin, Europartner, but the interaction with them was short as Europartner lost its license in 2008. Most of its clients, as well as the INP, have chosen to be certified by another Russian CB, called 'OOO Forest Certification', which was accredited by the ASI in May, 2009. In July 2009 they conducted their first surveillance audit of Lendrskii LC and the certificate was maintained.

As the INP had multiple frustrations working with SGS-Qualifor, which will be described in the next section, they decided to shift to working with NEPCon, a non-profit subsidiary of Smartwood, which is a program of the Rainforest Alliance in the USA. They started working with NEPCon on new certifications and in 2010 all areas certified by SGS-Qualifor were recertified by NEPCon.

In October 2008 the pre-assessment of the Muezerskii LC had been conducted by NEPCon. However, further certifications were postponed due to the severe economic crisis. In spring 2009, the decision was made to certify the Medvezhegorskii LC first, as the UK customers demanded certification of Medvezhgorskii saw mill products. The main assessment took place in October, 2010 and the FSC certificate was obtained in January 2011.[411] At the time of writing in April 2011, the Muezerskii LC was not yet certified and Kostomukshski LC passed the main assessment. The next section will focus in more detail on the interaction between the CBs and the company, and analyze the CBs as agents of institutional change.

8.4 Certification bodies as agents of institutional change

The CBs, as agents of institutional change, have many peculiarities. They not only verify compliance with the FSC standards in the country of implementation, but they also act according to their corporate policies, norms, rules and strategies. The procedures for the FSC audit itself are set up in a way so that step by step, the company is moved towards more sustainable forest management. At first, the company to be certified undergoes a pre-assessment. At this stage, the experts involved reveal and notify the company about inadequacies, if any, to the requirements of certification. This information is confidential. Generally speaking, this determines the company's remedial actions required in order to pass the main assessment in the future. The main assessment procedure envisages a certification body's possible reclamations as presented in the report, and the company is informed about the results after the assessment is finalized. Non-compliances can be either substantial or minor and the CB can accordingly request major or minor corrective actions

[410] Participant observation during the audits in 2006-2010.

[411] Interview with an Investlesprom certification manager, May 2010, public summary, http://www.fsc.org.

(major CARs or minor CARs). In case of major incompliance, the company is deprived of a certificate. In this case, it is granted a three-month period for any needed corrections before the next inspection is scheduled. During this period, the company has to submit a plan for remedying all the non-compliances and then correct them in accordance with the approved plan. As a rule, during the next visit to the company, auditors will be able to reduce the major corrective action requests to minor ones. In this case, the company may obtain a certificate. The certificate is valid for five years, provided that the annual surveillance audits are successfully passed. In the case that any major non-compliances are revealed again, the certificate should be suspended according to the regulations of certain CBs, while other CBs tolerate continuing mistakes. If auditors perceive the company's desire to fulfill all the requirements for certification, and if they see that it is really engaged in relevant activities to do so, they can skirt this rule and use other criterion for imposing CARs for the same breach, by wording the non-compliance in a different way. Thus, informal regulatory arrangements between the CB and the company often take place. One should remark that this plays a rather significant role in this process: assessments are being implemented by people of flesh and blood and – for all their honesty and aspirations to objectiveness – they can act following commonsense reasoning rather than the rigid following of formal norms.[412]

Much depends on the degree of mutual understanding between the auditors and the company. As in any legislation, the regulations and standards contain possibilities of different readings and interpretation. In addition, the attitude of auditors depends on subjective factors, such as their education, the country where they came from, their previous experiences and engagement, and the certification body's corporate policy (Maletz & Tysiachniouk, 2009). For example, a certification body itself can decide whether to use its own specialists or external experts in assessments. The ASI periodically checks the auditors' work during the assessment process. This ASI inspection usually takes place when claims against the CB or auditors' actions are raised. Sometimes it is a regularly planned periodic audit of the auditors. If a surveillance audit of logging operations is being held in the presence of a representative of the ASI, such an assessment is conducted in a more strict way, in full compliance with the prescribed standards. No informal arrangements will take place between the company and the CB when the CB itself is being audited by the ASI. For the ASI, the best auditors are the strictest and least compromising under any circumstances.[413] Since the auditors themselves can receive major and minor CARs in the course of such ASI checks, they tend to rigorously follow the formal rules. Such 'double' inspections often result in the suspension of either a company's certificate or a CB's license.[414]

The CB in the country of FSC implementation depends on the policy and strategy of its head office. A CB is naturally interested in being competitive, to expand their market and obtain as many certification contracts as possible. At the same time, they have to be legitimate actors with a good reputation; therefore they have to maintain the good quality and honesty of the certificates given to the companies. In general, when a CB has a good reputation, they attract fewer ASI audits and their relationships with the companies and the overall certification process goes more smoothly.

[412] Participant observation in 5 audits in the period 2006-2008.

[413] Interviews with SGS-Qualifor, GFA, and NEPCon CB representatives, participant observation during the audits.

[414] The presentation given by the ASI auditor at the auditor's meeting in Moscow, August 2008.

In the situation that many certificates given by a particular CB get suspended as a result of ASI audits, or the license of the CB gets suspended in one of the countries where it operates, the CB's general operation in the global certification market is at risk. The CB, in this case, must become a much stricter certifier and improving their reputation becomes the core of their strategy.[415]

CBs are impacted by the processes that take place at FSC-International and at the FSC offices in the countries of certification. In 2007-2008, environmental NGOs led by Greenpeace International were monitoring the quality of the FSC forest management certificates around the world. They wrote several motions and a report to the FSC General Assembly of 2008 requesting improvements in the certificate's quality.[416] The FSC allocated more funding for ASI audits, and the number as well as the strictness of the audits increased significantly.[417]

In November 2008, the Russian national FSC standard was approved by the ASI, which gave clearer and stricter requirements for both the social and environmental aspects of the FSC.[418] Since November 2008, all CBs are obliged to use this standard. The standard, as well as the quality, of FSC certificates were discussed at the joint meetings of CBs with the FSC Russian National Initiative in August 2008, May 2009, March and October of 2010. CBs were encouraged to improve the quality of certification, as this issue was a concern of the Russian National Initiative. These messages from the National Initiative were taken into account by the CBs operating in Russia.[419]

8.4.1 SGS-Qualifor as a certifier

As mentioned in the beginning of this chapter, the transnational CB 'SGS-Qualifor' has become engaged in certification of areas leased by the Segezha PPM. In the early 2000's, SGS-Qualifor was one of the world-wide leaders in the field of certification and dominated the FSC certification market in Russia. It was considered one of the most credible certifiers.[420] The main office of its Forest Department is located in South Africa, with its subsidiaries being located all over the world. In 2006, when the Segezha PPM's main assessment took place, SGS-Qualifor had lost its leading position in the world,[421] but its reputation in Russia was still good.[422] However, in 2007, they became a troubled company. In 2007, their license in Poland was suspended,[423] and the number of

[415] Interview with the head of SGS-QualiforNovember 2008, Cape Town, SAR (interview was conducted at the FSC general assembly).

[416] Participant observation at the FSC General Assembly, South Africa, 2008, informal communication with ENGOs including Greenpeace, Greenpeace report and motions.

[417] Interview with the ASI auditor, at the auditor's meeting, August 2008, Moscow.

[418] The author of this case is a member of the National Initiative in Russia and participated in the development of the standard.

[419] Participant observation at CB meetings, August 2008, May 2009, informal conversations with CB's representatives.

[420] Interview with the head of the FSC office in 2002, 2004, Moscow interviews with company representatives in 2004.

[421] Interview with the FSC office representative, 2006, Moscow.

[422] Interview with the Russian Forest and Trade Network representative, 2006, St. Petersburg.

[423] FSC newsletter, 2007.

complains and suspended certificates increased around the world.[424] In 2008, SGS-Qualifor took the initiative to impose a moratorium on all new certification contracts around the world in order to improve management inside the company.[425] This was a precautionary measure, originating from the fear of losing their license. They continued to work with their existing clients under contract and conduct surveillance audits (Tysiachniouk, 2010a: 223). In 2009, the Polish license was returned to SGS-Poland but the voluntary moratorium on certifications worldwide was not completely lifted, although a number of new certifications were allowed. Taking into account its troubles, SGS-Qualifor started to be extremely cautious on giving companies certificates in order to avoid questions and complaints related to the certificate's quality.[426] The critcisms that SGS-Qualifor experienced worldwide had an impact on their clients, including those in Russia. This CB became very slow in giving certificates and issuing reports. Some clients, including the INP, preferred to work with other certifiers on new certification projects.[427] However, the Segezha PPM subsidiary continued to work with them until 2010, as changing the certifier and re-certification was costly. Although the certification process of the Segezha PPM cannot be considered a success story, the company significantly moved toward the path of sustainable forest management.

8.4.2 Directing the company toward compliance with the FSC: ups and downs

Despite the fact that SGS-Qualifor, in general, was a troubled certifier, it played a significant role as an agent of institutional change in the FSC sites of implementation, and at woodlands in the Segezha PPM lease. Since SGS-Qualifor has a small-scale mission in Russia, international experts have been drawn into the assessment processes at the woodlands leased by Segezha PPM in all audits since July 2006. Pre-assessment was conducted with the participation of an expert from Ukraine, and the main assessment was conducted by a lead auditor from Poland. Participation of international CBs has intensified the transnational character of the assessment processes and has not allowed experts from Russia become immersed in Russian-specific reality, on the one hand, but the auditors could not always fully understand the Russian context on the other hand.

After making the decision to be certified, Segezha PPM chose a strategy which increased the risk of the certificate's suspension after re-assessment. The company did not prepare well for the pre-assessment or for the main assessment; they followed the course of superficially training their own experts. The latter did not have enough experience in certification proceedings; this entailed many mistakes and failures throughout the process. Such an approach was risky, with lots of major CARs being raised during the main assessment, and with a high probability of getting the certificate suspended as a result of repeated non-compliances for the same criteria.

FSC standards contain 56 criteria. It is quite natural that auditors often meet with difficulties in checking the company's compliance with all of these 56 criteria for certification. The choice of the criteria receiving the greatest attention during the assessment depends on the priorities of

[424] informal conversations with SGS staff during audits in the Russian Far East, October 2007.

[425] FSC Russia newsletter, interview with the head of SGS- Qualifor, November 2008, Cape Town, SAR.

[426] Informal conversations at the auditor's meeting in May 2009, Moscow.

[427] Informal conversation with the head of the FSC office in Moscow, and with SGS clients in Komi, St. Petersburg oblast and Karelia.

the experts involved. Main assessment at the Segezha PPM lease area in July 2006 was conducted by an expert team headed by a lead auditor from Poland. The team also included a specialist from the 'SGS-Qualifor' office in Russia who had a rich experience of collaboration with NGOs, environmental in particular.[428] This other expert was from CISR,[429] as according to the 'Qualifor' rules, assessments must be held with participation of an expert on the social aspects of certification. Inclusion of these two persons into the auditors' team resulted in a great number of CARs concerning the social and environmental aspects of certification.[430] The company got a clear understanding of the ways for improving forest management in its lease and has actively started working on them.

As a result, in early 2007 a certificate was obtained. The first surveillance audit in Summer of 2007 was held with participation of the above-mentioned expert from the Polish office and a trainee from the Russian office of 'SGS-Qualifor'. For the PPM, the audit went easily and smoothly. This 'easy success' has apparently blunted the enterprise's vigilance and lessened its efforts aimed at improving forest management. 'We thought everything was already on track, we can just keep it as it is'.[431] This also distracted the holding from understanding that certification is in essence a continual process, and not a number of measures to be taken at one time.

As it was mentioned, the company was faced with claims from the federal state forest agency (Rosleskhoz), which informed the FSC-Russia, who in turn informed the SGS-Qualifor. As a result, the surveillance audit held in May 2008 has become a precedent because of the number and quality of its participants. Instead of an ordinary annual company assessment, it turned into a four-layered procedure that involved audits of auditors themselves along with the audit of the company. In addition to the regular surveillance audit of the Segezha PPM forest management, SGS-Qualifor was audited by the ASI. Simultaneously, an audit of the Russian SGS office by the head of the South African 'SGS-Qualifor' office took place. At the same time, the ASI auditor was taking his qualifying 'exam' to become a lead ASI auditor. The exam was given by another auditor from the ASI, who had been working in the Amazon on a regular basis, and was visiting Russia for the first time. The surveillance was attended by observers from the federal Russian state structures, representatives of the agency that wrote the complaint, representatives from the Karelia Republic state bodies and from the local forest management unit Segezhskii. The audit was also attended by an observer from CISR.[432]

Because of such a peculiar situation, i.e. multilevel mutual inspections, all participants of the assessment process were compelled to strictly adhere to formal certification rules. The audit demonstrated that not all practices stipulated by the certificate were sufficiently institutionalized and implemented. Although, in respect to certain criteria, the Segezha PPM had used a creative approach to fulfill the certification requirements, it was, at the same time, lacking consistency in complying with others. It had not managed to achieve the introduction of new practices at all

[428] Later, this auditor worked in the Russian CB Europartner and since 2008 has taken a position as a certification manager at Investlesprom.

[429] The author of this case study.

[430] Participant observation, July 2006.

[431] Interview with an Investlesprom manager responsible for the FSC certification in Karelia.

[432] By the author of this case study, therefore during this audit participant observation was conducted.

subsidiaries; for example, in the situation of designating key biotopes, in some cases they were allocated properly, in other cases they were cut or damaged by mistake.[433] Another example of poor coordination of management efforts is a case with waste on the plots that was monitored by a satellite survey. Although the Segezha PPM was given an advanced notice about the complaint from the state bodies and was informed of the location of this waste, they only partly removed the waste at logging sites adjacent to the road. The waste was found again during the surveillance audit.[434] The mentioned mechanism of an informal approach to raising CARs on other indicators to avoid acknowledging repeated non-compliances might have been activated in an ordinary inspection, which implied an assessment of the situation by auditors in terms of common sense. In this case, the auditors might have considered non-compliances as primary issues and qualified them by another criterion. However, it was impossible in the situation of 'audit of auditors',[435] therefore; major CARs were raised, as is required by the FSC in cases of repeated nonconformities. The situation was aggravated by a close intertwining of national and transnational reviewers, as this assessment was attended by representatives of the state agency who had pointed out to the FSC's auditors that the company's activities should comply with national legislation. All these circumstances led to the suspension of the certificate.[436] Such a suspension normally lasts for three months and becomes an inducement for the company to take actions for the elimination of the defects. Otherwise, the certificate may be withdrawn. If the certificate is withdrawn, the company has to start the whole process from the very beginning and pass through overall evaluation of its activities, and not simply eliminate the incompliance.

The ASI's assessment of SGS's performance during the surveillance audit in Segezha acknowledged that this CB was working professionally. However, they found nonconformities and raised three major CARs with SGS-Qualifor. One was for the HCVF forest management: 'SGS had not raised CARs related to HCVF. ASI considers that identification, evaluation and management of HCVF are not fully implemented by the company, and thus need to be addressed by SGS.[437] The ASI report was finalized and made public in August, 2008. Therefore, the Segezha PPM could expect that in the fall 2008 audit SGS would raise the issues related to HCVF designation and management. However, the company took a minimalist approach and improved only the CARs raised by SGS in May, 2008. That is why the audit in the fall of 2008 resulted in a follow-up audit in February, 2009, where issues related to the HCVF were re-assessed, and a final decision to return the FSC certificate was made only in late June, 2009.[438]

As was mentioned, NEPCon became the new CB for the Segezha PPM. It was advantageous for INP to switch to a new auditor. Firstly, NEPCon/Smartwood has a good reputation as a strict

---

[433] Report prepared by a Yale University intern who spent two months at the NLC logging operation sites helping the company to designate a HCVF.

[434] Participant observation during the audit, May 2008.

[435] From auditing experience of the author, 2006-2008.

[436] Participant observation during the audit, May 2008.

[437] ASI-Accreditation services international GmbH, Annual Surveillance of SGS for 2008, Forest Management Audit to 'Segezha Pulp and Paper Mill', Russia, date of audit: May 19-22 2008, Public Summary (available on-line).

[438] Interview with the Investlesprom manager, responsible for certification, April 2009, e-mail communication in June.

auditor on environmental issues, both around the world and in Russia. As a nonprofit organization, they informally have more legitimacy in the FSC and environmental NGO community.[439] Secondly, NEPCon does not have a policy of analyzing the previous history of CARs given by previous CBs. Except for two minor CARs that still remained open from the SGS-Qualifor audit, NEPCon did not take into account the history of CARs during their assessment in January, 2010. This means that repeated mistakes, that would have required SGS-Qualifor to suspend the certificate, in the NEPCon audit appeared as new minor CARs, for most of which the company was given a period of a whole year for improvements, despite NEPCon finding the same incompliance's on HCVF, worker's safety, and gasoline leaks as had SGS-Qualifor in May, 2008. However, this did not result in any painful consequences for the company. It is important to note that NEPCon found many incompliances that were not previously detected by SGS-Qualifor and issued a total of 32 CARs; a significant number,[440] however the certificate was maintained during the surveillance audit in November, 2010.[441] NEPCon, however, paid much less attention to social aspects of certification and consultations with the local community members as compared with SGS-Qualifor.[442]

8.5 Transnational and local stakeholders in the process of FSC certification

Both transnational and local stakeholders are essential for the successful operation of the FSC (Table 8.1). Within transnational stakeholders, INP's buyers of FSC-certified products are the major drivers of the process. They represent an important part of the chain-of-custody, linking forest management sites of implementation with transnational spaces. Therefore, this section will start with the description of how buyers of INP products were involved and how they reacted when the Segezha PPM certificate was suspended. The section will then analyze the input of environmental and social NGOs, which were in this case combining the roles of stakeholders with the roles of experts, drivers and surveillance agents. Together with the industry representatives they were striving to institutionalize biodiversity conservation in the eyes of the state agencies. The section will show peculiarities of the interaction between the company and governmental agencies related to the FSC certification. Finally, the section will focus on local communities as stakeholders in the process and on the impacts of certification on local practices in the sites of implementation.

8.5.1 Drivers of FSC certification: the buyers of INP wood products

Segezha Packaging has sales offices in 24 countries. As was explained earlier, the INP units cover the whole chain from forest to finished packaging, including forest logging operations, mills, paper production and packaging. The NLC logs in the Segezha PPM leased territory, the pulp wood goes to the Segezha PPM and its affiliate Segezha Packaging in the town of Segezha in Karelia. Segezha Packaging produces paper, while paper sacks for the international market are made by Segezha

[439] Informal conversations with NGOs and FSC representatives.

[440] Public summary report for forest management audit of Segezha PPM 22-29 January 2010, http://www.info.fsc.org, viewed May 10, 2010.

[441] Informal conversation with INP FSC manager, December 2010, Moscow.

[442] Participant observation at SGS-Qualifor audit in May, 2008 and at NEPCon audit January, 2010.

Table 8.1. Major stakeholders in the spaces of place and transnational spaces.

Spaces	Stakeholders	Type	Interest
In the space of place	Rosleskhoz	GO	National forest management authority. Generally supports FSC certification. Did not develop regulatory norms for convention on biodiversity which became a barrier for FSC implementation in Karelia.
	Karelia Republic forest agency-Karelleskhoz	GO	Supervises the Leskhoz system in Karelia. Would like to act strictly according to Russian Forest Code of 2007, formally supports certification, but in practice does not believe in its values and acts as a barrier for FSC implementation.
	Leskhoz (since 2008 lesnichestvo)	GO	Manages forest on local level[1]. Leases land to the logging companies, monitors logging operations. Interested in responsible forest users that are in compliance with legislation, are engaged in thinning in young forests and in reforestation. Is not interested in compromising with FSC standards implementation.
	Karelia governments at regional centers	GO	In different places different reactions, mostly positive.
	Forest settlement and village administrations	GO	Supported the FSC and new practices to be implemented, however, does not believe in substantial improvements in community life due to FSC.
	NLC	Commercial company	Integrates logging subsidiaries in Karelia, has to maintain the old certificate and certify new territories. Has to implement the FSC standards. Does not have economic incentives and interests to implement the FSC, but has to comply with the INP requirement to be certified.
	Local subsidiaries of NLC	Business	Not interested in FSC certification, but has to comply with the NLC and INP requirements.
	SPOK	NGO	Interested in preserving all the virgin forests in Karelia, uses the FSC as a mechanism for implementing its mission, occasionally uses radical tools to promote sustainable forestry in Karelia.
	CISR	NGO	Interested in research on FSC certification, in strengthening the FSC scheme in Russia and beyond, implementing FSC social standards, development of methodologies that would help companies to implement FSC social standards, developing consulting services, in building the Model Forest Segozerie.
	Local libraries, house of culture, museum, schools	State institutions	Village intelligentsia, small grant recipients, interested in implementing their own ideas and initiatives, interested in funding to support their work, in revitalizing Karelian culture and traditional celebrations in traditional villages.

Spaces	Stakeholders	Type	Interest
	Community	Local villagers	Eagar to have an affordably priced fire wood supply, need sawn wood supply, in some villages concerned by roads and dust from the machines. Concerned about fewer jobs. Does not welcome the incorporation of Segezha PPM into INP and incorporation of small logging companies into NLC.
	Model Forest Segozerie	NGO	Interested to survive, to involve stakeholders in the process, to implement projects.
Transnational spaces	Investlesprom (INP)	Commercial company	Economic interest in developing integrated profitable business that includes Russian resource suppliers and buyers on the global scale. Interested in changing Russian legislation in a way that intensive forest management can be implemented. Interested in consistency, effectiveness and standardization of its business operations, interested in maintaining good relationships with Russian and international NGOs.
	CB SGS-Qualifor	CB, business	Expanding its services in Russia and worldwide, resolving issues and trends with the quality of FSC certificates worldwide, maintaining a good reputation worldwide.
	CB NEPCon/ Smartwood,program of the Rainforest Alliance	Not-for profit CB	Maintaining its services in Russia on a high level, to out-compete SGS on the global scale, to maintain good quality of FSC certificates.
	Buyers of wood products in the UK	Commercial companies	Both distributors and manufacturers demand FSC certification from Russian suppliers, however, in the case of short supply, some of them agreed to buy non-certified wood.
	WWF forest program	NGO	Promote sustainable forestry in the form of FSC certification, promote the FSC in both external and internal Russian markets, HCVF designation, protection and monitoring conducted by companies inthe right way.
	Greenpeace	NGO	To preserve all ancient (old growth, virgin) forests around the world and in Russia, to test Russian legislation implementation in Karelia.
	FSC International	NGO	To certify as many forests as possible, good quality certificates, interested that companies choose the FSC scheme.

[1] Only until 2008, after 2008 all reforestation, thinning and other forest management operations become the responsibility of the companies that lease the land, according to the new forest code.

Packaging plants in Europe (the former Korsnas packaging plant that was acquired by INP in 2006). The INP is considered the second largest producer of sacks in the world after Mondi Business paper. Five large Swiss, French and Mexican cement manufacturers are the major customers of Segezha Packaging. Each year they send questionnaires to the Dublin office of Segezha Packaging, requesting information on all the certifications that they have. The customers are interested in the traceability of wood supply and the FSC gives some advantage to the company,[443] according to the opinion of the top manager at the Dublin office. According to the INP Forest Resources Division's top manager, the FSC forest management certification is not required by customers buying sacks from Segezha Packaging, therefore the suspension of the Segesha PPM certificate did not play any role in these markets. However, the suspension of the FSC certificate had a significant impact on the sawn wood market and sales from the Segezha Saw Mill.[444]

Sawn wood is sold in the UK, the Netherlands, France, and Belgium. The UK buyers are established customers of sawn wood from the Russian Northwestern saw mills. They worked with them during Soviet times, when they became privatized during the Perestroika years and when, in 2006, they became part of INP. Between the late 1990's and early 2000's the demand for certified wood took hold and customers started asking Russian suppliers to get certified. In the period between 2004-2009 it became very difficult to sell timber that was not certified. One of the UK distributors said: 'The FSC established itself as a wholly grail ... the FSC is an alternate benchmark, the golden standard', although the PEFS is also acceptable, more and more customers are asking for FSC.[445]

Investlesprom owns five saw mills, and only two, including the Segezha saw mill – affected by the FSC certificate suspension, are certified. 'It is very bad for the company that the certification process is frozen because of the crisis'.[446] Because of old and stable connections, good quality sawn wood and appropriate sizes, INP still managed to sell non-certified timber. However, several customers were lost in the UK, the Netherlands and Belgium. Between 20-25% of Investlesprom's sawn wood goes to Egypt, where there are no requirements to certify forest management. But the prices for sawn wood sent there are significantly lower than in Europe. The most severe losses for INP were when the Heuvelman Ibis company in the Netherlands stopped buying sawn wood that it had used for construction of piers, sluices and shoring walls, because it was not certified.[447]

The economic downturn in 2007-2009 had a significant impact on markets, including the FSC niche market. With the mortgage crisis in the USA in 2007, the demand for wood used in construction decreased dramatically. The products from manufacturers that were previously sold to the USA stayed in Europe. Suppliers from Finland and Sweden, who are all FSC or PEFC certified, started to approach INP customers in the UK and created stronger competition between suppliers for construction project materials. Priority was given to those who were certified, although the quality of wood did not differ very much.[448] However, despite the crisis, the demand for garden

[443] Interview with the top manager of Segezha Packaging, Dublin office, October 2008, Moscow.

[444] Interview with the top manager of INP forest resource division, October 2008, May 2009, Moscow.

[445] Interview with the managing director, George Harrison Ltd, Hull 2009, UK.

[446] Interview with the Investlesprom top sales manager, April 2009, Moscow.

[447] Interview with the Investlesprom top sales manager, April 2009, Moscow.

[448] Interview with the Investlesprom top manager, head of Forest Resource Division, October 2008, Moscow.

sheds, outdoor decking, garden furniture and other DIY products continued to be strong in 2009, although merchants had fears about possible market declines in the winter of 2009.[449] Because of the crisis, saw mills were not operating for several months in the winter of 2009 and did not produce enough timber to supply the customers. Therefore, in June 2009 there was shortage of timber supply to the distributors. To some extent this facilitated sales of non-certified sawn wood, as buyers desperately needed wood. Therefore, all available wood from INP was successfully sold. For example, a distributor who normally bought only certified wood agreed to purchase a cargo from the Medvezhegorskii saw mill, even though the forest management operations there were at an early stage of certification.[450] However, since they could not accept the cargo at their regular terminal at the Hull harbor, the cargo went to another port and the wood was kept separate in order to avoid any questions from regular customers, such as B@Q, Home Base, or Travis Perkins, to keep the chain of custody pure.[451] Customers acknowledged that this opportunity for non-certified wood to get into the UK market was temporary.[452]

As was mentioned earlier, in the UK the major drivers for certification are DIY chain stores. Large ones, such as B@Q, or Home Base never accept non-certified wood. Smaller ones prefer certified wood, but when demand is high, they will buy non-certified wood. Another driver is governmental purchasing policy that accepts only FSC or PEFS certified wood. 'Joinery manufacturers and builders have pressure from local governmental authorities that timber related products should come with FSC or PEFC certification, should be from well managed forests, not from illegal sources'.[453] Originally, the UK government was against buying PEFS wood as its social standards are lower, but under the lobbying efforts of Finnish and Swedish suppliers, they agreed to accept PEFC wood as well.[454] The construction for the Olympic Games in London in 2011 is based on all certified wood and suppliers are obliged to pay a deposit of approximately 60,000 Euros. If any violations occur and non-certified wood gets into the Olympic's supply chain, the deposit will not be returned to the company.[455] The Olympics increased the incentives for both the distributors and manufacturers to filter out non-certified suppliers. Companies listed on the London stock exchange, such as Travis Perkins, are under close scrutiny from Greenpeace and other NGOs, and their shareholders, such as pension funds, are very sensitive to the company's reputation. This drives them to buy predominantly FSC wood.[456] Trade associations also play significant roles in promoting certification. If a manufacturer belongs to a trade association, it has to comply with the association's rules to buy only certified wood. This prevents manufacturers from making case by case decisions on purchasing certified or non-certified timber.[457]

---

[449] Interview with the managing director, George Harrison Ltd, Hull 2009, UK.

[450] The certificate was received only in January, 2011.

[451] Interview with the managing director, George Harrison Ltd, Hull, 2009, UK.

[452] Interview with the director, Arka Merchants, Ltd Sky Business Centers, June 2009, Hull, UK.

[453] Interview with the owner of Merchantile Timber Co, LTD, Timber and Plywood Importers, June 2009, Hull, UK.

[454] Purchasing manager, ArborForest Products ltd, June 2009, Hull, UK.

[455] Purchasing manager, ArborForest Products ltd, June 2009, Hull, UK.

[456] Interview with the managing director, George Harrison Ltd., Hull 2009, UK.

[457] Interview with the Investlesprom top sales manager, April, 2009, Moscow.

In the UK, most who buy from INP are in central England, in Hull. The company employed an intermediary person who was involved in the Russian markets for many years, working for a Russian owned distributor to the UK: Tsar Timber. This person became responsible for matching timber from the INP saw mills with UK customer demands. He was selling timber through two major distributors: George Harrison and Mercantile Timber, directly to a large number[458] of manufacturers (Figure 8.2). With such an organization, INP works without any agents or independent merchant's facilitating trades. The INP managers developed a sales strategy, in which certain saw mills are allocated to certain distributors and manufacturers. This allows them to be more customer driven, develop a brand and avoid price competition with one another. The Segezha saw mill, for example, supplies timber predominantly to Merchantile Timber, while the Onega saw mill supplies George Harrison.[459] The Segezha saw mill is considered within the UK by Merchantile Timber to be one of the top three suppliers in terms of quality of sawn wood. Therefore, if it holds the FSC certificate without suspension, it can supply a good share of the UK market.

Saw mills of INP, as do other Russian saw mills, have several disadvantages against Scandinavian mills, which prevent them from winning a greater market share in the UK. The first disadvantage is that they desperately need cash right away, and can not wait even for 30 days, which is common in trading with the UK.[460] Next, they do not have their own receiving terminals and therefore cannot develop 'markets on the ground'. Therefore, it is harder for them to be 'service driven' and penetrate the UK markets with greater depth. UK customers have to buy the whole cargo at once, while they often would prefer smaller loads with certain specifications. In some cases, UK customers are interested in buying specific lengths and the Russian saw mills do not produce to such specifications.[461] Above all, the biggest disadvantage is that only the Onega saw mill provided FSC certified timber, while the three other mills were not certified and the Segezha saw mill could not sell certified timber for almost a year because of the forest management certificate suspension.

The suspension of the FSC certificate at the Segezha PPM could have been tragic for the company, in the case the certificate had not been returned. The loss of the certificate could entail the loss of a good half of its clients, as over 50% of the company's sawn wood production was exported to socially and environmentally sensitive European markets.[462]

The Segezha saw mill produces 75,000 m^3 of pine sawn wood annually and 40,000 m^3 are sold to Western Europe; if the buyers stop buying, it would be a significant loss. If the UK customers rejected Segezha's sawn wood, this could even result in the closure of the Segezha saw mill, as it would become non-profitable.[463]

In 2009, in order to avoid damage to the reputation of the Segezha saw mill, major customers, especially the distributor Mercantile Timber, were keeping quiet about the fact that the FSC

[458] Around 50, on Figure 8.2 are Bayram Timber and Arbor Forest Products UK, I visited these companies and conducted interviews there.

[459] Interview with the director, Arka Merchants, Ltd Sky Business Centers, June, 2009, Hull, UK.

[460] Interview with the director, Arka Merchants, Ltd Sky Business Centers, June, 2009, Hull, UK.

[461] A paper developed by the Investlesprom representative in the UK, to the attention of Investlesprom managers in Russia, suggesting investment in a UK terminal, August, 2008.

[462] Interview with the certification manager of Investlesprom, April, 2009, Moscow.

[463] Interview with the Investlesprom top manager, head of Forest Resource Division, May, 2009, Moscow.

certificate was suspended and exhibited a willingness to wait for it to be renewed.[464] It took a long time for the partners to build the market, to match demand and supply, and to establish both formal and informal connections. None of the market partners was interested in destroying this long term trade relationship that had survived from Soviet times though Perestroika and the economic downturn. However, this willingness to wait for the certificate to be returned could not last for a time:

> *As an importer we supply a wide range of manufacturers. Slowly but surely they push for certified goods of FSC origin. Every saw mill without a certificate does not have a future. ... We were able to develop market share. With the loss of certification we lost our marketshare ... Because we can not provide the seal of the FSC, customers will not buy from us, they will look for alternative sources and we can lose our business'.*[465]

In the spring and summer of 2009, according to my informant, the monthly loss to INP originating from the FSC certificate's suspension was approximately 150,000 UK Pounds from Mercantile Timber alone. The assessment of losses of INP itself was approximately 200,000 Euros total (the estimate of the loss was lower because some timber that usually is sold to the UK was sold to Egypt and Turkey, and did not require certification).[466]

8.5.2 Environmental NGOs as agents of institutional change in localities

The involvement of NGOs in the processes of institutionalization of new practices in the course of certification may vary, from auxiliary to key roles. Every certification involves some interaction with the FSC International and FSC-Russia. Thus, the FSC certification manager at INP took an active role in the FSC National Initiative, served as a trainer for companies and auditors and in 2010 was elected to the FSC-Russia Board of Directors. Almost all the major holdings partner with the WWF in order to legitimize their activities. Compared to the other TNC's, the collaboration of the INP with the WWF was weaker and limited to consultations on HCVF and membership in the Global Forest and Trade Network (GFTN) that operates under the WWF (Figure 8.3).

However, expert NGOs played a key role in INP's certification. They performed dual functions: cooperative on one hand, and (potentially) 'supervisory-punitive', on the other. The NGO SPOK promoted the company's compliance with the environmental components of certification, while the CISR focused on social components. As we will see, both environmental and social NGO networks contribute to linking transnational spaces with the sites of implementation in the spaces of place and in institutionalization of new practices.

Since the early 1990s, Greenpeace has been actively working in Karelia, implementing one of its missions, i.e. protection of old growth forests against logging. At that time, in Karelia

---

[464] Interview with the owner of Merchantile Timber Co, LTD, Timber and Plywood Importers, June, 2009, Hull, UK.

[465] Interview with the owner of Merchantile Timber Co, LTD, Timber and Plywood Importers, June, 2009, Hull, UK.

[466] Interview with INP certification manager, May 2010, Moscow.

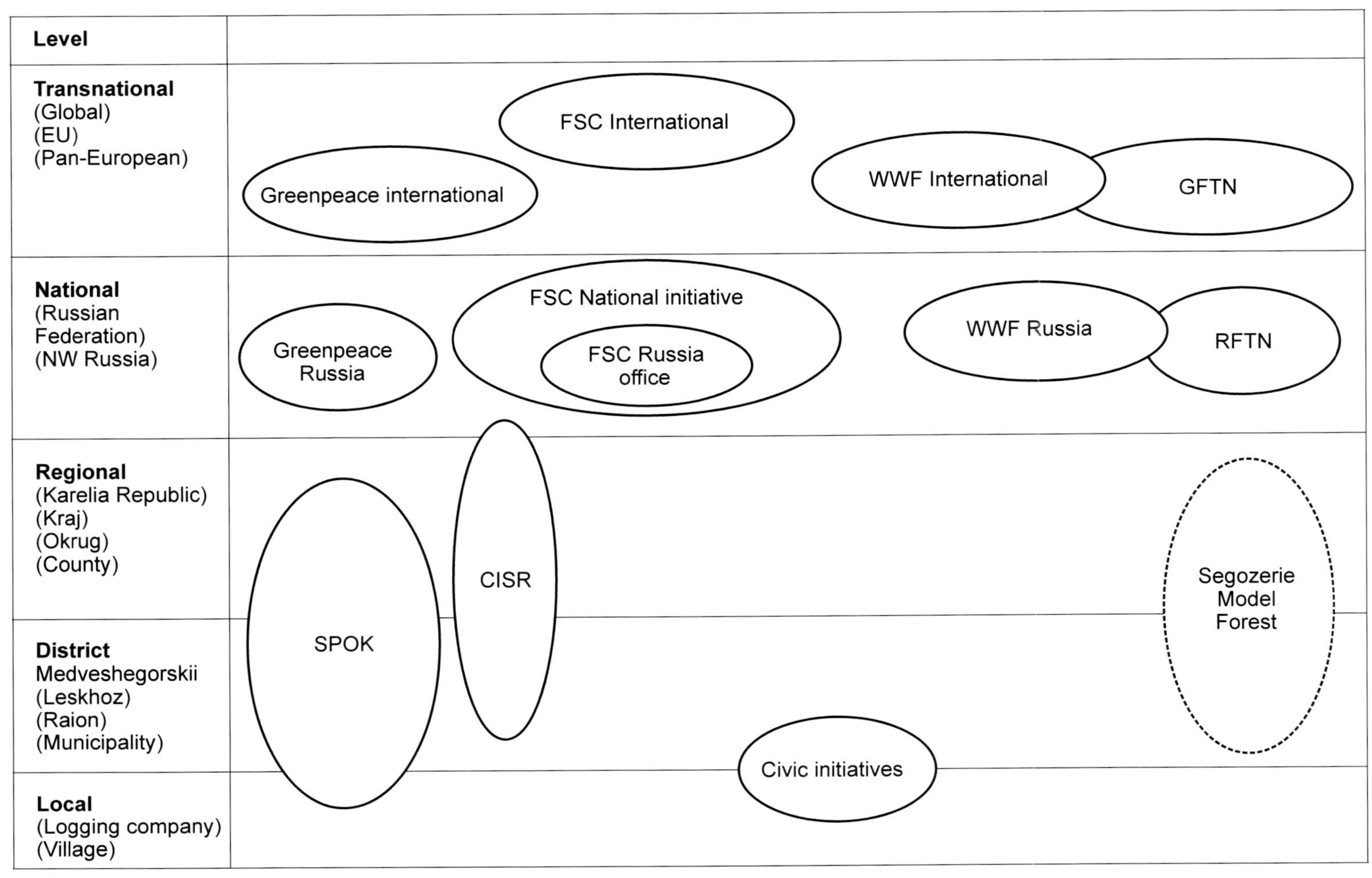

Figure 8.3. NGOs involved in the certification of territories leased by companies belonging to Investlesprom.

there was active cutting of forests, which according to Greenpeace's classification referred to the category of old-growth or intact forest landscapes, valuable for nature conservation. One should remark that it was Greenpeace and the NGOs belonging to the Forest Club[467] that initially applied the very notion of old-growth in relation to boreal forests in the North-West of Russia (Tysiachniouk, 2006a). For their conservation, it was necessary to conduct an enormous effort to institutionalize this concept, which at that time was recognized by neither businesses nor state institutions. Greenpeace, together with other NGOs, organized boycotts in Europe against the companies that bought Karelian timber.[468] By a joint effort, the NGOs mapped the Russian intact forest landscape. Fostered by Greenpeace, the regional NGO SPOK, located in Karelia, became a partner in promoting conservation of intact forests (Figure 8.3). Subsequently, along with confrontational actions, Greenpeace together with its partners relied on a more consensual based means for preserving old-growth forests (see Chapter 5). Forest certification has become an efficient mechanism for promoting activities aimed at preserving HCVF (Tysiachniouk, 2010b). That is why SPOK became involved in designating HCVFs for companies, leasing forests in Karelia as an expert consulting NGO.

For the holding company, it was a strategic partnership: the managers of the company preferred to have radical NGOs as partners, rather than as opposition. This particularly concerned SPOK, as by that time, this NGO had developed into a strong regional actor having a wide range of possibilities to influence businesses, which in turn had to take SPOK's interests into consideration. For the holding company, mobilization of SPOK as an expert has become a 'path of least effort'. Otherwise, there was a high probability that if any other expert, from the Karelian academy, for example, with its own priorities in nature preservation issues, played a role in designating HCVF, the radical SPOK would have sided with the opposition.

Effectiveness of the regional NGO SPOK at local levels and the high potential of it as an agent of institutional change, to a large extent, results from its deep connections with transnational actors: involvement in transnational NGO networks, such as Greenpeace and the Taiga Rescue Network, and also from its interaction with other networks. The Taiga Rescue Network is engaged worldwide in boreal forest issues, uniting many relevant western NGOs (Figure 8.3). This organization can, at any point, involve western socially and environmentally sensitive consumers in articulating the encountered problems.

The 'GAP-analysis' project[469] is an example of a very different project: grounded network uniting, alongside other NGOs, scientists and actors from other sectors. One of the purposes of this project is the creation of the 'Green belt of Fennoscandia', which creates a system of specially protected areas in Finland, Sweden, Norway and Russia. It attempts to prove that the Russian forests in the Karelian borderland are of particular interest and has become an object for study. SPOK conducted activities aimed to preserve old growth intact forests, working with the support

[467] The Forest Club belongs to Greenpeace, the Social Ecological Union, and the Center for Biodiversity Conservation (see Chapter 5).

[468] See the case study, described in Chapter 5.

[469] GAP-analysis – the analysis of biodiversity elements distribution contributes to long-term management through identifying and conserving its sensitive and representative elements in a community. In our case, it is the project on biodiversity maintenance in the European North through creation of PAs networks.

of respected international networking organizations and projects, i.e. with actors of transnational spaces. Such assistance raised its actions to a new level and increased its potential as an agent of institutional change. This greater potential allowed the organization to notably influence not only businesses, but even state agencies, as illustrated below.

For implementation of the FSC Principle 6 and Principle 9, holdings in Russia undergoing certification usually get NGOs involved as experts. These Principles regard protection of key biotopes, rare and endangered species, and maintenance of HCVF. One should mention, however, that different NGOs unequivocally interpret indicators to FSC criteria. Each of these organizations focuses on the component which has much in common with its own organizational views on what is a HCVF and their own understanding of nature conservation planning. That is why the INP, working in Karelia with the NGO SPOK, had to first face the task to designate and preserve intact forest landscapes. SPOK approached this issue in two ways: first, it demanded that the holding company sign a voluntary moratorium on the felling of old growth trees, using the general Greenpeace maps; second, it involved an expert for a field study to designate ecosystems of critical conservation value, and on the sites designated for logging, the preservation of key biotopes. In this way, SPOK set the stage and the INP had to implement it.

Although in this situation SPOK acted as the company's expert in forest management certification and as the FSC-Russia authorized consultant, conservation of intact forest landscapes (old growth forests) remained its mission and priority. Therefore, in many cases SPOK acts as a radical NGO, pressuring the company and not as a consultant working on the company's behalf.[470] Thus, the holding company itself had to deal with the other aspects of ecological requirements not concerning intact forest landscapes. For example, within the contract with the holding company, NGO SPOK, as an expert on certification familiar with modern GIS-technologies, could have executed mapping and described all HCVF types as required by certification. However, the company seemed to be too anxious about its intact forests and was interested only in coming to a compromise with NGO SPOK on that matter. It postponed appropriate mapping and reserved that for other contracts to be done when its financial situation would be better. Because of that, it repeatedly received CARs from the CBs.

As was descried above, the Segezha PPM became the first to obtain a forest management certificate in the NLC's territories of operation. During the first two years of certification in 2006-2007, the partnership with SPOK was developing rather smoothly, as not many intact forest landscapes were revealed in these territories and there were no problems with their designation. The Segezha PPM maintained contracts with SPOK and cooperation went smoothly.

The question of the designation of key biotopes became of concern for the company and, in 2007, the company, in cooperation with SPOK, developed the 'Field manual of key biotope designation in Central Karelia'. For both company managers and harvester-operators this manual, containing practical recommendations, became a useful tool on biotopes allocation. SPOK conducted training sessions for managers of the company. Simultaneously, as stipulated by the contract with the Segezha PPM, SPOK continued examining their territories for further designation of special value ecosystems. Forest sites with key biotopes are usually located in rocky areas or places hardly accessible by harvesters. There were no disagreements between SPOK and the company on the

[470] Interviews with the SPOK president in 2006, 2007, 2008.

exclusion of these sites from felling, since the company itself was not interested in conducting work in such hard-to-get-to places.[471] However, the company and SPOK sometimes had disagreements about the processes of designation and management of key biotopes. Thus, despite the trainings conducted and existence of the field manual, NLC apparently did not have enough managerial and organizational resources to control the key biotopes allocation and protection properly.[472]

In June 2008, during a training workshop organized for the company's staff, SPOK discovered a set of defects in designating key biotopes. Some of them were damaged during logging, others not allocated properly. SPOK's staff was strongly disappointed about such poor performance by the company. Yet before this, discrepancies in biotopes management were revealed during the surveillance audit held in May, 2008. The difficulties related to designation of key biotopes continued in 2009-2010. It was hard to implement new practices, even with the company's consent and the will to do so. The problem was insufficient organizational resources to introduce new practices at the level of each logging team. Institutionalization of practices assumes considerable additional work, creation of motivation for its performance and permanent internal controls with penalties for non-compliance.

In the process of further inspections of leased areas, the problems with valuable ecosystems became more acute. SPOK found a new old-growth forest territory near the village of Maslozero (in the Padansky forest management unit). The company intended to organize felling on this territory and had constructed a road. For both parties, the preservation of these territories became a topic of hard negotiations, which at once converted SPOK from an 'ideal partner' to 'acting opposition'. However, during the disputes and long discussions a reasonable compromise emerged.

As the company expanded its territories, more and more 'arguable' plots of land appeared. As it was mentioned above, the NLC operated on the leased lands of logging companies acquired by the INP. Forest management on these territories, as the latter became affiliated with the company, was becoming involved in the process of certification. Thus, the holding company acquired a logging enterprise in the Muezerskii region with an intact forest landscape. For many years SPOK fought on this point with the enterprise. After the holding company took possession of these territories, it also became SPOK's opponent on the issue. In this dispute, leaseholders argued that in Soviet times these forests underwent tapping, and consequently were doomed to die. SPOK opposed this, referring to the remoteness of events: if tapping somehow had damaged the forest, it would have died out by now.

Another example are the territories leased in the trans-Onega region (Zaonezhie) that have a special microclimate; they are geographically close to the Kizhi Island, which contains valuable historical and natural objects. The forests in these territories are referred to as HCVF. Some of these territories were to be designated as a regional Nature Reserve during state territorial planning. In 2010, some of these territories became a new arena for contestation and conflict between SPOK and the INP, as the company by a mistake – a lack of coordination between subsidiaries,[473] started logging forests in the area.[474] Finally, when the company declared that they will not cut forests

[471] Interview with the Segezha PPM director responsible for logging operations, October 2006, Segezha.

[472] Interview with the INP manager on intensification, November 2007, Moscow.

[473] Participant observation in Zaonezie, August 2010, villages in Zaonezie.

[474] In September, 2010, at the time when this case study was being written, the conflict between INP and SPOK was not yet resolved.

on planned reserves, the conflict was resolved which allowed the Megvezhegorskii LC to receive FSC certification in January, 2011.

These are only a few of the many complicated negotiations which drove SPOK's transformation from a company consultant to a radical NGO, ready to fight for the conservation of intact forest landscapes. However, when compromises were reached, the parties became partners once again.

For the past few years, such expert-consulting work has become one of SPOK's sources of financial stability. This was especially important under the conditions of curtailed financial support of Russian NGOs by foreign funders. However, one should note that to some extent this changed only the former radical image of the organization. In its current discussions with the company, SPOK succeeds in upholding its principle position as an uncompromising defender of forests. However, its role as a company adviser has somewhat moderated the organization's radicalism.

For the institutionalization of nature conservation practices, Greenpeace, in cooperation with the NGO SPOK, carried out an analysis of current legislation in special forums and on the internet. Through this activity, these organizations familiarize the public with novelties in the legislation and comment on these laws. To check how well the laws are implemented or, more precisely, to prove that in practice they do not work as intended and need improvement, Greenpeace, together with SPOK, carried out inspection raids in the Karelian forest areas. Their reports on detected infringements were sent to the Procurator's Office of the Republic of Karelia. Although these measures were aimed, first of all, at cooperating with state structures and improving the legislation, representatives of business may be affected as well if these infringements on their leased territories are revealed. This caused additional tension in the relationship between the company and SPOK. However, there are also commonalities where the business sector and the third sector act together in full consent, e.g. regarding the legislative enactment of practices of biodiversity preservation on the logging plots. Despite the fact that Russia ratified the Convention on Biological Diversity many years ago, these practices were not institutionalized in Russia until 2009. On the contrary, necessary measures for biodiversity conservation as required by forest certification contradict current Russian legislation. Companies complying with the FSC were often penalized by state forest management units. This especially concerned the Republic of Karelia, as certified companies in other regions had already found relevant opportunities to compromise with current legislation. Actually, the Republic of Karelia has remained the only region where authorities took a conservative standpoint and refused to meet international requirements. Therefore, both SPOK and the Segezha PPM were interested in the institutionalization of practices of biodiversity conservation and often jointly participated in different workshops and other relevant events.

As I showed, SPOK played not only an auxiliary role of expert-consultant, but also used the opportunity of transnational spaces to bring social change into the spaces of place. It acted an agent of institutional change as far as intact forest conservation is concerned, both in partnership with the holding company and independently. One should remark that despite all the difficulties of this complicated partnership between two such different agents – a multinational holding and a radical NGO – and perhaps even owing to the difficulties of this interaction, the achieved results proved to be more than the sum of the individual efforts. One could assert that there can be a certain cumulative effect in such partnerships.

8.5.3 Socially-oriented NGOs

Cooperation with the CISR developed somewhat differently compared to cooperation with SPOK, although the CISR participants also played multiple roles in the FSC process. Starting as an auditor, the CISR representative switched to an expert-consultant and a stakeholder, combining all these roles with their research mission. The same researcher is a member of the FSC International, part of the Russian national initiative and, since 2010, serves on the board of directors in the FSC Russia. All these multiple affiliations and functions facilitated the role of being an institutional entrepreneur, connecting transnational and local actors.

The CISR specialist was involved in the main certification assessment held in 2006.[475] After the assessment, in the same year, this expert agreed with the company to conduct research at both the Segezha PPM and in local communities adjacent to the leased territories. The enterprise's management system, as well as the effects of post-Soviet enterprise reorganization on the local population, became a central subject for the study.[476] At the time of the research, the company was not paying the CISR for its consultations, because of the conflict of interest. After a year, the former auditor declared that they had stopped participating as an FSC assessor of the company and became a leader of the CISR research team that combined their study with the consulting.

When visiting villages and settlements located near the Segezha PPM's leased territory, researchers identified key stakeholders who could participate in consultations within the framework of the certification processes. They also explored local initiatives and projects which could help the company comply with the FSC requirements through financial support in the form of small grants. Researchers gathered information on those social value forests[477] that have to be designated and protected by the company to comply with the FSC requirements. The enterprise rendered feasible assistance to researchers in organizing their study. Since then, cooperation between the CISR and the Segezha PPM has continued.

If the mission of SPOK was to preserve the intact forest landscape, the CISR focused on the promotion and realization of FSC certification in Russia, in particular by focusing on its social aspects. The CISR acts as an agent of institutionalization of new practices; it transmits global standards to the sites of implementation through available channels. Playing the role of 'transmitter' of international standards from transnational spaces to local spaces, the organization also serves as a feedback connection from local spaces to a node of global design (in this case – the FSC's international center).

As asserted above, the organization uses different channels to transmit global standards to localities. The group of CISR researchers involved in the project formed a special forest certification unit and registered with the FSC as authorized consultants. As members of the FSC national initiative, they participated in compiling the Russian national FSC standard regarding social aspects of certification. Therefore, the group was interested in testing and working out methods of implementing standards in local places. It gave recommendations to the company about the most

[475] The author of this case study.

[476] Material was being collected as part of the project related to natural resources management in the Northwest Russia supported by the Finnish Academy.

[477] HCVF 5-6, required to be designated by Principle 9.

efficient way to organize working with stakeholders, carrying out public hearings or consultations, and designating social HCVF. In its cooperation with the company, the NGO CISR ensures in-depth implementation of the FSC requirements on the ground, combining support from their own grants to communities with financing from the company interested in conforming to the requirements of certification.

In 2007, by order of the FSC-Russia, the group developed short recommendations on the social aspects of certification as previously tested at the Sehezha PPM leased territory, placed them on the FSC website and made these available for public discussion. Thus, the organization transmited standards 'top-down' through three different channels, including participation in the preparation of the FSC national standard; cooperation with and consultations for business; and elaborating general methodical recommendations for introducing best practices at enterprises and the dissemination of these practices among the enterprises directly or through the FSC official channels.

Feedback from the locality to the node of global governance design – the FSC International – is as follows. Participants of the forest certification group at the CISR are members of the Social Chamber of the FSC international. This provides them with an opportunity to make proposals to change generic international standards if current standards prove to be inefficient in the sites of implementation. In 2008-2010, the CISR participated in the revision of global standards through the FSC Social Chamber consultations and the HCV network technical panel input.

Among the above-mentioned channels for transmitting international standards and practices, the company's leased areas (in this case, with the INP holding) functioned as experimental sites for implementing new practices and standards. The most successful implementation techniques were intended to be applied at other places. In 2007, on the eve of the inspection of the Segezha PPM leased land, a number of meetings with the participation of local residents were held in villages located on these territories. Consultations with the population included a wide range of questions to be discussed, in order to cover, during one session, as much criteria of conformity to the certification requirements as possible. For example, one public hearing was devoted to the conformity to Principle 9 of the allocation of HCVFs of social significance, and also to Principle 4, related to company-community interaction and the respect of worker's rights.

Small grant programs have become another successful measure to familiarize the local people with their rights, as provided by the FSC and to implement new practices. These tactics are based on revealing the interests of local initiatives and combining them with the interests of adhering to the certification requirements.

In contrast to SPOK's approach to cooperation with the company, the CISR group has distinguished itself, in particular, in its desire to provide the company with a maximum of FSC documentation support and programs which could facilitate compliance with the FSC. It aimed to direct the company towards maximum conformity with the expectations of the auditors engaged in the certification process. The group conducted so-called 'aggressive institutionalization' of new practices within localities.

The CISR group continued to apply such 'combined' and, whenever possible, integral approaches to issues related to certification. Involving other networks and funding agencies was important in using transnational space agents' resources for facilitating the institutionalization of new practices on the ground. Although, generally the CISR and SPOK belong to different networks and contributed in different ways to linking transnational actors to local ones, the intersection of these networks – by working on the same project – was important for contributing to institutional change in the sites of implementation. In 2008-2009, the CISR involved SPOK as its partner in a practically-oriented project supported by the European Union. Participation of both NGOs in the project ensured a combination of the project's social and environmental constituent parts. At the same time, it was a combination of the two channels of institutionalization of practices, which was amplified by the fact that the European Instrument for Democracy and Human Rights (EIDHR) grant in itself was also a channel of institutionalization of international practices in localities. This channel of global standards transmission to localities was used to strengthen democratic trends in the certified territories by promoting public participation in forest management.

In February 2008, a conference within the project was organized in Petrozavodsk with the participation of SPOK, to which all the stakeholders were invited. During the discussions, the parties defined their roles in promoting certification for fulfillment of environmental and social requirements, and worked out ways of implementing new practices. Implementation of global rules assumes a mass of work aimed to present new concepts, and discourses at national, regional and local levels. This process needs locations for negotiations where stakeholders would meet, exchange opinions and discuss new practices. At this conference, which was such a place, the organizers brought the discussion to the regional level. It was very important to do this in Karelia, where the concept of certification was barely understood or recognized by the representatives of state authorities.

Another example of the joint effort of social and environmental NGOs implementing the FSC on the ground was their cooperation in the Muezerskii district. In this case, the environmental and social aspects of certification were interconnected, as allocation of intact forest areas from the areas intended for felling resulted in job destruction and hence the infringement of the local population's social rights. After SPOK had selected intact forest landscape areas, it was necessary to convince the local people that such measures were undertaken for protection of their own rights in a long-term perspective, and the CISR became engaged in such explanatory work. In 2008, both NGOs undertook a joint expedition to Zaonezhje to prepare for the certification of newly leased territories belonging to the holding company. A new cycle of public hearings and consultations, as well as HCVF designation, was more effective, because the CISR and SPOK used a methodology developed and tested earlier, which proved to be efficient, and also relied on working with the other timber companies. However, in 2010 relations between the NGOs and the company changed: the CISR continued building company-community relationships in Zaonezhie, while SPOK entered a cycle of confrontation with the company after they logged a HCVF on one of their plots by mistake. This made joint a CISR-SPOK expedition impossible.

The CISR was also involved with the company in building a Model Forest 'Segozerie', where implementation of the FSC standards went deeper. The model forest served as another channel for transmitting and institutionalizing new standards and practices from transnational spaces to localities. Certification was only one of many activities linked to the model forest.[478] The Segezha PPM forest resource directors decided to build the Model Forest 'Segozerie' after discussions with the CISR and international experts in sustainable forestry. Besides SPOK and the CISR, the

[478] See case study 2-3.

organizers involved the Karelian NGO SOCIO-logos in the Model Forest working group. This NGO was engaged in research on the sustainable development of forest communities in Karelia. Representatives of all the forest ministries and departments of Karelia, research institutions, and local governments which could provide a close connection with locality were invited into the working group. In addition to adjusting global practices in accordance with the local situation, the organizers of the Model Forest have had another task: to give particular attention to preservation and restoration of the local historical and cultural heritage.

The discussion of the Model Forest Segozerie was launched in 2006, all preparatory works were completed in 2007, and in 2008 a noncommercial foundation was officially registered. The NLC, working in the territory of Kareliya, became a founder of this organization on behalf of the INP. The Model Forest working group transformed into a permanent resource for negotiations on elaborating the Model Forest strategy and the plan of concrete action within the project. Each actor of the project had its own interests. One of the INP goals in the model forest development was experimentation with logging practices aimed at the intensification of forest management and legitimating this kind of activity in the eyes of the regional authorities. The INP was interested in lowering the age of the trees at felling and to have this approved by the federal and republic departments.

The CISR was interested in developing a model that ensured the greatest conformity of business practices to the FSC's certification criteria. SPOK had a major interest in nature conservation planning and training. Before Model Forest 'Segozerie' had been officially registered, projects were initiated in localities, i.e. institutionalization of the new forest management practices developed by the working group. In a wider sense, the purpose of these projects was to implement sustainable forest management into practice. In this case, however, its concept had a wider interpretation and contained a deeper approach to the matter than was required by the FSC certification. For example, an ecological trail was created for the demonstration of key biotopes and the principles of their designation. In 2007, 'Karelian House' had become the most important project, intended to preserve the cultural and historical heritage of the Karelian village Yukkoguba. The project represented a master training program for builders of traditional Karelian houses, using traditional building technologies. Afterwards, the local people who had been trained built a house for a private owner in the village of Yukkoguba, which preserved the appearance of a traditional Karelian dwelling. The third project was devoted to the revival of age-old Karelian holiday events in the villages of Selgi and Yukkoguba. All of these projects were financially supported by the company. The creators of the Model Forest regarded the projects with great enthusiasm. However, at the end of 2008, the INP made a decision to suspend financial support for the project due to the financial crisis in the country and world. The Model Forest has slowly started its revitalization again in 2010.

Concluding the section, it is important to note that the tasks of outside agents (CISR and SPOK in this case) were to undertake sociological intervention, to stimulate the forming institutional field, and to introduce new rules and norms. It is supposed that the major agent (the company in this case) should conduct the routine everyday efforts. However, the setting up of new practices has not always been a success, for the following reasons: inertia and resistance in the established system, and the lack of organizational, time, personnel, and financial resources in the company. Despite effective work in building relationships between the logging companies and communities, as facilitated by CISR, the activities initiated by the group decreased or even faded away when the CISR researchers had to leave. The CISR experienced the same frustrations as had SPOK when the company was left alone with the task of designating key biotopes. As was explained at the beginning of the chapter, the restructuring and permanent reorganization of the holding company weakened its managerial capacity, combined with the economic difficulties along the way, weakened its commitment to sustainable forest management.

8.5.4 Interaction with government during forest certification

Despite the Russian government's declaration that it generally supports FSC certification, this section will demonstrate the barriers that the INP faces in implementing global rules in Russia. Resistance to the new global rules is characteristic in all federal, regional and local governmental agencies; however, at each level this resistance is expressed differently. It is important to mention that the companies effectively cooperate with NGOs in order to overcome governmental resistance to non-state regulation. When the FSC rules contradict Russian legislation, the related issues can be resolved through special agreements officially signed between the companies and governmental agencies, or through informal agreements, on regional or local levels. Some local state forest management units (FMU) do, but others do not, get engaged in informal agreements unless they have specific orders from the upper level of state hierarchy. In this case, state forest management units might penalize the companies for violating Russian legislation in order to comply with the FSC. State FMU's also may violate legislation and get engaged in conflicts with a company on both legislative and economic grounds, with these conflicts being resolved through lawsuits. The INP made an effort in establishing good relationships with state structures at federal, regional and local levels, while simultaneously insisting on its own needs and interests. The INP office in Moscow dealt with the federal Rosleshoz, while the office in Petrozavodsk worked mostly with the republic government. At the local level, the NLC worked with the FMU involved in controlling their lease.[479] All these reciprocal actions, however, were carried out across the hierarchy of scales, and were concentrated around specific problems to be solved, involving multi-level structures when necessary. The most important issue to be resolved with the various governments was the getting of special permission for measures that allow forest management intensification. Intensive forest management is not a requirement of the FSC, however it is important both for economics and good ecology, as it allows for the maximum economic outcome with fast replanting and re-growth from the areas closest to the PPM without moving deeper and deeper into the forests. For these purposes, the INP was interested in initiating an experimental project called 'Segezha Pine' that stipulated intensive forest management in the territories leased by the Segezha PPM. Within the project, it was planned to lower the felling age while simultaneously intensifying restoration of forests for commercial exploitation. This project was planned to be implemented together with the above-mentioned 'Model Forest Segozerie', which ensured that attention would be paid not only to economic innovations, but to environmental and social aspects as well. Without such an integrated economic-environmental approach it would have been difficult to achieve the approval of the 'Segezha Pine' project by both the authorities and civil society.

[479] Leskhozes, that later in 2008 were transferred to Lesnichestvo according to the new Forest Code.

To establish constructive relationships with the state, the holding worked on an agreement for cooperation with the authorities. Initially, the agreement was supposed to be signed between the INP and the federal agency Rosleshoz. At the same time, many problems concerning compliance with the FSC certification standards in Karelia were connected with the republic's authorities' stubborn unwillingness to welcome innovations.

In 2008, a tripartite agreement between Inveslesprom, Rosleshoz and the government of the Karelian Republic was signed. The initial tasks of the agreement were to lobby for intensive forest management, the elimination of discrepancies between certification requirements and Russian legislation, and the promotion of the model forest where economic, environmental and social aspects of forest management would be combined. The agreement originally specified mutual obligations of all three parties. However, lengthy revisions have resulted in a document calling for the holding company to take on the obligations of implementation of planned measures, while the state structures limit their own activities to the providing of assistance. The model forest's promotion has disappeared from the document altogether. The situation was described in one of interviews as follows: 'Officials tried to evade responsibility completely. They have emasculated everything. There is no mention that they have to do something – not a whit. They assist – that is all that remained ...'[480] The agreement, as it is now, has not yielded any practical results. The process of forest management intensification was especially difficult. In general terms, Rosleskhoz approved the idea of intensification, but official permission to launch the project was never received (at the time of writing, April 2011).

At the level of the Karelia Republic, the relationship between the company and different state agencies and departments developed differently. Most of the conflicts and problems were related to Karelleskhoz, which supervises all the FMUs in the Republic, with the INP's activity being substantially dependent upon this agency and its subordinated units. Many of the tensions and long lasting conflicts have occurred between the INP and the Segezha FMU. In 2005, the Segezha PPM came out against the Segezha FMU, which was thinning in its leased lands and transforming this procedure into a profitable business: they were cutting down the best trees for their subsequent commercial utilization instead of caring about the forest's sustainability. At that time it was permitted, according to the legislation, and such practices were widely used over the whole republic.[481] The FMU actions caused a notable economic loss to the company. SPOK and Greenpeace supported the Segezha PPM by putting the relevant information on the internet, and helping with its dissemination. It caused a federal investigation of the activities of the Segezha FMU, which resulted in a decision to recognize the Segezha PPM's complaints as justified. The incident has aggravated relations between the holding company and Karelleskhoz. The company, however, has used any possible means to overcome the tensions. For example, in 2006, together with SPOK and Greenpeace, it held a joint field workshop and got officials involved in the event. During the workshop, representatives of the company demonstrated their skill in thinning using harvesters. After the workshop, the company was allowed to carry out thinning in its leased areas – which was what it had wanted to achieve. At that time it was a victory, as the existing legislation did not provide for such a case, until the new Forest Code came into force in 2007.

[480] Interview with an INP manager, August 2008, Petrozavodsk.

[481] Interview with a Greenpeace representative, June 2007, Moscow.

In 2008, the discussions of the forest utilization plan became an arena of basic contradictions between the regional authorities and the company. As was mentioned above, according to the new laws, the Karelian Republic had to develop and submit a forest utilization plan for approval by federal bodies. This plan needed to be compliant with both the new Forest and Water Codes and other regulatory documents. The regional authorities delegated the working-out of the plan to a contractor. In Karelia there are many wetland areas; all of them had to be taken into consideration by the plan's developers. As a result, water protection zones and production forests were arranged in such a way that the annual allowable cut was reduced from 8.8 million to 5.6 million cubic meters. Considering the fact that 2 million m^3 of the remaining amount are economically unfeasible, the plan envisaged an allocation of 3.5 million m^3 of annual felling. Such an unexpected course of events was also aggravated by the fact that the construction department made the new scheme of territorial planning without preliminary consultations with the timber producers. Moreover, on the initiative of SPOK, this scheme was also considerably supplemented with a number of new specially protected areas. This would have additionally reduced the allowable cutting areas. As a result, the INP found it self dealing with two new documents, acceptance of which would threaten the company with great economic losses.

For the NLC, such a reduction of the annual allowable cut was absolutely unacceptable, since the resulting volume would be insufficient for the Segezsha PPM's needs and the company would need to increase the volume of wood they buy from territories that are far from the mill. The Karelian timber producers and the INP's department for intensification focused on a critical analysis of the forest plan and paid great attention to negotiations with the republic's authorities on the issue. Managers of the holding company were of the strong opinion that this plan did not provide for profitability of the timber enterprise; it would provoke a shortage of raw material and lead to the closure of one third of the republic's logging enterprises. They substantiated their criticism by predicting mass unemployment, social tensions and sharp decreases in payments into the budget. They asserted that such a plan was in conflict with the Long-term Strategy of Forestry Development in the Russian Federation leading up to 2020. Interestingly, the republic's state authorities did not want to witness such consequences. On the contrary, they reckoned on industry development and investments; they advocated for intensification of forest management, job growth and increasing budget payments. They also understood that the idea of rapid development of tourism and recreational businesses as a means of increasing the republic's prosperity is ephemeral. Thus, because of confusion in the legislation and the ongoing reforming of power, the decision-making process had been passed over to organizations and people who lacked relevant experience.

The republic's government did not manage to correlate the decision taken with the possible consequences, and approved a document which contradicted their own interests. At the federal level, the plan as presented was not ratified. After revision and adjustment, a new version of the plan was approved; one more acceptable for the company. The given example is indicative to understand where the desired reform of forest legislation and forest management does not match reality. This also created difficulties with certification: according to certification standards, the enterprise's forest plans should be embedded in the general plan of the republic, and be in compliance.

The issue of biodiversity conservation remained a stumbling block in relations with the regional state structures for many years. The holding company had to be accountable for biodiversity to the FSC, and Rosleskhoz has promised to provide only for methodological guidelines. Therefore, another agreement with the Karelian government was necessary. The INP managed to sign an agreement with the Karelian agencies regarding biodiversity conservation by November, 2008. For several years, since 2005, the company was ineffectively fighting for approval of their concept of biodiversity conservation and for an agreement with Karelleskhoz. There was precedent for this in the Arkhangelsk region for FSC certified companies. The agreement there was reached with the help of the WWF. However, in Karelia, these guidelines on biodiversity conservation took much longer to accomplish. The company involved officials in various educational activities: for example, they were invited to attend courses on key biotope's allocation in the summer of 2007, where the company demonstrated an eco-trail they had created. However, despite the participation of representatives of state structures in actions such as that, their approval of biodiversity conservation remained uncompleted. In practice, they were not going to change their course by building relationships with the company.

Especially large problems concerning key biotopes designation were encountered in the places where the state forest management units didn't wish to make informal agreements. The Segezha FMU, whose relationship with the company was, as described above, spoiled in 2005, has become the most 'principled' fighter against the allocation of key biotopes. They were judging key biotopes before logging operations were finalized and forced the company to pay penalties. Finally, the Segezha state FMU agreed with biotopes conservation, provided they are registered in the license for allocation of forest plots. In practice, however, enterprises usually do not thoroughly explore allotments beforehand, since this action requires additional time and human resources. That is why it often happens that a biotope is identified at later stages, when working crews come in to pursue logging operations. According to formal rules, in such a case it is necessary to receive a new FMU endorsement, and to list the biotope in the forest license and thereafter to protect it from felling. One cannot just leave a biotope, as the absence of respective prior permission would be regarded by the FMU as undercut. The FMU workers traditionally consider undercuts as a serious violation of felling regulations, which are punishable by fines. The basic accusation from the Segezha FMU was as follows: in its opinion, the company left inconvenient sites (with unfavorable conditions for felling), having referred them to as biotopes. Here is a revealing comment from an informant: 'They bring an application to me and say – `we ask to recognize this undercut as key biotope'.[482] There were lots of fines, and the company attempted to assert its rights in court. The company initiated legal proceedings seven times, and seven times they failed. One should note that in defiance of all the obstacles and fines, the company continued to allocate biotopes to comply with the FSC. As was mentioned, the battle ended when the agreement on biodiversity conservation was signed with the INP in November, 2008. In 2009, Krellesprom allowed other companies, certified by the FSC, to receive special permission for biodiversity conservation. Therefore, the INP and SPOK served as agents of institutional change; shaping and later creating a new path for non-state rules to be semi-officially institutionalized in the sites of FSC implementation. They used different methods, including multiple forums of negotiation, in order to persuade the state agencies to accept the FSC's requirements.

[482] Interview with a senior forest manager of the Segezha FMU, May, 2008, Segezha.

8.5.5 Global rules and locality: interaction with local people in forest settlements

One of the FSC's requirements is that local communities must participate in forest management. The certificate holder, according to the FSC rules, has to be proactive and ensure public engagement in the FSC processes, e.g. public consultations, allocation of HCVF, etc. There are several barriers for successful public participation in the decision making processes. The first barrier is disempowerment and a lack of belief among the local population that their suggestions will be taken into account, originating from the lingering effects of the Soviet past. The second barrier also originates from the Soviet past – it is a high expectation that the company can play the role of the State and provide for significant community infrastructure. Thirdly, the modernization and restructuring of the forest industry reduces employment needs and brings additional tensions to the local communities. All these difficulties in engaging the public in decision making can be clearly seen in this case study.

Incorporation of the formerly independent logging companies into the NLC and, hence, their newly established affiliation with the INP inevitably entailed a changed relationship with the local population. During the Soviet past, state logging enterprises formed the infrastructure of a settlement. At that time, temporary logging settlements were constructed for servicing the enterprises and people came for work from all around the country and settled down there. During the period of Perestroika, many of the state logging companies were privatized; others were ruined or re-registered and handed over to new owners. At the same time, these enterprises traditionally had to maintain the infrastructure of the settlements, although some of them managed to be released from their obligations as a result of bankruptcy or re-registration. The affiliation of these former individual logging enterprises with the holding company has ruptured the regular course of relationships between the enterprises and the local citizens. The holding is focused on maximizing profits and carrying out modernization of production. This inevitably entailed dismissal of some of the workers from the enterprise, athough the living standard of those who remained was raising. Settlements where enterprises were utterly ruined and closed were the worst hit by such reorganization: the holding company in such cases employed contractors, and the local population was left without their former support for infrastructure maintenance. The situation in settlements where local people still worked at the operating enterprises was somewhat easier. Larger settlements, which had the status of regional centers, had more opportunities to both defend their rights through the higher state structures, and to require support from the holding company. Villages, which existed on these lands even before the logging companies were founded, underwent less noticeable changes because of traditional rustication: due to well-developed personal subsistence plots people were, to a lesser extent, dependant on the enterprise.

The supply of firewood and sawn timber for the local population was one of issues traditionally arranged by the logging enterprises. After the merging of logging enterprises with the holding company, almost all of the woodlands of the region were included in the leased area of the company's structural divisions. By that time, there were no more small businesses who would supply the population with fire wood and sawn timber. Such a situation is typical for this very case, for as in other regions, as a rule, a number of smaller companies operated in the marketplace alongside the larger timber companies, allowing citizens to choose where to buy wood. Meeting a demand of the population, these small companies provide for competitive conditions and regulated the prices for firewood and sawn timber. In Karelia, however, the holding – due to circumstances related to lease – became a monopoly in the wood business. The company was not interested in supplying the population with firewood and sawn timber. As a result of modernization and technical re-equipping of production means, the NLC used more widely assorted harvesting, in which timber is cut to pieces of required size on the spot and is hauled directly to the final destination, omitting local timber storage.[483] Because of this, local timber storage operations were closed and torn down. This resulted in reducing the number of jobs and hindered solving the problem of the local population's needs for firewood and sawn timber.

Sawn timber supply was an especially difficult problem for the holding company. It was inexpedient to allow private saw mills in the area because all the woodlands were leased by the Segezha PPM. It was possible to sell wood from the Segezha saw mill, but in this case transportation would be costly and make sawn wood unaffordable for local villagers. Therefore, providing the local population with firewood and sawn timber has become an expensive task, demanding additional efforts and resources. The holding company would probably refuse to become involved in settling these questions. However, traditional Soviet practices, generally annihilated in the recent past, began to revive in the form of global practices proceeding from global spaces to local places through certification and the holding company's activity.

Certification requirements stipulate that an enterprise has to maintain consultations with local communities and to acknowledge their rights, including those connected with access to forest resources in the places of residence. Moreover, Russian certification standards requires annual consultations to be organized in all settlements where the public can obtain information on logging plans and access issues, whenever logging activity would hamper other kinds of forest utilization. During these consultations, the company should involve the local community in the designation of areas which are to be defined as having historical, cultural, and archaeological values. The places available for hunting, the gathering of mushrooms and berries may also be specified, provided that they are traditionally used by the population as a means of subsistence and recreation. In the course of consultations, the company has to report to the local community on its forest management activities and also collect complaints and requests from the population, in order to develop the mechanisms and procedures for resolving problems and eventually, to resolve them. As was mentioned above, in 2006 and 2007 the holding company initiated such consultations together with the CISR experts; in 2008 and in 2010 the consultations were conducted on its newly acquired lease territories. Consultations were organized to establish a constructive relationship with the company. During such consultations, key stakeholders were identified; including such institutions as schools, libraries, administrations as well as concerned people with whom it would be possible to continue the work once begun.

In 2006, with the intent of more actively involving local citizens in cooperation with the company and supporting HCVF 5-6, e.g. areas of social significance, a small grant program was launched and continued into 2006-2007. Small grants were a means to support local initiatives and were focused on the supporting of creative activities and participation, rather than direct aid to the population. The projects supported by the company were as follows: the above-mentioned revival of a traditional Karelian village and folk-festivals in Segozerie; a historical museum of the Valdai logging company; the partisans' trail; an eco-trail and others.

[483] Transit timber yard at proximity of a settlement, where timber is being accumulated for further transportation.

Another tool for establishing a relationship between the company and the local population was a social fund allocation voluntarily undertaken by the company. Initially, they contributed 10 rubles per one cubic meter of harvested wood. The funds were received by the local municipal government, which then had to decide how to distribute them. Such actions promoted the population's confidence in the company's activities and evoked a wave of public participation, which was a necessary condition for certification.

In 2008, due to the economic crisis in the forestry sector and the crash of prices for sawn timber, social funds allocations were cut to 2 rubles per cubic meter of wood cut, and later totally canceled. The small grant program was closed, with the decision to revitalize it not made until 2010. Local communities were somewhat disappointed, especially because the situation of firewood supply worsened. However, as a result of actions aimed to re-establish the company's relationship with the population, the latter became more self-confident: for example, in 2008 the population of the settlement of Padany approached the company with a letter concerning fears for the safety of a small spawning river located close to the settlement where the holding company was permitted to execute thinning in water protection zones. In response to the letter, the company suspended felling and suggested that the population itself should find an expert who could examine the area and submit an expert opinion on whether thinning is harmful for the small river. The company promised to consider the results of the inspection and to resume felling only in the case of it being harmless. One should remark, however, that the local community did not find an expert and the initiative came to a standstill.

An intensive interaction by the company between the local community and other stakeholders took place in 2009-2010 around the proposed logging near the village of Tunguda. The Company allocated logging plots prior to the public hearings and the process of designation of social HCVF. The Karelian social NGOs Trias and Young Karelia and seasonal residents from Moscow, who had bought a house in Tunguda, joined with the village residents to protect the forest near the village. The conflict situation could have grown in size; however, all the issues were resolved. Stakeholder interaction was mediated by an expert from the CISR; social HCVFs were designated, and ultimately, the stakeholders started cooperation with the company in forest management near the village.

Proceeding from an overall assessment of the FSC rules' impact on the local community, one should recognize that for the local community no significant change of practices was observed, except for some minor democratization in rural settlements. To a greater extent, changes affected the employees of the enterprise, as the worker's safety improved, as required by certification. For the local population, certification has not provided any particular material welfare or measurable benefits, except for the right to participate in forest management. Certain stimulation of local participation and intersectoral dialogue has emerged in local settlements. The population understood that it has a right to participate in forest management and to become aware of the opportunities which participation can ensure. For all that, the situation is still far from being a genuine on-site democracy, since civil society and local self-government remain underdeveloped. However, due to the actions undertaken by the holding company in compliance with the requirements of certification, the local population, in participating in social value forests allocation and consultations, becomes a participant in forest relations; formerly it was a prerogative of the experts alone.

8.6 Discussion and conclusion

8.6.1 Interplay of actors of transnational space and the space of place in the framework of FSC certification

In this chapter, the interaction of actors from transnational spaces and the spaces of places in the process of FSC implementation by the holding company have been analyzed. Using empirical findings, I construct a more generalized model of the FSC certification of a TNC which involved NGOs as experts (Figure 8.4). Actors that play a role in changing norms and practices in localities affected by the FSC standards are referred to as agents of institutional change. Companies being certified, their experts, state agencies, and environmentally and socially-oriented NGOs also became such agents, providing different kinds of inputs into the process. Therefore, they all became part of the FSC-GGN (Kortelainen *et al.*, 2010). In each specific case, certain actors became major agents of institutional change, depending on their agency and capacity for shaping the institutional field, while other actors contribute to the process by catalyzing or blocking institutional change. If the FSC certificate holder is a company and there is no outside funding for certification, as was described in this chapter, the company is the major implementer of the FSC (Figure 8.4: TNC). In the described case, however, the role of the holding company as an agent of institutional change is not limited to certification. It is also related to the impact of its growth, restructuring and modernization. Therefore, the holding company acts not only as a channel for institutional changes in the certification processes, but also as a generator of institutional change that comes from the internal processes and dynamics occurring in this corporation. Although in general, the vertical structure of a holding company has a good potential to implement change in practices of its subsidiaries at the local level, the current case demonstrates multiple challenges that the company faces because of the institutional turbulence of the surrounding social and institutional environment, as well as the consequences of their own restructuring.

There are always other actors involved in implementing new institutions or changing existing ones (Table 8.1). UK chains selling DIY products as well as governmental purchasing policies play critical roles in providing economic incentives for companies to get their products certified (Figure 8.4: buyers). Enterprises belonging to the holding company which are located in Western Europe are another channel of interaction between the holding company and the global space (TNC Europe).

As was demonstrated in the chapter, experts from NGOs participating in the certification processes are characterized by attempts to promote their own agenda in addition to the FSC's requirements, which can be either an environmental or a social agenda (Figure 8.4: NGOs). In general, the role of NGOs in such cases differs from those cases in which NGOs are the major implementers of the governance design on the ground, as described in Chapters 5-7.

By imposing their demands and adjusting them to the local context, CBs are one of the major agents promoting institutional change in local places. Peculiarities of the subsequent changing practices depend on the expert community involved: its composition, experience, members' educational level, etc. (Maletz & Tysiachniouk 2009). In cases of ordinary FSC certifications, in which NGOs are not playing the role of major implementers of any model of sustainability, the role of the CBs increases because there is a tendency in the business community to comply

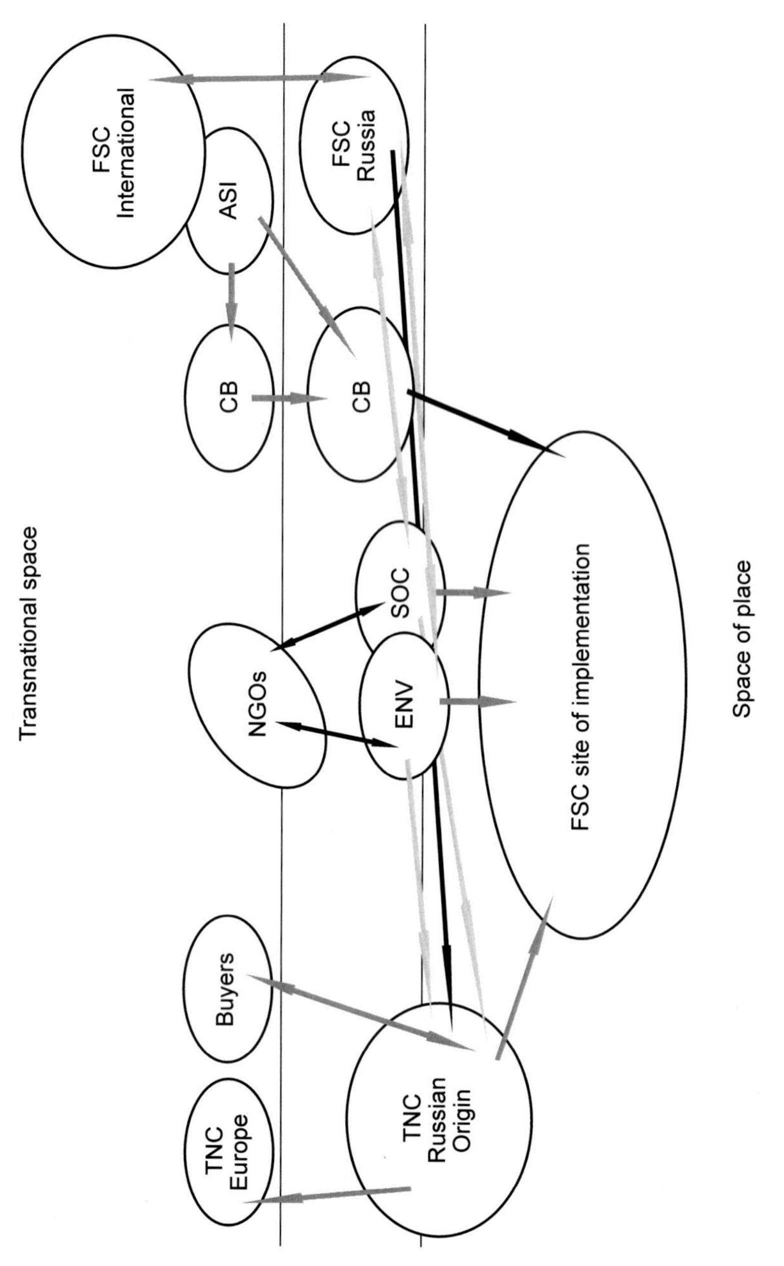

Figure 8.4. Interplay of actors of transnational space and the space of place in the framework of FSC certification.[1]

[1] *Lighter errows are used between FSC-Russia and other actors as they were not directly involved in the project.*

with the FSC rules in a minimal way. Although officially they are not allowed to consult the company, they provide a direction for it, by imposing CARs and monitoring compliance. As was demonstrated in the described case, the ASI also influences implementation of the FSC rules on the ground by fostering stricter audits by the CBs. The CB in turn becomes more demanding in their interaction with the companies (Figure 8.4: CB, ASI). As one can see, the role of CBs as an agent of institutionalization of new practices is to urge the company to adopt new practices, to adjust their actions, and to redirect their energies. This can be seen as 'soft pressure' exerted by an actor of transnational space on local actors. Sometimes, as in the case of the suspension of a certificate, this role changes and a CB becomes a sort of 'alarm bell'. Such a signal prompts the company to undertake actions for changing the institutionalization of practices and warns it against the possible negative effects of its inertness. This signal very actively penetrates into the institutional field and reshapes it. Therefore, a rather rigid pressure from transnational space to locality is trilateral: it comes simultaneously from the certification bodies, the UK market, and the company striving to comply with FSC and please its customers.

Certification does not involve state structures directly in its process as it is a non-state and voluntary mechanism of forest governance (Cashore *et al.*, 2004). However, in practice, certification cannot be implemented without the participation of state institutions, in one form or another; the fact that land in Russia belongs to the state and is leased to the companies is one reason (Tysiachniouk, 2006a). Another is that legislation can support or contradict the FSC rules. Hence, state structures cannot be interpreted as a transnational agent of institutional change, although they play the role of a change agent at the national level by regularly intervening in this transnational process and influencing it.

In addition, the other parties involved in forest management, i.e. stakeholders, can also be agents of institutional change. The role of stakeholders as agents of institutional change depends on the degree of their activity and participation in certification, and on the mode of their interaction. The latter can include both confrontation and conflict or partnership relations; it can promote or impede change.

8.6.2 Application of the FSC-GGN concept to the case study

The FSC-GGN involved in the INP certification, with the peculiarities of its node of design, forums of negotiation and sites of implementation is visualized on Figure 8.5. The global FSC standards were adjusted by the CBs (after 2008, Russian National FSC standards came into force and replaced the CB interpretations) and adopted by the TNC of Russian origin, Investlesprom. The TNC and social and ecological NGOs played a subordinate role in interpreting the standards and developing the strategies for the company to achieve compliance (Figure 8.5: nodes of design, FSC, adopted by TNC, with NGO experts). The research showed that expert NGOs had different priorities in interpretation of certification standards. For example, social experts developed specifically the social part of certification. Ecological NGOs (SPOK) stimulated development of the methodologies for determining of the key biotopes, which further were spread throughout Karelia. Thus, combination of experts and auditors determined the interpretation of the FSC standards and their institutionalization at the site of implementation.

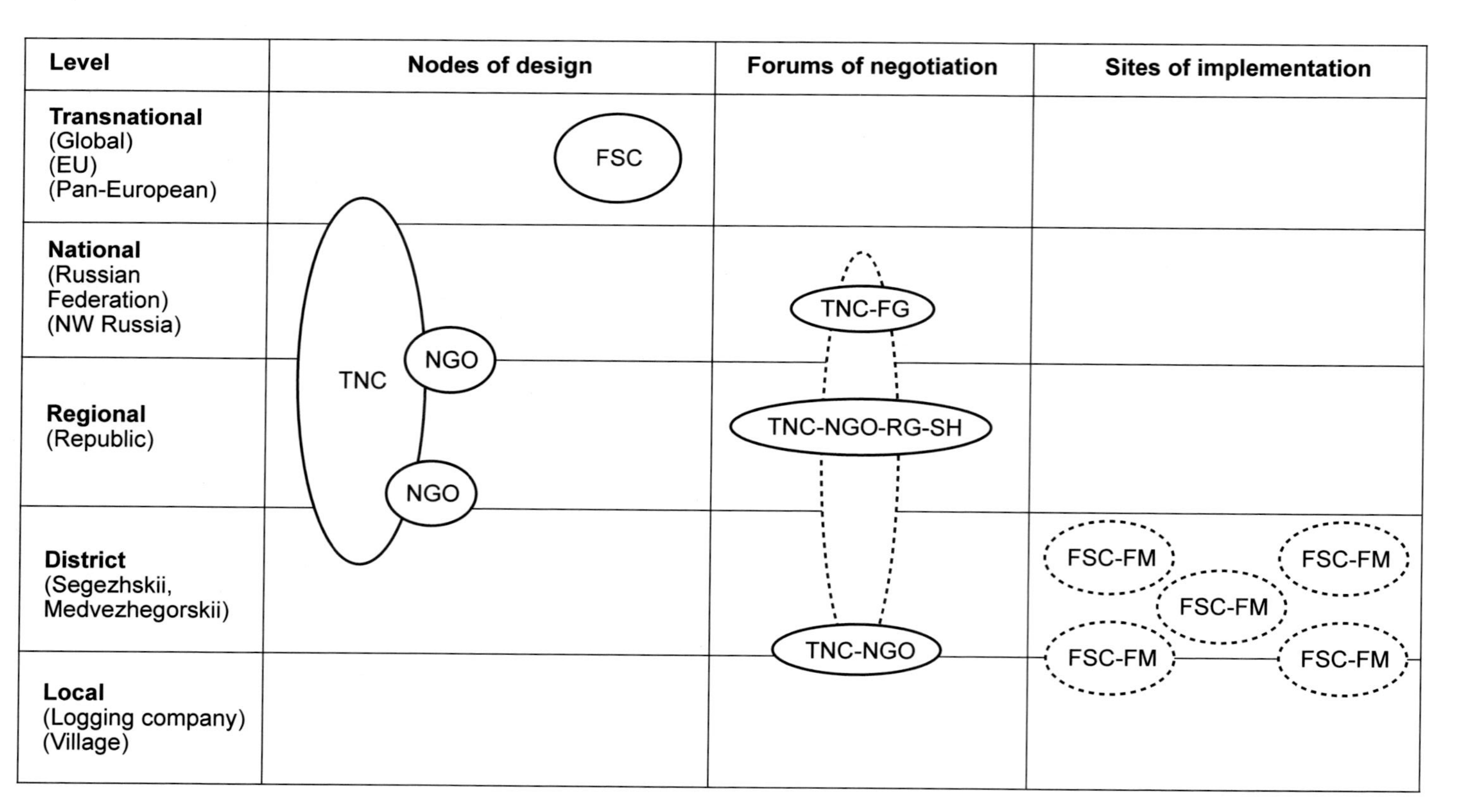

Figure 8.5. Analysis of FSC certification of the Northern Logging Company.

The process of interpretation of the FSC requirements stimulated forums of negotiations on federal, regional and local levels. On the federal level, there were very few negotiations organized by the company with federal agencies around intensification of forest utilization and harmonization of the FSC standards with the national legislation (Figure 8.5: forums of negotiation: TNC-FG). The most difficult forums were on the regional level and local levels. On the regional level, forums were much more intense on the issues of protection of biodiversity and contradictions between the FSC and forest legislation and involved negotiations with regional state agencies, NGOs and the whole variety of stakeholders (Figure 8.5: forums of negotiation: TNC-NGOs-RG-SH). In the sites of implementation, the forums were conducted with local people living in forest settlements and villages. Intensive negotiations took place with the local FMUs on key biotope preservation issues that contradicted existing forest legislation (Figure 8.5: forums of negotiation: TNC-NGO-FMU-SH). In spite of the fact that employees of the FMU are forest specialists who understand the necessity of an environmental approach to forest exploitation, established long-term practices prevented them from recognizing the new global practices of commercial forest use. These forums contained little co-design, as the main negotiations were organized between stakeholders to agree about understanding of certification requirements and to decide exactly what they have to conserve in forests. Most often, the forums were organized in the form of conferences, seminars, and stakeholder training with the discussions following. The only exception was a short period of time when the TNC had the intention to construct a model forest for developing innovations for sustainable forest management. When a new initiative springs up beyond the frames of certification, new types of the forums arise. Thus, the trial of the model forest resulted in a more intensive dialogue of stakeholders, which involved all the interested parties and led to the implementation of a few environmental and social projects.

Two major types of barriers in the process of institutionalization of the FSC rules have been identified. The first is related to the turbulent institutional environment in Russia as a whole, and the Karelia Republic in particular. The second type of barrier was related to the institutional turbulence within the holding company itself, as it was acquiring and restructuring former Soviet enterprises and simultaneously modernizing them. This process became especially painful for both the companies' subsidiaries and the local communities during the economic downturn in 2008-2009. It was problematic for the company to spread corporate policy and the FSC's innovations to its own structural sub-divisions. That is why the FSC rules often were not transferred into concrete local practices. As a result, the FSC certificate was suspended and returned only after the necessary improvements in local practices.

The policy of the INP was to certify all forests in their lease and to unify the approach toward certification in order to increase efficiency. However, research demonstrated that mimetic institutional isomorphism was only partly working for the company, despite the considerable efforts. The logging companies that the INP was acquiring and incorporating into the NLC were at different stages of their economic development, the territories that they leased had different contexts and different levels of resistance to new owners and new rules. The forest areas to be certified were also very different, as some of them had large territories with virgin forests, which the FSC rules require to be preserved. For example, in the early stages of certification at the Segezha PPM lease, few old growth forests were found and the company developed the algorithm of their designation in consultation with the NGO SPOK. Later, more valuable virgin ecosystems were found and large territories were on the Greenpeace maps in the newly acquired Muezerskii LC with their lease. These peculiarities of territories fostered deeper and sharper forums of negotiation around the new rules at sites of implementation. Originally, the approach to FSC certification was developed by INP at the Segezha PPM lease territory, and was planned to be disseminated to the other sites. However, when the FSC certificate at the Segezha PPM was suspended, it became harder to use it as a model; therefore the other subsidiaries had to focus on an analysis of the mistakes and failures in the FSC implementation at the Segezha PPM lease site. In addition, the Russian national FSC standard was accredited by the ASI in November, 2008 and fostered new approaches to be implemented. The company had to work with different CBs, which had different approaches and interpretations for the FSC standards implementation. This also made it harder to implement an isomorphic approach to all certifications of the holding.

Coercive isomorphism was the key in the institutionalization of new practices in the site of implementation. In sociological literature, the term coercive isomorphism is usually used in the analysis of state policy implementation that involves new regulations that force the changing of practices and behavior. In this case, the agents fostering coercive isomorphism were mostly private. The FSC itself imposed new rules of the game to which companies have to comply by institutionalizing change. Although the FSC is a voluntary certification system, companies get involved in the FSC process as they are forced to by their western customers and markets. Therefore, the INP was forcing its subsidiaries to become certified. Environmental NGOs also acted as coercive agents. The case demonstrated the coercive role of SPOK that was fostering the company to implement new rules for HCVF preservation, management and monitoring. The CISR acted more softly with the company, playing mainly the role of a consultant, and less as an NGO demanding change. Therefore, the CISR was an agent of mimetic isomorphism with few elements of coercive power. The state in this case also played some role in implementing coercive isomorphism, but not in a standard way. The state used the FSC organizational infrastructure to foster enforcement of its own regulations by writing a complaint to the FSC office regarding the operation of logging enterprises at the Segezha PPM lease.

The normative pressure to institutionalize change can be illustrated in the case of preservation biodiversity on logging sites and the designation of HCVF. The 'norm negotiation' on biodiversity took place over several years with the efforts to implement the FSC rules contradicting Russian legislation. Slowly, the biodiversity preservation norms, as stated by the FSC and the convention on biodiversity that Russia had ratified, were winning the battle. As follows from the above, with the lack of state support for introducing global practices springing from global spaces, the most difficult task was to institutionalize these practices in the spaces of places. If, at national and regional levels institutional turbulence impedes institutionalization of new practices, at local levels their introduction is complicated by the inertia of a well-established institutional field.

Despite the difficulties, due to the large dispersion of the company's leases throughout the country, changes in forest management practices could extend to other districts in other regions. Therefore, the overall institutional change turned to be highly isomorphic, yet poorly enforced. This constitutes the difference in this particular holding company from other cases, in which corporate agents had better managerial abilities for institutionalizing their own policies in the sites of implementation.

Chapter 9. Conclusions

9.1 Introduction

Russia has no major deforestation, although poor forest management is a major cause of forest degradation (Zagidullina & Romaniuk, 2011: 61). Large intact forest landscapes of global significance are under threat as many of them are situated in territories designated for logging and state authorities do not recognize them as having great value. Companies that lease the land, due to Russian legislation, are allowed to cut these valuable forests with high biodiversity. However, there are moratoriums on logging, following NGO pressure (Tysiachniouk, 2009). There is not an adequate state forest policy in Russia. Constantly restructuring and changing legislation not only accelerates the pace of forest degradation, but also creates an institutional void in governance. This leaves a window of opportunity for non-state actors to cope with poor forest management through introducing new forms of global governance, such as governance through FSC forest certification.

Since the collapse of the Soviet Union in 1991, Russia has experienced transnationalization of governance arrangements that are characteristic of the current globalized world. As in other countries, new governance schemes, involving new actors such as international private and state aid foundations, NGOs and TNCs have appeared in Russia. In order to protect the intact old growth landscapes from logging, transnational NGO networks have organized market campaigns, fostering Western retailstores to boycott companies involved in old growth forest cuttings (Tysiachniouk, 2009, Chapter 5). TNCs operating in Russia became interested in pursueing adoption of voluntary forest management standards as a license to operate in environmentally and socially sensitive markets of the West. Both private and state aid foundations facilitated creation of new institutional infrastructure for adoption and implementation of forest certification and sponsored experimentation and modeling of innovative forest management tools on the ground (Chapters 6-7). These new actor's involvement fostered development and institutionalization of the FSC in Russia.

Therefore, the central objective of this study was to understand the operation of the FSC voluntary forest certification scheme as transnational non-state governance institutions, both globally and within the nation state. Three research questions were formulated:

1. How to conceptualize the process of institutionalization of transnational private governance through certification schemes?
2. What is the role of global and local non-state governance actors, especially NGOs, in the process of installing new sustainable forest management practices in Russia?
3. How and to what extent do FSC governance arrangements transform forest management practices in Russia?

The focus of this volume is on highlighting the mechanisms of how the FSC creates new governing institutions, affecting not only forest management but communities and workers. Using several case studies, this study assesses to what extent governance through the FSC has a transformative character and can fill an institutional void, by giving voice to private actors and enabling them to foster sustainable forest management in Russia.

Voluntary regulatory schemes, such as the FSC, foster market transformation toward more environmentally friendly and socially responsible wood production. The forces of the market are used by actors to provide incentives for companies to enter ecologically and socially 'sensitive' niche markets, in which they can get rewards and benefit from demonstrating their responsibility towards the environment and local people (Cashore *et al.*, 2004). Governance through the FSC uses consumer power for providing market incentives for companies to get engaged in this voluntary scheme. It appeals directly to companies by allowing them access to environmentally sensitive and more lucrative markets of the West. Adoption of corporate social and environmental responsibility is the price that companies have to pay for their market benefits. Therefore, the global aim of the FSC as a regulatory instrument is to improve production practices in countries where the laws are weak, social protections enjoyed by the population are low and ecological issues are not treated as a priority by the majority of citizens. 'Responsible' purchases have become a method to connect one's voice to the demand that producers of goods in developing countries and in countries with transitional economies abide by a certain social justice and ecological responsibility (O'Rourke, 2005: 116; Tysiachniouk 2010b,f, 2012c).

These new modes of governance that have appeared in Russia, and almost in every corner of the world, do not fully replace the state's top-down approaches of policy making, but enter into complicated relationships and co-existence with them. Non-state actors have formed private authorities in governance, e.g. voluntary certification schemes, that co-exist with both state initiated international treaties and national regulatory systems. Contemporary governance can be characterized by a multiplicity of actors, scales and rules of the game. Therefore, global voluntary governing arrangements change the role of the state in governance. In particular, as was shown in this volume, the FSC bypasses governments on the stage of global policy design and standard setting, but cannot avoid working with state agencies as stakeholders on the stage of policy implementation in a particular country.

Such global governance arrangements cannot be explained through old theoretical state centered concepts. Therefore, in order to explain the phenomena of governance through the FSC, this volume contributes to scholarly discourses and theorizations on global/local interplay in multi-level, multi-stakeholder governance of natural resources. It provides a new analytical framework for understanding the mechanics of transnational environmental governance. In particular, an effort has been made to provide a new theoretical and empirical lens for analyzing transnational governance through certification and labeling schemes, such as the FSC, MSC, fair trade and many others.

This chapter describes both theoretical and empirical contributions of this study to the field. It starts with revisiting the major analytical concept of the book (the governance generating network), theorizes further and explains how it works in practice. It also compares the GGN concept and its explanatory power with another conceptualization of certification schemes (Section 9.2.) The chapter proceeds with revisiting the conceptual framework for the study involving different types of networks, summarizes empirical findings and their diversion from the previously identified ideal types. Moreover, it explains what aspects of sustainable forest management can be achieved using the FSC certification scheme and suggests why other aspects fail to be institutionalized (Section

9.3). The chapter concludes by highlighting perspectives for future research and scholarship in the field (Section 9.4).

9.2 Revisiting the concept of a governance generating network

In this volume the concept of a governance generating network (GGN) was developed, both theoretically and empirically. It was developed empirically as a grounded theory using the case of the FSC and theoretically built on the basis of the sociology of transnational processes (Castells, 1997, 2009; Sassen, 2003, 2008). Following Castells, the study distinguishes between the spaces of places, in which day to day practices take place in specific territories, and in transnational spaces, disembodied from concrete geographical settings, in which global rules and standards are being created. The new category of networks, the GGNs, link transnational and local spaces and generate new governance arrangements in both spaces. The volume demonstrates how GGNs are produced, maintained and extended.

Following Giddens' structuration theory, the GGN concept is built taking into account the balance between agency and structure within the network (Giddens, 1990). GGN actors develop global regulatory tools, products, or standards to be implemented in different parts of the world. Different kinds of governance actors became agents of institutional change and contribute to the development of new standards and adoption of the existing standards to different natural and social contexts. GGNs involve three structural elements: nodes of design, forums of negotiation, and sites of implementation; and policy is conceptualized through four processes: abstraction, standardization, operationalization (translation into national standards), and implementation-institutionalization in concrete practices on the ground. The processes of abstraction and standardization are transnational and non-territorialized, while the processes of operationalization, implementation and institutionalization are territorialized and mediated by the local conditions, e.g. the social, political, economic and environmental contexts of forests in different regions of the world.

The nodes of global governance design are formed by different kinds of transnational stakeholders, drawn from all over the world and representing strategically operating groups and individuals. Sites of implementation are physical place based territories, where new rules are institutionalized. Forums of negotiation are part and parcel of standards development by the node of design, adoption by national stakeholders and implementation on the ground in different parts of the world (Kortelainen *et al.*, 2010; Tysiachniouk, 2006b, 2010g). The standards are developed predominantly in the nodes, where abstraction from local issues takes place, next adopted and adjusted to local environments that constitute operationalization predominantly on the national level and transferred to the sites of implementation. In the sites of implementation the standards become rules in use that regulate forest management practices.

FSC-GGN represents a form of private authority in governance; therefore power distribution changes both in transnational and local spaces from governments to governance arrangements. The increased participation of transnational actors does not make national domestic processes fully transnational, but it gives national arrangements additional actors, some of whom are transnational. This situates domestic processes in the analytical borderline between the national and the global (Sassen, 2003).

9.2.1 FSC-GGN and its policy circle

Chapter 3 explained the operation of the FSC certification scheme from the perspective of a GGN. It examined how new policy and standards are being developed transnationally in the node of global governance design, which is comprised of FSC members and staff from all over the world. The chapter explained how new transnational policies are set up and negotiated at the regular FSC forum, the FSC General Assembly. NGOs, companies, indigenous people groups and trade unions, representing economic, social and environmental interests, can participate in working groups, and/or send their suggestions for policy change in the form of motions to the general assembly. At the general assembly, members participate in negotiations and vote. FSC policies and standards are implemented and transformed into concrete practices in multiple sites of implementation, which are represented by FSC certified forest management territories.

Therefore, it seems that the process of policy change to some extent resembles the conventional policy circle with the major difference that it involves non-state actors as agents of institutional change (Figure 9.1: external circle) and governments are not directly involved in the transnational part of the FSC-GGN. In the spaces of places they, to some extent, affect the process of institutionalization of global rules through legislation and power structures. In such a way the policy circle involves processes both in the transnational spaces and the spaces of places. Non-

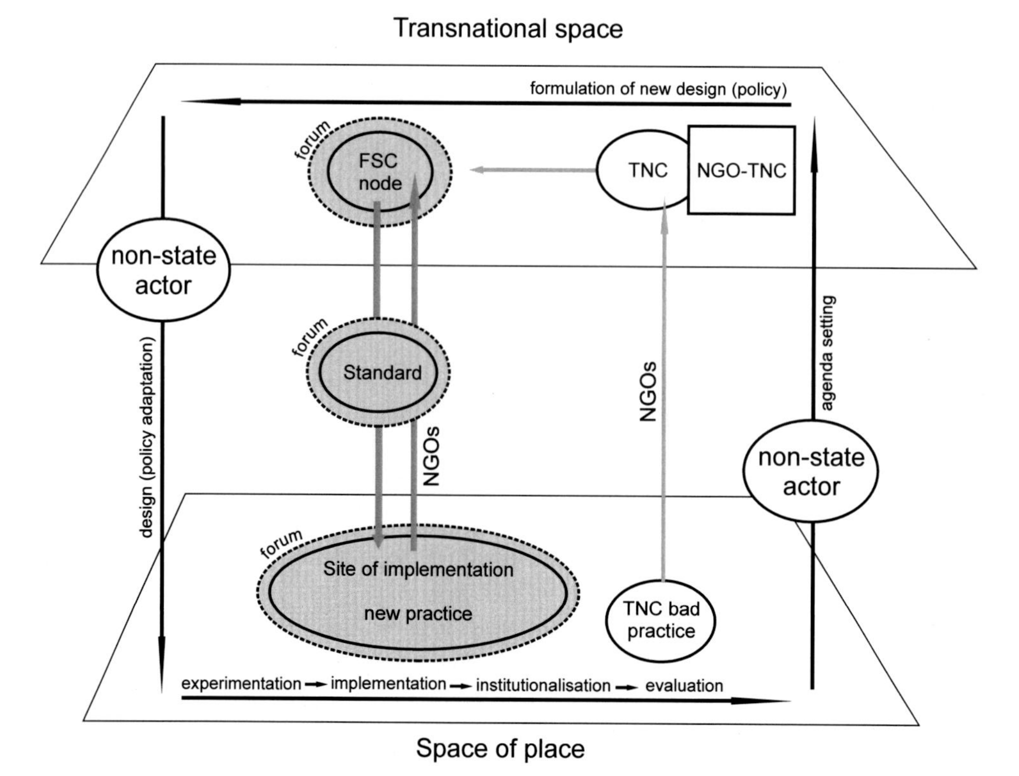

Figure 9.1. Policy process in the FSC-GGN.

state actors, FSC members and stakeholders from around the world participate in agenda setting. Some standards are harder to implement than others; some contradict local legislation and/or do not fit local social, political and environmental contexts. Therefore, different interest groups request standards and/or policy change and send motions to the general assembly and by doing so are setting up new agendas for the entire network. Motions provide a feedback loop mechanism between local and global, in which local actors can contribute to changing global agendas.

In transnational space, new policy is negotiated and formulated. Next it is adapted and adjusted to national environments. The space of place represents a type of interpretative platform in which stakeholders negotiate the ways the standards can be implemented. Here, in the sites of implementation, a lot of experimentation takes place. In certain sites, models are built by FSC implementation with the aim to later reproduce best practices and lessons learned (see model forests developed by NGOs in Chapters 6 and 7). Holding companies that adopt the FSC and have forest management areas in different countries and/or in different regions of one country experiment with and model the implementation of FSC rules as they strive for efficiency and standardization of their practices on the ground (see Chapter 8). Institutionalization of new rules takes place in FSC certified territories, and over time sites of implementation come to represent islands of sustainability in different nation states. In certain cases the new standards are adopted by national legislations, like in Bolivia, Belarus, Ukraine and Sweden (Elbakidze *et al.*, 2010), in others they are not, as in Russia where it was explained in this volume. FSC policies and standards are evaluated by national stakeholders in each particular country through different means and both grievances and best practices from the sites of implementation contribute to the process of non-state actor's agenda setting. Every three years at the general assembly new policy strategies and targets can be set up and the actors involved in GGN's node of global governance design formulate new policies and standards. The new designs and policies then again enter the stage of adaptation and adjustment to multiple contexts (Figure 9.1: external circle).

A closer look at the FSC-GGN shows that its operation is not limited to the conventional policy circle, because there is a combination of several policy making arrangements taking place simultaneously, with top-down and bottom-up connections between actors in transnational spaces and the spaces of places. In particular, in the transnational node of design, actors from all over the world are setting up and negotiating new policies and standards that are based on global discourses on sustainable development. They transfer wider discourses on sustainability into new regulatory arrangements for sustainable forest management (Arts *et al.*, 2010). There is a time lag between transnational negotiations of new global Principles and Criteria and their adoption and re-negotiation in the regions. Thus, for example, new FSC Global Principles and Criteria were voted by the membership in January, 2012, and since then global generic indicators are in the process of development (see Chapter 3). Simultaneously, on the level of concrete nation states, members and stakeholders are involved in adoption and adjustment of policies which were set up on the transnational level much earlier. National standards developed under old principles and criteria (approved in 1993) are still in use in different parts of the world. In Russia, for example, in February 2012 members and stakeholders were still negotiating the 7th version of a national FSC standard, which is based on old global Principles and Criteria, despite the new global Principals and Criteria which were approved by members in January, 2012. Simultaneously, companies and stakeholders on the ground in different regions are interpreting and implementing the 6th version of the Russian national standard (see Chapter 4). Multiple forums with local stakeholders are set up in this process of institutionalization of the standards at the sites of implementation (see Chapter 8).

FSC members and experts can participate simultaneously in both global and national standard development across a hierarchy of scales, bringing to the global arena concerns from local sites of implementation, and vote directly for the global and/or national standards and policies, vote for the FSC International board of directors (Figure 9.1: FSC-GGN is shown vertically in grey color, policy setting and forums of negotiation are organized by the FSC node, on the national level by FSC national offices for policy adaptation and at the sites of implementation during the institutionalization process).

The FSC-GGN cannot totally avoid top-down hierarchical arrangements, but the overall system seems to be very democratic. A closer look demonstrates that on the one hand, it is more of 'direct democracy' than in a regular policy process, on the other hand it is 'managed democracy' based on membership and not on overall society, where transnational actors play a superior role. Transnational stakeholders operating in the node of design and forums organized by the node are global rule makers that have more freedom compared to national and local stakeholders, which have to comply with already established rules of the game. When rules come to the sites of implementation, they are still top-down, as are state laws and regulations (the same as legislation-policy), although much more flexibility is allowed for interpretation in the process of institutionalization as compared to legislation.

One of the differences of a GGN with the conventional policy circle is its enforcement structure. In addition to Certification Bodies, who verify compliance with the FSC rules in the sites of implementation and are comparable with enforcement agencies, NGOs represent an additional voluntary surveillance agency. In case these companies are FSC certified, NGOs operate within the FSC network, following those procedures. They also may charge non-certified companies with harmful practices on the ground, confront their head offices in transnational places, negotiate with them, and/or organize a consumer boycott (see Chapter 5). As a result companies are converted toward a more sustainable way of doing business (Cashore *et al.*, 2004), join the FSC-GGN and contribute to the policy circle, often on several levels (Figure 9.1: internal thin-grey arrow pointing up, right side-TNC bad practice in the space of place-NGOs affecting TNC office in transnational space, NGO-TNC negotiation fosters TNC to adopt FSC). This surveillance agency, with its threat of consumer boycott, is a necessary component of FSC operation (Conroy, 2001, 2007). It introduces a certain tension in FSC-GGN and, to some extent, affects the policy circle.

9.2.2 FSC-GGN and institutionalization of new rules on the ground

Section 9.2.1 dealt with how a GGN works in theory, this section will highlight how it works in practice. Although the global FSC Principles and criteria are the same around the world, the practices that one can see on the sites of implementation on the ground are very different. This happens partly because of how the FSC-GGN is set up. Global standards are developed transnationally; they have to be general enough to be relevant to all kind of social and environmental contexts. Therefore, the process of their development involves abstraction and standardization (Kortelainen *et al.*, 2010). Global principles and criteria cannot be changed nationally. However, as was explained in Chapters 3 and 4, a national standard can be developed. This constitutes the process of operationalization. Members of social, environmental and economic chambers of national offices can develop indicators and verifiers which, together with the global forest management standard, constitute a national standard that has to be accredited by Accreditation Service International (ASI). Various actors get involved in negotiation in order to pursue with their own interests through exchange and cooperation with others, yet without compromising their autonomy (Herting, 2007: 50). Therefore, the national indicators and verifiers differ from the global generic indicators and verifiers in their strength, depending on the composition of the experts involved in the national standard development (Figure 9.2: global standard (round shape) and national standard (square shape)). If the country is big and has diverse economic, social and environmental features, sub-national standards can be developed for different regions of one country. These standards cannot contradict or be significantly different from the national standard, but can have specific regional indicators. They also differ in different countries. For example, both environmental and social requirements are stricter in the 6th version of the Russian national standard than in the Swedish standards in use (Elbakidze *et al.*, 2010).

As explained earlier, the CBs are third party auditors, who verify compliance with the FSC standards on the ground. They are very important agents of institutional change as they have the ability to direct activities of the company within the framework of the FSC. They also have the right to develop indicators and verifiers to the standards in countries in which a national standard has not yet been developed and/or not yet accredited by Accreditation Service International (ASI). If this is the case, different CBs can adjust the global standard to a national context. The process of standard adaptation depends on the backgrounds and composition of the team involved in the CB (Malets & Tysiachniouk 2009). Standard's interpretation and assessment of company compliance also depends on the composition of the team involved. If different CBs are certifying different territories belonging to the same holding company, the auditing approaches on the ground can vary significantly, because different CBs can give more or less attention to social and environmental aspects during the certification process. This results in different practices that the company implements on the ground (Figure 9.2: CB-1 or CB-2-result in different configurations (in grey color) on the sites of implementation). Although CBs are accredited by ASI using the same scheme (Figure 9.2: CB-1 and CB-2-circles), in the field they use different local experts (Figure 9.2: CB-1 (square), CB-2 (triangle)) which affects the auditing process: differently composed teams perform both standards adjustments and assessment on the ground in different ways. Even when different CBs are using the same national and/or sub-national standard, practices that are institutionalized in the sites of implementation can differ due to varying interpretations by team members in different CBs (see Chapter 8 on the example of forest management by the Northern Logging Company). It is important to acknowledge that CBs are not the only important actors of institutional change affecting practices in the sites of implementation. Local practices depend on what kind of experts the company employs during the certification process, what kind of company staff is involved in certification process and what kind of stakeholders are involved in certification process. Although the standards are the same, the practices implemented differ from one site to another (Figure 9.2: standard implementation, square and circle configurations – grey color).

Therefore, institutionalization of the FSC rules in the sites of implementation depends upon which actor has the larger agency in the FSC-GGN network. The composition of actors, as shown in the study, influences the process of implementation-institutionalization of the standards in concrete

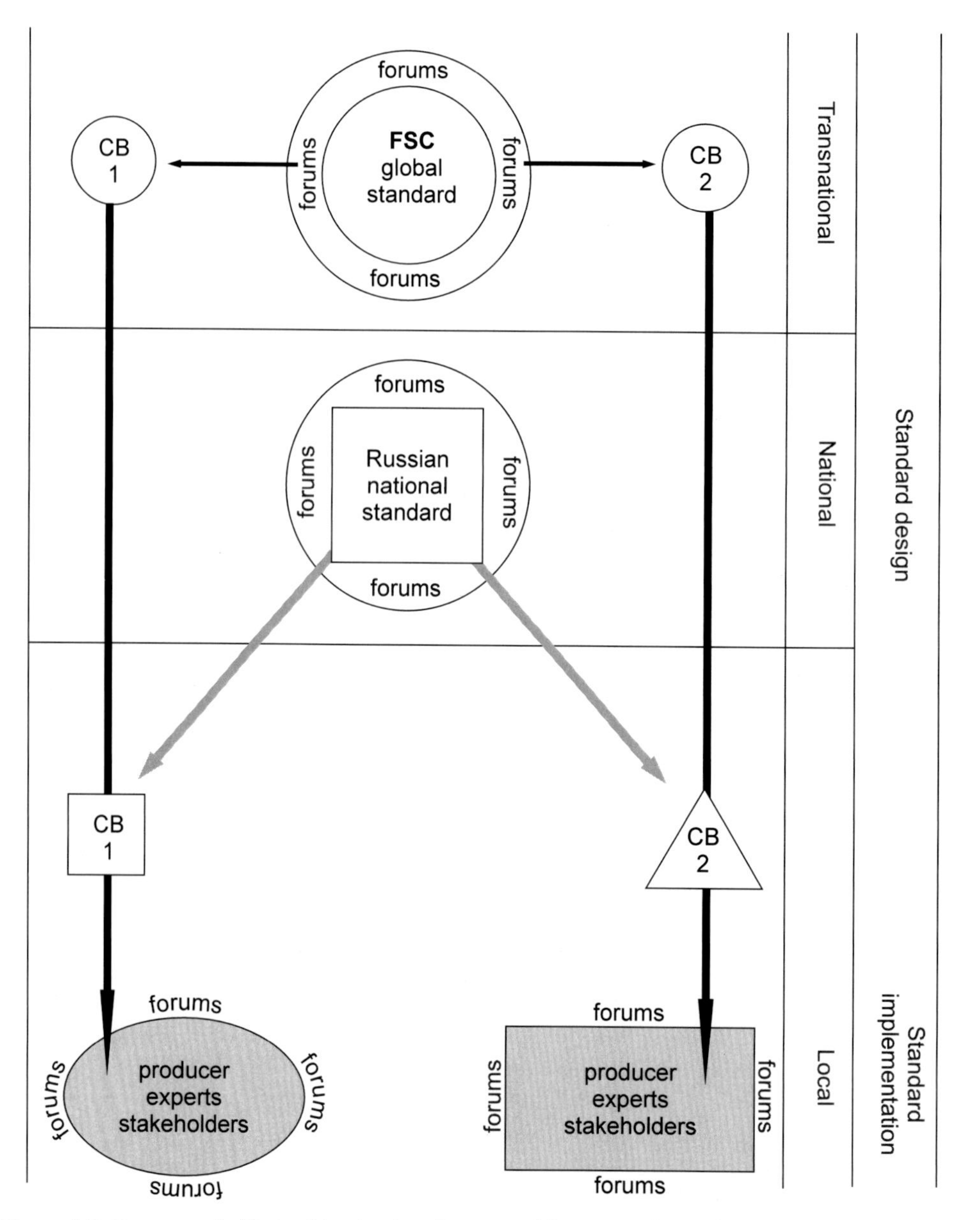

Figure 9.2. Causes and effects of institutionalization of the FSC rules at the sites of implementation.

practices on the ground. In particular, it depends on the companies (certificate holders), on experts involved in the implementation process, on governments of different level, on local communities and other stakeholders. For example, if the major implementer is the company, involving experts in some particular issue, such as intensification of forestry and nature conservation planning, those standards related to the company's priorities are more deeply implemented than are others (see Chapter 6, Pskov Model Forest). If an NGO partners with governmental agencies in building a model of sustainable forest management, it experiences fewer difficulties in the process of FSC implementation. In a case in which an NGO implementer has external foundation money, it can invest both time and money for more profound implementation of both social and environmental standards and even for over-compliance with the FSC rules (see Chapter 7, Priluzie Model Forest). In a region where there is a radical environmental NGO which monitors designation of key biotopes and preservation of old growth forests, the FSC certified company will take implementation of related standards more seriously and more significant changes in forest management practices can be observed compared to other regions (see Chapter 8, the case of the Northern Logging Company).

As demonstrated in the study, the FSC standards are not prescriptive and straight forward. There is a lot of space for negotiation and interpretation of the standards during their implementation. Practices that are implemented depend on consensus building between stakeholders, which contributes to additional value of the product related to the democratic process of decision making, which is termed the consensual value of the product. The flexibility that the FSC-GGN allows in interpretations and negotiations at all levels helps to adjust the standards to local social, political and ecological environments in the sites of implementation. It also allows activating the feedback loop at the transnational level and changing the standard in case it does not fit the local environment.

9.2.3 The role of NGOs in the institutionalization of new norms, designs and practices on the global and local levels

The FSC-GGN involves in the process of standard setting adoption and implementation many strategically operating actors, such as TNCs, experts-consultants, etc., that are constrained by previously established rules of the network. However, the main governance agencies in all described cases were NGOs, facilitating the global/local interplay and connecting transnational and local spaces while installing and institutionalizing new practices. They turned out to be effective actors and agents of institutional change and managed to change rules of the game, power distribution, and discourse (Arts *et al.*, 2006) in both spaces. The new governance arrangements, as well as lessons learned, were possible to reproduce, at least partly, in the framework of the FSC. Therefore, the institutional change turned to be isomorphic.

For understanding, explaining and theorizing the role of NGOs in institutionalizing new norms and practices, I combine the approach of sociologists of transnational processes (Castells, 1997, 2009, 2011) with the more conventional approaches of neo-institutional sociology (DiMaggio, 1991; Fligstein, 2002; Powell *et al.*, 2005) to bring to light the mechanism of generating new governance. Recent studies show that it is possible to expand the foundations of neo-institutional sociology, based on the nation state, to the global, transnational level (Dingwerth & Pattberg, 2009b) and by doing so to explain the mechanisms of institutionalisation of new norms and practices and to highlight the role of NGOs in these processes.

NGO-led market campaigns against companies involved in harmful logging practices, or buying wood from unsustainable and/or illegal sources, as widely described in the literature (Cashore *et al.*, 2004; Conroy, 2007) and in this book (Chapter 5), provide coercive pressure both on TNC transnational offices and their local subsidiaries and suppliers in the space of place (Figure 9.3: NGOs arrows up and down – coercive). It forces companies to change their own global environmental and social policies and standards, to adopt or revise their codes of conduct, to

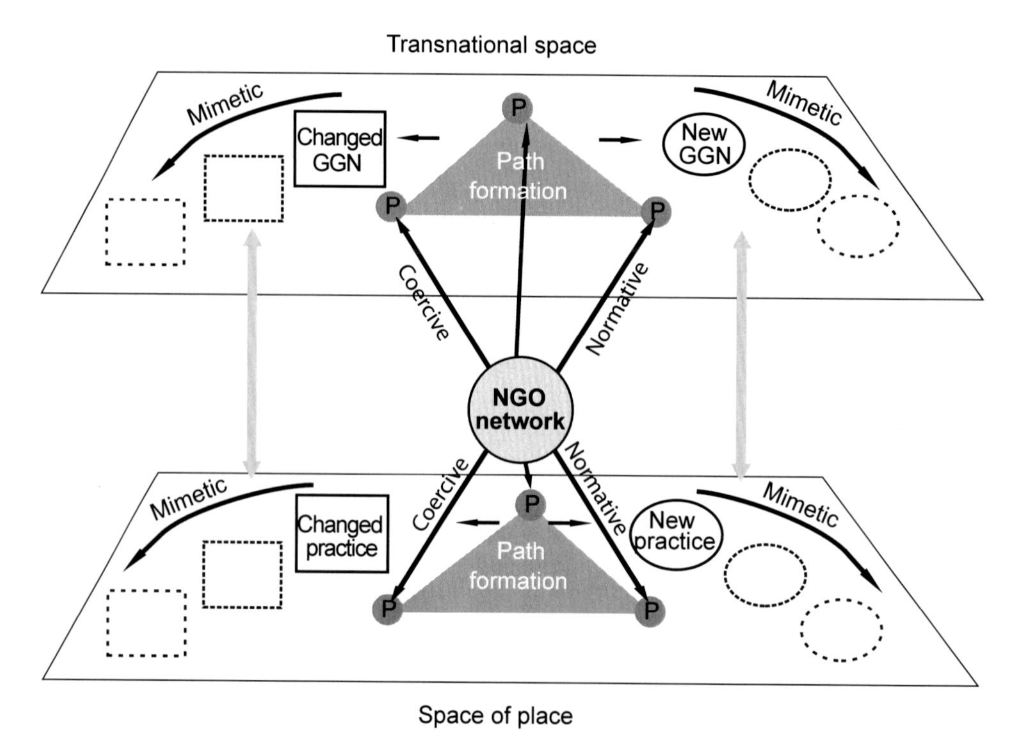

Figure 9.3. NGOs institutionalizing new practices in transnational spaces and the spaces of places.

create new 'taboos', such as logging old growth forests (see Chapter 5), to adopt ILO conventions for their workers, and to respect the rights of local communities and indigenous peoples.

Simultaneously with market campaigns, NGOs build models of sustainable forest management on the ground, advertise such models and educate both transnational and local stakeholders about the advantages of sustainable forest management (see Chapters 6 and 7, Figure 9.3: NGOs, arrows up and down – normative). By exercising both coercive and normative influence, NGOs foster new discourses that help to institutionalise new practices, to break from the old and form a new path: a path towards sustainable forest management (DiMaggio, 1991; Powell *et al.*, 2005). They foster creation of new institutional fields (Bartley, 2007a) with their institutional infrastructure, including expertise, surveillance agencies and new rules of the game. In this way they contribute to a market transformation towards sustainability. Facing both coercive and normative pressures, TNCs adopt FSC rules and become part of FSC-GGN (Figure 9.3: transnational spaces – changed GGN (square)). In the case in which a large holding company adopts FSC certification, it fosters sustainable forest management practices of the subsidiaries and suppliers on the ground (see Chapter 7, the example of Mondy Business Paper and Chapter 8, the example of companies belonging to Investlesprom, Figure 9.3: spaces of places, changed practice – squares). Power relationships change along with the formation of new institutional fields and new path dependencies, both in transnational spaces and the spaces of places (Figure 9.3; triangles with P in the corners).

The FSC is capable of mimetically reproducing its sustainable forest management practices as well as its transnational policies and standards (Figure 9.3: on the left side – squares in transnational spaces and spaces of places). Sustainable forest management in the framework of the FSC is gradually becoming the norm. In other sectors, NGOs along with other governance actors, foster new governance arrangements to mimetically reproduce the FSC global governance model (Figure 9.3: on the right side). New GGNs, such as the Marine Stewardship Council, the Tourism Stewardship Council and others, based on private authority in policy making, multi-stakeholder engagement and similar to the FSC's institutional set up, havecome into being, institutionalized their practices and created new institutional fields (Dingwerth & Pattberg, 2009b: 723-724) both in transnational and local spaces (Figure 9.3: right side, new GGNs – circles in transnational spaces, which bring along new practices – circles in the spaces of places). These new governance arrangements, with their institutional fields, created new islands of sustainability around the globe and constituted new environmentally and socially sensitive niche markets. New governance is generated in such way.

9.2.4 Certification and labeling scholarship: what this study contributes?

The booming scholarship on certification and labeling involves, but is not limited to, neo-institutional economic studies (Bartley, 2007a; Dingwerth & Pattberg, 2009b), governance literature (Cashore *et al.*, 2004; Gulbrandsen, 2010), and policy evaluation and impact research (Newsom & Hewitt, 2005), involving value chain analysis (Auld *et al.*, 2008b; Blackman & Rivera, 2011; Vagneron & Roquigny, 2011). There are no clear boundaries between these research streams, and this volume contributes to all of the above mentioned literatures.

One of the approaches of economic scholarship, used in this study as a complimentary-explanatory theory, is the neo-institutional organizational analysis of business regulation through certification and labeling. Scholars analyze why firms choose to participate in voluntary certification and labeling schemes in addition to, or instead of, using self-regulatory instruments or simply complying with the existing state regulations, and how the incentive structure is built in one or another labeling schemes. Economic studies focus not only on firms, but on niche markets involving sustainable production and consumption in which certified companies and environmentally and socially sensitive consumers are engaged. They analyze how negative externalities affecting communities in free markets are eliminated in niche markets constructed through certification and labeling schemes. Neo-institutional organizational analysis (Bartley, 2007a; Dingwerth & Pattberg, 2009b; Powell & DiMaggio, 1991) is used in some studies for explaining how firms are able to reduce transaction costs by engaging in certification and labeling, how old path dependencies are replaced by new ones; how new institutional fields involving sustainable markets are formed; and to what extent they can be institutionalized and reproduced (see Bartley, 2007a; Chapters 2, 5-8,). This study, along with Tim Bartley's research on certification and labeling, is contributing to neo-institutional economic sociology by extending its foundations beyond the nation states to the transnational level (see this Section 9.2.3).

The major contribution of this study, however, is to governance literature on certification and labeling, in particular on private authority in governance. Governance by private authority through certification schemes was previously conceptualized by Benjamin Cashore as a 'non-state market driven governance' (NSMDG). The theory of NSMDG was primarily developed for the FSC certification of wood products (Cashore, 2002) and later expanded to other certification systems, in particular to the apparel industry, toys, cosmetics, production of coffee, tea, cocoa, sugar, bananas and other agricultural products, the construction industry, fisheries, mines and the emerging tourism certifications (Auld *et al.*, 2009; Bernstein & Cashore, 2007; McNichol, 2006: 349). The theory of NSMDG (Bernstein & Cashore, 2007; Cashore, 2002; Cashore *et al.*, 2004) explains the mechanisms lying behind the sustainability of global supply chains.

For the analysis of the same phenomena, the concept of GGNs is used, which captures certain points that cannot be explained by NSMDG. The term NSMDG itself does not precisely reflect the core of the phenomenon underlying certification systems as global regulatory mechanisms. In fact, governance through voluntary certifications is not driven by the market, but by actors, predominantly NGOs who use market forces and incentives, to make impact on companies' production processes and practices on the ground.

The theories of NSMDG and GGNs originate from different schools of thought. NSMDG arises from political science, while GGN, to the larger extent, from the sociology of transnational processes. Both have great explanatory power for analyzing the emergence of sustainable niche markets in which the consumer's desire for environmentally and socially friendly products can influence corporate policy. Both approaches acknowledge the significant role of NGOs as agents of social change that are able to transform the conventional markets and foster development of the ecologically and socially sensitive markets. As an added benefit, GGNs, as compared to NSMDG, capture processes and practices that take place in the sites of implementation, involving local actors and stakeholders that react and adjust to the new global standards. The GGNs concept also, to a greater extent, focuses on the importance of multi-stakeholder forums of negotiation in the development, adaptation and institutionalization of global rules.

Another approach in governance literature is on the certification and labeling standard setting processes. For understanding transnational governance research, effort is made to study how non-state rule setting authority gets established at the transnational level, how the standard setting process through transnational multi-stakeholder involvement is organized and how it gains legitimacy in the market and the transnational policy arena (Hallstrom & Bostrom, 2010). The literature on standard setting processes also analyzes political rationalities that the standards promote and assesses to what extent the newly developed standards can resolve potential conflicts between economic, social and environmental aspects of sustainability (Djama, Fouilleux & Vagneron, 2011). For example, ISO, FSC and MSC scholars analyze complicated processes, such as power struggles, power asymmetries, mutual adjustments and mutual learning in multi-stakeholder meta-organizations in which members represent different types of organizations, involving firms, NGOs and other entities. Although scholars argue that multi-stakeholder involvement is necessary at all stages of the certification and labeling process, they talk less about the network structure of certification and labeling governance arrangements and look at networks through the lenses of sociology of organizations. The feedback loop of a particular certification scheme and processes on the ground in particular countries is left out from the analysis. Compared to the GGN concept, the multi-stakeholder organization approach is limited to studying the node of design and forums of negotiations organized by the node in the process of standardization and labeling. Such studies bring many insights to the governance literature, however, they leave out the processes taking place on national, regional and local levels, in which standards are adjusted, renegotiated and transferred to concrete practices on the ground. The GGN concept, compared to the multi-stakeholder organization approach, allows capturing relationships between the standard setting processes and standards implementation processes in specific territories; it allows comparing transnational standard negotiations with negotiations in specific regions and in states where political regimes and contexts differ. In this respect, the GGN approach is most important in that it is able to integrate structures and processes located in different spaces and places and is able to provide a lens for analyzing the functional unity of actors involved in a particular certification and labeling network. It can also analyze relationships between GGNs: their familial relationships, cooperation, alliances and competition (see Chapter 3).

There is little research in governance literature that, in line with this study, analyzes certification and labeling using the sociology of transnational process, in particular Castell's network society theory as conceptual tool (Oosterveer, 2007: 45-48; Oosterveer & Spaargaren, 2011: 100). In a similar way as in this study, following Castells and Urry, these scholars acknowledge time compression coming with globalization and distinguish between spaces of places where production and consumption of goods takes place and spaces of flow (transnational spaces) where transnational governance arrangements are settled (in this study in GGN nodes). Using certification schemes in the agro-food business as an example, Oosterveer explains how local social and environmental concerns get incorporated in new transnational private governance arrangements, which mediate between the places of production and places of consumption (Oosterveer, 2007: 194-216). Both approaches can be easily combined and the GGN concept can benefit from incorporating spaces of consumption into its framework and emphasizing consumers as actors and agency in these spaces. Generally, Oosterveer's approach is broader than the GGN concept that allows following each particular certification project from the standard setting process through standard implementation in a specific territory.

Generally, the GGN approach, compared to other approaches, can capture how the certification scheme is adapted to changes in global policy priorities, such as incorporating carbon into the standards for coping with climate change (by studying negotiations organized by the node of design) and at the same time can monitor local responses and consequences of new standard implementation in specific certified territories (studying sites of implementation). It can follow the discourses, grievances and interests of multiple stakeholders involved in the different levels of forums of negotiation. It can assess to what extent the feedback loop is working between global and local dimensions within a particular GGN.

The GGN concept can also bring phenomenological, explanatory power to the impact studies. There are different ways which allow evaluating the implementation of any certification scheme. The most common is the quantitative approach, for example, how many old growth forest landscapes are preserved with FSC implementation, or how many farmers in a particular country benefit from fair traded coffee or bananas. The GGN approach allows for assessing the transformative character of implementation of a certification scheme on the example of a specific case study in a particular country and region. As it is actor oriented, it can explain why the same standard is implemented differently in different sites of implementation.

Therefore, empirical study of the FSC-GGN offers analytic leverage because it allows scholars to discern agency within the network and to determine which actors generate and accumulate power, and therefore are more important. When major agents of institutional change are identified within a particular certification project, it then becomes possible to analyze how they change authority and power distribution in the sites of implementation.

9.3 The role of actors in shaping global/local interplay in the process of installing new rules and practices

Empirical study of GGNs, with their nodes of design, sites of implementation and forums of negotiation requires multi-sited research, which was conducted for this volume. The study of several cases in Russia revealed how different configurations of networks make for differences in practices on the ground. Distinct cases demonstrate where power, resources and authority lies in a network. It became possible to analyze the combinations of actors and agency that matter for the eventual effectiveness of environmental governance. In using GGN as an analytical framework, it is still important to look at actors and agency; in my cases the interplay of NGOs and TNCs that clash with the rigidity of the Russian state.

In this study, three types of networks involved in installing new sustainable forest management practices in Russia were analyzed. The intention was to understand how governance actors are shaping global/local interplay and what outcomes it brings in fostering sustainable forest management in specific sites in Russia. All selected types of networks are related to the Forest Stewardship Council (FSC) certification and were selected according to the conceptual model outlined in Chapter 2.

9.3.1 Revisiting the conceptual model and the case studies

Three types of NGO networks, selected in accordance with the ideal types, have been analyzed (Figure 9.4). The first type, the Bottom-up Market Campaign is directed from the space of place to transnational space, in this study a market campaign for saving old growth forests in Karelia. This market campaign served as a prerequisite for certification, as it threatened business as usual and encouraged companies to adopt the FSC. A bottom-up consumer boycott was an important means of institutionalizing new norms, creating a new 'taboo', and convincing companies to have proof of environmentally friendly operations. The second type, the Top-down Guided Standard Implementation, consisted of two cases of networks, directed from transnational space to the space of place, namely the Pskov Model Forest and the Model Forest Priluzie. Both cases represent a model of sustainable forest management build with foundational support, in which the FSC took part. Top-down models with best practices on the ground served as sites of experimentation. Funding from foundations enabled negotiation, elaboration and testing of different kind of innovations in specific territories. The cases differed immensely in the FSC implementing agents; in the first case, a NGO partnered with the company and in the second case, with state agencies. The outcomes were different. Finally, the third type, the Hybrid Glocal Governance, involved a two-way network, constantly linking transnational spaces with the spaces of places represented by 'pure' FSC certification of a particular company, namely the Northern Logging company. Despite these broad variations in directionality, field research showed that in practice all networks diverged from the suggested ideal types in different ways.

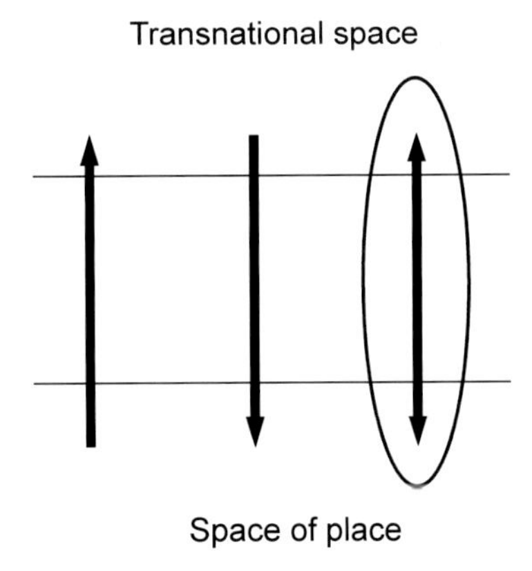

Figure 9.4. Conceptual Model: ideal types of networks installing forest management practices. The oval circle indicated the new way of decision making, being sustainable.

Market campaign (case 1)

Market campaigns represent a special type of advocacy campaigns, which use the force of the markets and impose market-based pressures on TNCs in order to foster changes in their practices in an environmentally and socially friendly direction in both production processes and trade relations (Conroy, 2001, 2007; Friedman, 2006). Therefore, the distinctive feature of market campaigns, which differ from other protest movements, is the use of market mechanisms as a powerful tool of impacting corporations.

In the late 1980s-early 1990s, NGOs involved in a market campaign to protect the old-growth Karelian forests attacked European buyers of Russian wood, in my framework it was a 'bottom-up network'. Directing its grievances toward corporations in transnational spaces by naming and shaming them, the campaign not only achieved its conservation goal in the space of place in Karelia, but it helped to institutionalize the new practice of preservation of old growth forests in Northwest Russia. It opened the window of opportunity for the FSC. Companies realized that they need some instrument to prove their dedication to the preservation of the old growth forests, as the logging of high conservation value forests became a non-tariff trade barrier with European markets. Such an instrument was provided by the FSC certification; hence almost all large holding companies joined the certification process (see Chapter 5). As in other countries, (O'Rourke, 2005; Conroy, 2007) in Russia the campaign reached the multiplier effect and fostered the FSC in Karelia and other parts of Northwestern Russia.

The bottom-up market campaign was not the only network responsible for installing new sustainable forest management practices in Russia. Several other networks were involved in the same issue of preservation of old growth forests in Karelia. For example, there was a European Union project involved in a top-down network providing resources for designating specially protected areas in Karelia. Another network was a trans-local one, connecting the Finnish and Russian NGOs interested in preserving the Kalevala cultural folk singing villages and surrounding old growth forests on both side of the border. Both networks pursued similar goals, using less confrontational measures. They actively softened the stressful effects of the market campaign on local actors. The same NGOs that were involved in the market campaign received foundation grants to designate a new specially protected area, the Kalevala National Park. NGOs participated in all three networks simultaneously, facilitating the overall positive outcome (Figure 9.5: case 1, bottom-up (black), a major ideal type network, top-down (grey), supplementary network (a small black arrow in the bottom, a supplementary trans-local network connecting Russian and Finnish NGOs).

Model forests (case 2 and 3)

The model forest cases (both projects in early 2000s) represented networks in which the NGO-implementer used resources from industrialized countries (granting agencies) for implementing sustainable forest management, in my framework labeled 'top-down' networks. Model forests were an arena in which global ideas and designs of sustainable forest management were tested and implemented on the ground in a transitional country. In these networks, FSC certification was an essential part of building a model of sustainability on the ground, which went hand in hand with the top-down model forest network (Figure 9.5: case 2 and case 3, in addition to the top-down (black), FSC supplementary network (grey), see Chapters 5 and 6). In this case the model of

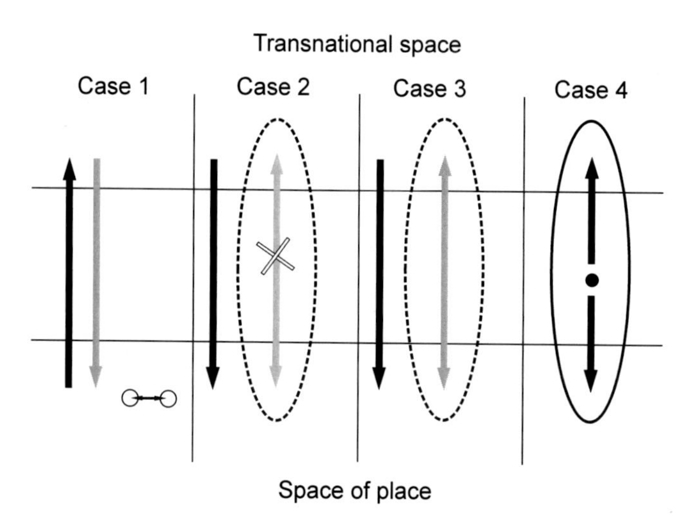

Figure 9.5. Divergence from the ideal type. Black, thick arrows represent ideal types of the networks; grey arrows represent networks, additional to the ideal types, involved in concrete case studies. They supplemented and fostered implementation. Grey arrows represent empirical findings. The cross through the supplementary network in case 2 shows that the network (FSC certification of the area) no longer exists. The dotted circles around the supplementary networks (case 2-3) show connections between the spaces of places and transnational spaces coming with governance through certification. The combination of black and grey arrows represent the divergence of the empirical findings from the ideal types. Two small circles with horizontal arrow (case 1) mean a translocal network, see more explanations in the text.

sustainability was supposed to go beyond FSC requirements. In such cases, FSC certification was facilitated by the availability of outside funding, coming through the model forest project. In both cases, the NGOs were using the resources coming from transnational spaces for implementation of practices at specific spaces of place, which were the sites of implementation of the models. In certain ways, this outside funding allowed the FSC requirements to be exceeded. The main purpose of the NGOs in those cases was to transfer the design and resources of transnational space into expert and human resources at the space of place. This transfer included the organization of the forums of negotiations, empowerment of the stakeholders and local communities, and stirring up public participation. As project organizers, NGOs strived to achieve consensus between the stakeholders and moderated their negotiations. In contrast to consumer boycotts, their activity was rather conflict-free, and, if tensions appeared, they solved them through negotiations.

The two model projects studied in this book had different outcomes, as the network agencies were different. It turned out that partnership with state agencies was essential for the success and reproduction of lessons learned in installing sustainable forest management practices in Russian regions. Partnership allowed the institutionalization of certain practices of sustainable forest management through the Komi Republic (case 3). The Pskov Model project implementing team, working on behalf of NGO-TNC (WWF-Stora-Enso) partnership, developed many innovations in intensive forest management which go beyond the FSC requirements; however, it could not disseminate them to other regions because these innovations contradicted existing Russian legislation. It was only possible to spread nature conservation planning practices that are part of FSC requirements, although some of them also contradict Russian legislation. Although the TNC-NGO partnership developed many innovations in forest management, they were not able to reproduce outcomes of their work. This difference of network outcomes depended on the involved agents of institutional change, e.g. NGO-state partnership versus NGO-TNC partnership. The two model forest projects differed in their sustainability. The Stora Enso subsidiary STF-Strug was sold and when the funding ended the project failed to continue and a new company operating in the area did not maintain the FSC certificate (Figure 9.5: grey-crossed, see Chapter 6). The WWF-State agencies partnership in the Model Forest Priluzie allowed implementers to overcome multiple contradictions with Russian federal legislation of innovative sustainable forest management practices. Special permissions and exclusions that the Komi regional government received from the federal government allowed implementers to overcome legislative barriers and to disseminate some of the innovations throughout the Komi republic. After the end of the project, the TNC Mondi Business paper, which is a key player in the Komi Forest sector, became interested in supporting the NGO Silver Taiga, which then became the company's FSC consultant. With major support from Mondi, Silver Taiga continued testing different kinds of innovations on the territory of the Model Forest Priluzie and continued to foster FSC certification in Komi. This new NGO-TNC partnership ensured sustainability of the project; however, in recent years activities have not gone beyond the FSC's requirements. Certification in Komi is growing and no longer needs the outside funding it relied on in the early stages (see Chapter 7). Therefore, the network gradually morphed into a regular certification project, described in case 4.

Certification of the company lease (case 4)

FSC certification, in which outside funding is not used, is a relatively stable long-lived entity as it is built on incentives for producers to implement sustainable forest management practices at the logging sites. The case of certification of the Northern Logging Company shows how the FSC operates in practice in Russia, demonstrating ups and downs as well as difficulties related to the process. The sustainability of the network is maintained by market dynamics. Despite drawbacks, new rules and practices were institutionalized. The NGOs involved in the certification of the holding company took on several roles simultaneously. They were stakeholders in certification, and played the roles of the business' opponents as well as that of its consultants and experts. This certification project turned out to be a complicated process that coincided with holding company's formation, modernization and transformation, processes involving painful impacts on both company workers and local communities. The FSC, to some extent, softened the negative impacts of the holding company's transformation on local residents by fostering a new type of corporate social responsibility towards communities in the leased territory. The case study illustrates the complicated process of FSC institutionalization, with constant interplay of local and global actors going through interruptions and breakdowns, in which the FSC certificate was suspended and later returned (Figure 9.5, case 4 interrupted two ways networks with constant global/local interplay of actors).

In contemporary Russia, the bottom-up NGO networks that previously organized consumer boycotts (case 1) have become informal surveillance agents for the FSC certificate holders in Russia. As the majority of the holding companies in Northwest Russia have become FSC certified, NGOs are now able to operate through the FSC framework. The role of NGOs has changed from blaming and shaming towards working in the framework of the FSC. NGO representatives turn the attention of CBs towards poor certificates and organize monitoring trips to the FSC certified forest management areas. The threat of a NGO-lead market campaign still exists and provides the necessary tension, which keeps the TNCs involved in improving their compliance with the FSC standards. The official surveillance agency, in the form of CBs whose aim is to monitor compliance with the FSC standards on the ground, is not enough to ensure the high quality of new sustainable forest management practices on the ground.

The top-down networks, in which development agencies sponsor implementation projects in Russia, have diminished immensely as Russia now is perceived by the international community as a country with a developed market economy and a high inflow of 'petro-dollars'. The interests of development agencies have shifted from the countries with transitional economies to countries in the global South with tropical forests. Therefore, the top-down cases described in this book were important in setting up FSC institutions, in experimentation and modelling with adjustments of the FSC to the Russian context. The top-down models shifted to a more sustainable, marked-based instrument of sustainable forest management: the FSC. Based on market incentives, FSC certification, despite multiple challenges and drawbacks, has potential in installing new sustainable forest management practices in Russia. Actors involved in FSC certification constantly link transnational and local spaces, providing a new way of participatory decision making and new governance arrangements through multiple stakeholder involvement.

9.3.2 FSC implementation: transformation of forest management practices and unintended consequences

This study assessed the capacity of different actors involved in FSC-GGN in transferring general rules expressed in the FSC Principles and Criteria to concrete forest management practices. It also analyzed how actors behave and negotiate their interests when the FSC rules coincide with other regulatory streams in forest management coming from transnational spaces, e.g. international conventions and TNC policies, coming from the space of place, e.g. Russian national legislation.

As mentioned, the FSC certification scheme has 10 guiding principles and 56 criteria that are supposed to be implemented in concrete forest management territories. The case studies demonstrated that not all criteria were fully implemented and forest management practices were not fully transformed toward sustainability in the sites of the FSC standard implementation. Some FSC rules were transferred into new practices while others were not fully transferred. This can be partly explained by the network agency with their implementing actors involved in a particular certification project, such as NGOs, and TNCs. The context of the nation state also turned to be important. On the stage of implementation of the FSC standards, implementing agents were obliged to take into account state agencies and their policies and regulatory instruments. Differences in state actor's reactions to the FSC rules determined many differences from one site to another, as well as regional differences in standards implementation.

NGO actor's engagement at all levels of the FSC-GGN turned out to be a determining factor (see Section 9.2.3). Major transnational NGOs, such as the WWF and Greenpeace, as well as their networks in Russia, are strongly involved in preservation of intact old growth forest landscapes worldwide and in Russia, in particular. The first steps were consumer boycotts of companies logging old growth forests (see Chapters 2 and 5). The second step was working in the framework of the FSC. NGOs helped to formulate FSC Principle 9 and the criteria belonging to it (old growth forests belong to the HCVF-2 category). They got involved in enforcement and monitoring compliance of old growth preservation as FSC surveillance agents (see Chapter 3) and achieved compliance both globally and in Russia. Although Russian legislation does not require old growth forest preservation, they are all preserved in FSC-certified territories in Northwest Russia under voluntary moratoriums signed by NGOs and the logging companies (see Chapters 5-8). In many districts, attempts are being made to exclude the old-growth forests from logging via creation of specially protected natural areas. Designation of new nature reserves, however, is a very long and challenging route; therefore by the year 2012 this issue remained unresolved.

The FSC social chamber is weaker than the environmental and economic chambers, both on the transnational level and in Russia. The former involves small social NGOs, a few trade union representatives and indigenous peoples organizations (Hallstrom & Bostrom, 2010, Chapter 4). Social NGOs in Russia do not play the same role as environmental NGOs. In Russia, most of them contribute to FSC standards adjustments, standards negotiations on the national level and are consultants for both companies and communities in the sites of implementation. Occasionally, they are involved in FSC standards implementation as surveillance agents, but this role is marginal, compared to environmental NGOs. They can be credited, however, for developing the standards and fostering the implementation of the designation of forests of social value (HCVF 5-6 category). Despite the fact that FSC certification in Russia has already existed for a long time, the issue of designation of social value forests (HCVF 5 and 6) came to more or less satisfactory implementation only in the years 2007-2009 (see Chapter 8) due to social expert's involvement in the process. Before that time, this issue was not raised in most cases because of the absence of methodology of designation of the HCVF 5 and 6, or it was implemented only partly, as the term 'forests of a high social value' was narrowly understood by the companies and the FSC auditors. For example, in the Komi Republic, only the places for gathering berries and mushrooms were designated (see Chapter 7), and in the Karelia HCVF 5 and 6 were fully designated much later, after certification was established there (see Chapter 8).

Social NGOs and experts fostered new forms of intersectoral dialogue with civil society institutions in Russia, thus enhancing development of democratic initiatives in rural communities and democratic decision making processes dealing with forests. The FSC requirements and NGO pressure forced companies to adopt a new type of corporate social responsibility with closer interaction with local communities, as the standards involved contributing to community infrastructure, social impact assessment, and local employment.

In FSC certification cases, TNCs were important agents of institutional change as they adopted and implemented this voluntary standard on the ground. TNC's of Western origin, who had previously adopted into their internal policies the requirements of the ILO convention (case 2), showed better performance than a TNC of Russian origin (case 4) in implementing standards related to workers safety (see more details below).

As mentioned, the state actors, state policy and legislation became important in the FSC-GGN at the national level, especially at sites of implementation, where the standards need to be transferred into concrete practices. In Russia, the co-management of forests by state and non-state actors turned out to be a complicated process, because of the constant institutional turbulence. The almost constant, and incomplete, reform of the forest sector which started during perestroika times and continued into the 2000's is one of the factors that shaped the institutionalization of global rules. The reform itself substantially influences relationships between state and non-state actors as jurisdictions change, state authorities shift agencies and positions, and legislation is poorly implemented. This is especially relevant to this study because the Forest Code of 2006 partly contradicts the FSC rules and is not satisfying any of the stakeholders. In general, although the Russian government accepts the idea of certification, it still has difficulties in communicating with private sector organizations, especially with NGOs. Despite the efforts of NGOs and companies, many negotiations on the national level went in circles and did not achieve results in harmonizing national legislation with the FSC rules (see Chapter 4). It was exceptionally complicated for the government to incorporate private processes into their activity, in view of the conservatism of the state agencies in Russia. Despite contradicting Russian legislation, the FSC standards were implemented and institutionalized. Nevertheless, interaction between state actors and non-state actors differed from case to case (see Chapters 5-8). Despite regional differences, some generalizations can be drawn from the case studies.

As mentioned, some FSC requirements contradict while others are in harmony with Russian legislation. In the situations where the FSC rules do not contradict Russian legislation, the FSC demonstrates a relatively high enforcement capacity compared to conventional state surveillance. This was clearly seen regarding forests that provide basic services, e.g. watershed protection, erosion control, etc. (HCVF-4 category in FSC standard). The same tendency was observed with gasoline leaking on to the ground, which state agencies do not pay significant attention. State enforcement agencie's work was reinforced by additional surveillance by the FSC certification bodies.

Another slightly different example is workers' safety. The ILO convention in Russia is officially harmonized with national legislation concerning safety measures, such as uniforms for loggers, special protective equipment, medical supplies, accident prevention measures, fire equipment and training. However, the enforcement is poor; logging companies manage to avoid buying expensive equipment and never pay enough attention to safety measures. Strict FSC requirements obliged logging subsidiaries to obey the rules and regulations on safety. Regular FSC audits forced companies to take measures, implement internal policies and finally improve worker's safety (see Chapters 6-8). However, in this case not only the FSC requirements, but also the TNC's corporate policy is important for successful implementation of the standards. The compliance with these FSC requirements on worker's safety went smoothly in cases where the TNC, in addition to the FSC requirements, imposed its own policies on worker's safety and was strictly monitoring for compliance (see Chapter 6, the case of Stora Enso). Companies in the process of certification were constantly receiving corrective action requests from FSC auditors and improved their performance from time to time (see Chapters 7 and 8). In such a way, a combination of FSC certification and TNC policies can help to enforce national legislation and Russian obligations towards the ILO international convention.

Very different dynamics happen when the FSC requirements contradict Russian legislation. For example, leaving critical habitat areas (key biotopes) untouched contradicted Russian legislation. Companies that did not cut all the wood on their leased territories were fined by the state forest management unit (see Chapter 8). This offers an incentive, quite popular in Russia, to get around the laws through informal person-to-person agreements. Therefore, the fulfillment of the pertinent FSC requirements became strongly dependent on relations with state structures and on the role and identity of an auditor (Tysiachniouk, 2008a, see also Chapter 8, the case of the Northern Logging Company). In another case, key biotopes were framed as in non-exploitation zones, which were allowed to be left on the logging site (see Chapter 6, subsidiaries of Stora Enso). The issue with key biotopes was resolved only in 2008, when at the regional level permission was granted to the FSC certified companies to preserve biodiversity on their logging sites.

There are also some unintended consequences of setting up and implementing FSC certification, observed both at the national level and in the sites of implementation, in every case studied (see Chapters 4, 6-8). Findings presented in this volume showed the positive externality of the global regulation of the supply chains: building up deliberative democracy, especially in rural areas. For Russia, it was demonstrated that the efforts of regulating transnational supply chains and the process of standardization and labeling in late 1990s and early 2000s fostered inter-sectoral dialogue and overall democratic development, especially in rural areas. The FSC's policy and standards require the national and regional FSC initiatives to resemble the global three chambers, economic, social and environmental, with equal representation of each group of stakeholders. This requirement fostered the FSC office in Russia to attract and encourage companies' participation, to attract social NGOs, indigenous peoples associations, trade unions, despite the fact that inter-sectoral dialogue was not well established in the society. This top-down FSC requirement promoted more democratic ways of decision making on national and regional levels.

Deliberation processes observed at the sites of implementation are even more apparent. Generally, the FSC standards are formulated in such a way that they allow negotiations, interpretations and re-interpretations between stakeholders; therefore, they foster negotiations between the different interest groups. The FSC site of implementation becomes an arena with contradicting rules coming both from transnational spaces and the spaces of place. All of the participating parties, e.g. the companies, environmental NGOs and the local authorities, were negotiating contested issues and trying to take advantage from the unclear rules. Forums of negotiation involved informal agreements, special permission, approved instructions, law suits and fines. Forums arose around issues of how to interpret the rules, what areas are considered to be of high biodiversity, to what extent the forest areas used for non-commercial purposes can be labeled as intact forest landscape, and what areas can be labeled as key biotopes.

The FSC standards require engagement of local stakeholders and regular consultations with local communities on such issues as designation of high social value forests: places of historical, cultural, religious and archeological significance (the above mentioned HCVF 5-6). In addition, companies organize community meetings to decide on maintaining community infrastructure, e.g. maintaining roads, repairing bridges, supporting local events and celebrations. Negotiations related to community infrastructure are very intense, due to the lingering effects of the Soviet past as communities are accustomed to paternalistic relationships with the state and companies. The companies resist paternalism in relationship with local communities as they are market actors and cannot even partly substitute for the state's roles as it was during socialism. In addition to the benefits that communities receive from the companies as a result of negotiations, they are empowered through the process and learn how to participate in the decision making processes. Therefore, in such a way the FSC-GGNs foster democratic engagements.

9.4 Epilogue: perspectives for future research

The GGN approach can be used for comparative studies of FSC implementation outside Russia in different countries. The FSC-GGN node of design is the same for the whole world, but actors and agencies of the network, forums of negotiations and processes at the sites of implementation differ in varied national contexts. The FSC-GGN model can provide a useful framework for comparative analysis of FSC institutionalization and implementation because global Principles and Criteria originating from the same node of design are the same, yet political, environmental, social and economic contexts are different in various countries. A comparative analysis can involve both the implementing actors and processes, such as national standards setting, standards adaptation, standards institutionalization and other types of stakeholder forums of negotiation. For example, assessment of the FSC standard and its impact on biodiversity on the ground was shown to be stronger in Russia than in Sweden, despite the fact that the FSC rules contradicted Russian legislation on biodiversity conservation, a factor that can be a barrier in standards implementation (Elbakidze *et al.*, 2010). By comparing which stakeholders were engaged in the forums of negotiation related to national FSC standard setting in Russia and Sweden, it would be possible to explain why there is such a difference.

The role of implementing actors, such as NGOs, and companies and their relationships with state agencies in the process of FSC implementation can be compared and assessed. Even in countries that seem to be similar, differences can be observed. In post communist countries with transitional economies, implementing actors and their interactions with the state and other stakeholders can be different as, for example, it is in Russia and Byelorussia.[484] In Byelorussia there is much less freedom for NGO involvement because of the Lukashenko authoritarian regime. Yet the state itself is interested in certification, and effectively harmonized national legislation with the FSC requirements and fostered implementation of the FSC rules on the ground. Although democratic decision making processes are poorly developed in Byelorussia, citizen involvement in forest management was accomplished through still existing Soviet institutions, such as the Young Communists and Hunting and Fishing Societies.

The GGN concept also provides a framework for comparing different certification projects in one country, where the national standard is the same, but local stakeholders, forums of implementation and institutionalization, as well as regional contexts, are different. The GGN concept allows having a closer look at the process of institutionalization of new rules in places and analyzing the social, environmental and economic impacts on the ground.

Differences in FSC implementation can be seen in different regions of one country. Market incentives for forest producers are the most important aspect of FSC regional development. For example, in the Russian Far East, FSC certification is much less developed than in North Western Russia, because of the geographical closeness of the Asian markets, which do not require certification (Tysiachniouk & Reisman, 2006). The Russian Far East situation changed recently[485] because Russian holding companies have built new processing plants and became interested in selling their products to sensitive European markets, which require certified products. Comparison of the stakeholder forums of negotiation in Northwestern Russia and the Russian Far East shows significant differences. This can be explained partly in the time differences in the introduction of the FSC in Russia. The FSC is relatively new for the Far East, while in Northwestern Russia it took hold more than 10 years ago. Delay in participation in the process of certification brought complications. The Russian national standard is seen by both companies and governments in the Far East as top-down and not applicable to the context of Russian Far East. Companies were not interested in the Russian FSC standard setting process at that time. In 2011-2012, the Far Eastern companies criticized the standard, made efforts to soften it through involving regional governments in Far East on their behalf, and wrote complaints regarding the Russian FSC standard to the FSC international.[486] This blocked and/or delayed the certification process of the Far Eastern companies.

Comparison of the institutionalization of the FSC rules in sites of implementation of those companies that received the FSC certificates in the Far East also shows differencesfrom those in Northwestern Russia. The FSC implementation of the Terney Les Company, which was certified 6 years ago, shows that many more issues arise with indigenous peoples in the Far East, as their

[484] The author has not done formal research in Byelorussia, but worked in a team with the certification body SGS-Qualifor.

[485] Research of the author conducted in the Fall 2011 on the FSC in the Russian Far East in the framework of the project 'National Interests and Transnational Governance: Russia's Changing Environmental Policy', funded by the National Council for Eurasian and East European Research (results are not yet published).

[486] Research of the author conducted in the Fall 2011 on the FSC in the Russian Far East in the framework of the project 'National Interests and Transnational Governance: Russia's Changing Environmental Policy', funded by the National Council for Eurasian and East European Research (results are not yet published).

lifestyle is more deeply embedded in nature and their hunting grounds are larger than those in Northwestern Russia. It was harder for the company to implement environmental standards, because the Far Eastern forests are richer in biodiversity and endangered species. This certification project was more complicated than were the certification projects in Northwest Russia described in this study because of the different organizational and institutional environments in the Russian Far East and in Northwestern Russia.[487] While the FSC rules of the game were easily accepted by the companies in Northwestern Russia, the same rules were widely resisted in the Russian Far East. A GGN approach can give many insights in understanding the dynamics of FSC implementation, by analyzing actors, network agencies and processes within the network.

The GGN concept can give similar insights to the study of other certification schemes; some of them, such as the Marine Stewardship Council, the Tourism Stewardship Council, etc., were building on the basis of the FSC example. Other schemes, which can have greater differences with the FSC, but are also based on the multi-stakeholder global standard setting processes, can be investigated using the GGN concept (see Chapter 2). They have a transnational node of governance design. They negotiate the global standards, then adjust them to national environments and implement them. Therefore, they have forums of negotiations. The standards are transferred into concrete practices on the ground, at the sites of implementation. It seems that a multi-sited qualitative study of each particular GGN can give most interesting results; however, studying certain segments of the network with their actors and agency can also be useful.

In a similar way it is possible to study other GGNs that generate new global standards, but are not related to certification schemes. This can prevent social sciences from acquiring a too narrow understanding of local processes on the ground and undermining the global span of what is supposed to be local (Sassen, 2003). It can also encourage globalization scholars to look at concrete processes of institutionalization and in-depth consequences of the application of global rules to particular places.

[487] Participant observation at the FSC audit of Terney les in 2010.

References

Aksenov, D., Dobrynin, D., Dubinin, M., Egorov, M., Isaev, A., Korpachevskiy, M., Lestadius, L., Potapov, P., Purekovskiy, A., Turubanova, S. & Yaroshenko, A. (2002). *Atlas of Russia's Intact Forest Landscapes*. Moscow, Russia: Global Forest Watch Russia.

Anheier, H. & Katz, H. (2005). Network Approaches to Global Civil Society 2004/5 Yearbook. In: H. Anheir, M. Glasius, M. Kaldor & F. Holland (eds.), *Global Civil Society* (pp. 206-221). London, UK: Sage Publications.

Arquilla, J. & Ronfeldt, D. (2001). *Networks and Netwars: The Future of Terror, Crime, and Militancy*. Santa Monica, CA, USA: Rand Corporation.

Arts, B. (2011). Forests Policy Analysis and Theory Use: Overview and Trends. *Forest Policy and Economics*, 16, 7-13.

Arts, B., Appelstrand, M., Kleinschmit, D., Pülzl, H., Visseren-Hamakers, I.J., Eba'a Atyi, R., Enters, T., McGinley, K. & Yasmi, Y. (2010). Discourses, Actors and Instruments in International Forest Governance. In: J. Rayner, A. Buck & P. Katila (eds.), *Embracing Complexity: Meeting the Challenges of International Forest Governance (A Global Assessment Report)*. Vienna, Austria: International Union of Forest Research Organizations (IUFRO).

Arts, B. & Buizer, M. (2009). Forests, Discourses, Institutions: a Discoursive Institutional Analysis of Global Forest Governance. *Forest Policy and Economics*, 11(5-6), 340-347.

Arts, B., Leroy, P. & Tatenhove, J.P.M. (2006). Political Modernisation and Policy Arrangements: A Framework for Understanding Environmental Policy Change. *Public Organization Review*, 6(2), 93-106.

Auld, G., Balboa, C., Bernstein, S. & Cashore, B. (2009). The Emergence of Non-State Market-Driven (NSMD) Global Environmental Governance: a Cross-sectoral Assessment. In: A. King & M. Toffel (eds.), *Governance for the Environment: New Perspectives* (pp. 183-215). Cambridge, UK: Cambridge University Press.

Auld, G., Bernstein, S. & Cashore, B. (2008a). The New Corporate Social Responsibility. *Annual Review of Environment and Resources*, 33, 413-435.

Auld, G., Gulbrandsen, L.H. & McDermott, C.L. (2008b). Certification Schemes and the Impacts on Forests and Forestry. *Annual Review of Environment and Resources*, 33, 187-211.

Autio, S. (2002a). *Soviet Karelian Forest in the Planned Economy of the Soviet Union, 1928-37 in Rise and Fall of Soviet Karelia*. Helsinki, Finland: Kikimora Publications.

Autio, S. (2002b). *Forests and the Ecological Dimensions of Industrialization of the Soviet Union in the 1930's*: BASEES.

Bächtiger, A., Niemeyer, S., Neblo, M. & Steenbergen, M.R. (2010). Disentangling Diversity in Deliberative Democracy: Competing Theories, Their Blind Spots and Complementarities. *Journal of Political Philosophy*, 18(1), 32-63.

Backstrand, K. (2006). Democratizing Global Environmental Governance? Stakeholder Democracy after the World Summit on Sustainable Development. *European Journal of International Relations*, 12(4), 467-498.

Bartley, T. (2003). Certifying Forests and Factories: States, Social Movements, and the Rise of Private Regulation in the Apparel and Forest Product Field. *Politics & Society*, 31, 433-464.

Bartley, T. (2005). Corporate Accountability and the Privatization of Labor Standards: Struggles over Codes of Conduct in the Apparel Industry. *Research in Political Sociology*, 14, 211-244.

Bartley, T. (2007a). How Foundations Shape Social Movements: The Construction of an Organizational Field and the Rise of Forest Certification. *Social Problems*, 54(3), 229-255.

Bartley, T. (2007b). Institutional Emergence in an Era of Globalization: The Rise of Transnational Private Regulation of Labor and Environmental Conditions. *American Journal of Sociology*, 113(2), 297-351.

Bartley, T. (2010a). Certification as a Mode of Social Regulation. In: D. Levi-Faur (ed.), *Handbook of the Politics of Regulation* (pp. 441-453). Cheltenham, UK: Edward Elgar.

Bartley, T. (2010b). Transnational Private Regulation in Practice: The Limits of Forest and Labor Standards Certification in Indonesia. *Business & Politics,* 112(3), 7.

Bartley, T. & Child, C. (2008). *Shaming the Corporation: Globalization, Reputation, and the Dynamics of Anti-corporate Movements (Working Paper).* Bloomington, IN, USA: Indiana University.

Bartley, T. & Child, C. (2011). Movements, Markets and Fields: The Effects of Anti-Sweatshop Campaigns on U.S. Firms, 1993-2000. *Social Forces,* 90(2), 425-451.

Bartley, T. & Smith, S. (2008). *Structuring Transnational Fields of Governance: Network Evolution and Boundary-setting in the World of Standards (Working paper).* Bloomington, IN, USA: Indiana University.

Baxter, P. & Jack, S. (2008). Qualitative Case Study Methodology: Study Design and Implementation for Novice Researchers. *The Qualitative Report,* 13(4), 544-559.

Beck, U. (2005). *Power in the Global Age: A New Global Political Economy.* Cambridge, UK: Polity Press.

Bernstein, S. (2001). *The Compromise of Liberal Environmentalism.* New York, NY, USA: Columbia University Press.

Bernstein, S. (2011). Legitimacy in Intergovernmental and Non-state Global Governance. *Review of International Political Economy,* 18(1), 17-51.

Bernstein, S. & Cashore, B. (2004). Non-State Global Governance: Is Forest Certification a Legitimate Alternative to a Global Forest Convention. In: J. Kirton & M. Trebilcock (eds.), *Hard Choices, Soft Law: Voluntary Standards in Global Trade, Environment and Social Governance* (pp. 33-66). Aldershot, UK: Ashgate Press.

Bernstein, S. & Cashore, B. (2007). Can Non-State Global Governance be Legitimate? An Analytical Framework. *Regulation and Governance,* 1(4), 347-371.

Blackman, A. & Rivera, J. (2011). Producer-level Benefits of Sustainability Certification. *Conservation Biology,* 25(6), 1176-1185.

Börzel, T. (1998). Organising Babylon: On the Different Conceptions of Policy Networks. *Public Administration,* 76(2), 253-273.

Bostrom, M. & Gartsen, Ch. (2008). *Organizing Transnational Accountability.* Cheltenham, UK: Edward Elgar.

Bouteligier, S. (2011). Cities and Global Environmental NGOs: Emerging Transnational Urban Networks? In: M. Amen, N. J. Toly, P. McCarney & K. Segbers (eds.), *Cities and Global Governance* (pp. 151-175). Surrey, UK: Ashgate.

Bozzi, L., Cashore, B., Levin, K. & McDermott, C.L. (2012). The Role of Private Voluntary Climate Programs Affecting Forests: Assessing their Direct and Intersecting Effects. In: K. Ronit (ed.), *Business and Climate Policy: The Potentials and Pitfalls of Private Voluntary Programs.* Tokyo, Japan: United Nations University.

Bryant, D., Nielsen, D. & Tangley, L. (1997). *The Last Frontier Forests.* Washington, DC, USA: World Resource Institute.

Callon, M., Meadel, C. & Rabeharisoa, V. (2002). The Economy of Qualities. *Economy and Society,* 31(2), 194-217.

Campbell, J.L. (2004). *Institutional Change and Globalization.* Princeton, NJ, USA: Princeton University Press.

Carlson, L. (2000). Towards a Sustainable Russian Forest Sector. *Natual Resources Forum,* 24(1), 31-37.

Carlson, M., Wells, J. & Roberts, D. (2009). *The Carbon the World Forgot: Conserving the Capacity of Canada's Boreal Forests Region to Mitigate and Adapt to Climate Change.* Seattle, WA, USA: Boreal Songbird Initiative and Canadian Boreal Initiative.

Carlson, M., Chen, J., Elgie, S., Henschel, C., Montenegro, A., N., Roulet, Scott, N., Tarnocai, C. & Wells, J. (2010). Maintaining the Role of Canada's Forests and Peatlands in Climate Regulation. *The Forestry Chronicle,* 86(4). Available at: www.borealcanada.ca/documents/03-2009-063_HR.pdf.

Cashore, B. (2002). Legitimacy and the Privatization of Environmental Governance: How Non-State Market-Driven (NSMD) Governance Systems Gain Rule-Making Authority. *Governance,* 15(4), 502-529.

Cashore, B., Auld, G. & Newsom, D. (2004). *Governing Through Markets: Regulating Forestry Through Non-state Environmental Governance.* New Haven, CT, USA: Yale University Press.

Castells, M. (1996). *The Information Age: Economy, Society and Culture. Volume I: The Rise of the Network Society.* Oxford, UK: Blackwell Publishers.

Castells, M. (1997). *The Information Age, Economy, Society and Culture. Volume II: The Power of Identity.* Oxford, UK: Blackwell Publishers.

Castells, M. (2009). *Communication Power.* Oxford, UK: Oxford University Press.

Castells, M. (2011). A Network Theory of Power. *International Journal of Communication,* 5, 773-787.

Charmaz, K. (2006). *Constructing Grounded Theory: A Practical Guide Through Qualitative Analysis.* Thousand Oaks, CA: Sage Publications.

Checkland, P. & Holwell, S. (1998). Action Research: Its Nature and Validity. *Systemic Practice and Action Research, 11*(1), 9-21.

Chernova, Ye.B. (2010). *Intensivnoe Lesopol'zovanie Dlya Rossii: Opyt Innovatsii Proekta «Pskovskiĭ model'nyĭ les».* St.Peterburg, Russia: GreenForest.

Chriakenen, K. (2005). First Steps of the Kalevala Park. In: P. Tervonen & K. Harkonen (eds.), *Kalevala Parks, Life on the Border and Pristine Nature* (pp. 202-213). Kainuun Sanomat, Finland: Kajaani.

Conroy, M.E. (2001). Can Advocacy-led Certification Systems Transform Global Corporate Practices? Evidence and Some Theory. *Working Paper Series,* 21, 1-25.

Conroy, M.E. (2007). *Branded!: How the Certification Revolution Is Transforming Global Corporations.* Gabriola Island, BC, Canada: New Society Publishers.

Counsell, S. & Terje Loraas, K. (2002). *Trading in Credibility: The Myth and Reality of Forest Stewardship Council. Report with Case Studies.* London, UK: The Rainforest Foundation UK.

Delmas, M.A. & Young, O.R. (2009). Introduction: New Perspectives on Governance for Sustainable Development. In: A. King & M.W. Toffel (eds.), *Governance for the Environment: New Perspectives* (pp. 3-11). Cambridge, UK: Cambridge University Press.

DeVaus, D. (2001). *Research Design in Social Research.* London, UK: Sage Publications.

DiMaggio, P.J. (1991). Constructing an Organizational Field as a Professional Project: U.S. Art Museums, 1920-1940. In: W.W. Powell & P.J. DiMaggio (eds.), *The New Institutionalism in Organizational Analysis* (pp. 267-292). Chicago, IL, USA: University of Chicago Press.

DiMaggio, P.J. & Powell, W. (1983). The Iron Cage Revisited: Institutional Isomorphism and Collective Rationality in Organizational Fields. *American Sociological Review,* 48(2), 147-160.

Dingwerth, K. & Pattberg, P. (2009a). Actors, Arenas and Issues in Global Governance. In: J. Whitman (ed.), *Global Governance* (pp. 41-65). Basingstoke, UK: Palgrave Macmillan.

Dingwerth, K. & Pattberg, P. (2009b). World Politics and Organizational Fields: The Case of Transnational Sustainability Governance. *European Journal of International Relations,* 15(4), 707-743.

Djama, M., Fouilleux, E. & Vagneron, I. (2011). Standard-setting, Certifying and Benchmarking: A Governmentality Approach to Sustainability Standards in the Agro-Food Sector. In: S. Ponte, P. Gibbon & J. Vestergaard (eds.), *Governing through Standards – Origins, Drivers and Limitations* (pp. 184-209). London, UK: Palgrave Macmillan.

Douglas, J.D. (1976). *Investigative Social Research.* Beverly Hills, CA, USA: Sage Publications.

Efimova, N., Kopylova, E., Kulikova, E., Newell, J., Pankratov, V., Shmatkov, N., Shuvaev, Y. & Voropaev, A. (2010). *Keep it Legal Country Guide: Practical Guide for Verifying Timber Origin Legality.* Moscow, Russia: WWF Russia.

Elbakidze, M., Angelstam, P. K., Sandström, C. & Axelsson, R. (2010). Multi-stakeholder Collaboration in Russian and Swedish Model Forest initiatives: Adaptive Governance Toward Sustainable Forest Management?. *Ecology and Society,* 15(2), 14.

Elliot, K. & Freeman, R. (2003). *Can Labor Standards improve under globalization?* Washington, DC, USA: Institute for International Economics.

Food and Agriculture Organization of the United Nations (2011*). The State of the World's Forests*. Available at: http://www.fao.org/docrep/013/i2000e/i2000e.pdf.

FERN, Greenpeace & Inter-African Forest Industry Association (2008). *Precious Woods and Tropical Forest Trust. Regaining Credibility and Rebuilding Support: Changes in the FSC Needs to Make to Ensure It Regains and Maintains Its Credibility*. Available at: http://www.fern.org/node/4297.

Field, B. (2001). The Development of Global Forest Policy: Overview of Legal and Institutional Frameworks. *Mining, Minerals and Sustainable Development,* October(3), 2-22.

Fligstein, N. (2002). *The Architecture of Markets: An Economic Sociology of Twenty-First-Century Capitalist Societies*. Princeton, NJ, USA: Princeton University Press.

Fligstein, N. & McAdam, D. (2012). *A Theory of Fields*. Oxford, UK: Oxford University Press.

Follesdal, A. (2006). Political Consumerism as Chance and Challenge. In: M. Micheletti, A. Follesdal & D. Stolle (eds.), *Politics, Products, and Markets* (pp. 3-20). New Brunswick, NJ, USA: Transaction Publishers.

Forest Stewardship Council (2012a*). Global FSC Certificates: Types and Distribution*. Available at: http://www.fsc.org/fileadmin/webdata/public/document_center/powerpoints_graphs/facts_figures/2011-11-15-Global-FSC-Certificates-EN.pdf.

Forest Stewardship Council (2012b*).* Rossiiskie Predpriyatiya, Imeyushchie Sertifikaty na Upravleniye Lesami po Skheme FSC. Available at: http://www.fsc.ru/.

Friedman, M. (2006). Using Consumer Boycotts to Stimulate Corporate Policy Changes: Marketplace, Media, and Moral Considerations. In: M. Micheletti, A. Follesdal & D. Stolle (eds.), *Politics, Products, and Markets* (pp. 45-82). New Brunswick, NJ, USA: Transaction Publishers.

Giddens, A. (1990). *The Consequences of Modernity*. Cambridge, UK: Polity Press.

Gille, Z. (2006). Detached Flows or Grounded Place-Making Projects. In: G. Spaargaren, A.P.J. Mol & F.H. Buttel (eds.), *Governing Environmental Flows. Global Challenges for Social Theory* (pp. 137-156). Cambridge, UK: MIT Press.

Glaser, B.G. (1992). *Basics of Grounded Theory Analysis*. Mill Valley, CA, USA: Sociology Press.

Glaser, B.G. & Strauss, A.L. (1967). *The Discovery of Grounded Theory: Strategies for Qualitative Research*. Chicago, IL, USA: Aldine.

Grigor'ev, A., Pakhorukova, Y., Shmatkov, K. & Ryeĭting, N. M. (2012). *Informatsionnoe Soderzhanie Internet-saĭtov Organov Gosudarstvennogo Upravleniya Lesami Sub'ektov Rossiĭskoĭ Federatsii 2011 goda* Moscow, Russia: Vsemirnyĭ fond dikoĭ prirody (WWF).

Gulbrandsen, L.H. (2003). The Evolving Forest Regime and Domestic Actors: Strategic and Normative Adaptation. *Environmental Politics,* 12(2), 95-114.

Gulbrandsen, L.H. (2005). Mark of Sustainability? Challenges for Fishery and Forestry Eco-labeling. *Environment,* 47(5), 8-23.

Gulbrandsen, L.H. (2008a). Accountability Arrangements in Non-state Standards Organizations: Instrumental Design and Imitation. *Organization,* 15(4), 563-583.

Gulbrandsen, L.H. (2008b). Organizing Accountability in Transnational Standards Organizations: the Forest Stewardship Council as a Good Governance Model. In C. Garsten & M. Bostrom (eds.), *Organizing Transnational Accountability* (pp. 61-80). Cheltenham, UK: Edward Elgar.

Gulbrandsen, L.H. (2009). The Emergence and Effectiveness of the Marine Stewardship Council. *Marine Policy,* 33, 654-660.

Gulbrandsen, L.H. (2010). *Transnational Environmental Governance: The Emergence and Effects of the Certification of Forests and Fisheries*. Cheltenham, UK: Edward Elgar.

Guthman, J. (2004). *Agrarian Dreams: The Paradox of Organic Farming in California*. Berkeley, CA, USA: University of California Press.

Guthman, J. (2007). The Polanyian Way? Voluntary Food Labels as Neoliberal Governance. *Antipode*, 39(3), 456-478.

Hallstrom, K.T. & Bostrom, M. (2010). *Transnational Multi-Stakeholder Standartization*. Cheltenham, UK: Edward Elgar.

Haufler, V. (2009). Transnational Actors and Global Environmental Governance. In: A. King & M.W. Toffel (eds.), *Governance for the Environment: New Perspectives* (pp. 119-143). Cambridge, UK: Cambridge University Press.

Haufler, V. (2010). The Kimberly Process Certification Scheme: An Innovation in Global Governance and Conflict Prevention. *Journal of Business Ethics*, 89(4), 403-416.

Henry, L. (2010a). *Red to Green*. Ithaca, NY, USA: Cornell University Press.

Henry, A. (2010). Between Transnationalism and State Power: the Development of Russia's Post Soviet Environmental Movement. *Environmental Politics*, 19(5), 756-781.

Hertting, N. (2007). Mechanisms of Governance Network Formation – a Contextual Rational Choise Perspective. In: E. Sorenson & J. Torfing (eds.), *Theories of Democratic Network Governance* (pp. 43-76). New York, NY, USA: Palgrave Macmillan.

Hong, P. & Song, I. (2010). Glocalization of Social Work Practice: Global and Local Responses to Globalization. *International Social Work*, 53(5), 655-669.

Jaffee, D. (2007). *Brewing Justice: Fair Trade Coffee, Sustainability, and Surviva*. Berkeley, CA, USA: University of California Press.

Kawulich, B.B. (2005). Participant Observation as a Data Collection Method. *Forum: Qualitative Social Research*, 6(2), 43.

Keck, M. & Sikkink, K. (1998). *Activists Beyond Borders: Advocacy Networks in International Politics*. Ithaca, NY, USA: Cornell University Press.

Keskitalo, E., Sandstrom, C., Tysiachniouk, M. & Johansson, J. (2009). Local Consequences of Applying International Norms: Differences in the Application of Forest Certification in Northern Sweden, Northern Finland, and Northwest Russia. *Ecology and Society*, 14(2), 1.

Khanna, M. & Brouhle, K. (2009). The Effectiveness of Voluntary Environmental Initiatives. In: M.A. Delmas & O.R. Young (eds.), *Governance for the Environment: New Perspectives* (pp. 144-182). Cambridge, UK: Cambridge University Press.

King, A. & Toffel, M.W. (2009). Self-regulatory Institutions for Solving Environmental Problems: Perspectives and Contributions from the Management Literature. In: A. King & M.W. Toffel (eds.), *Governance for the Environment: New Perspectives* (pp. 98-115). Cambridge, UK: Cambridge University Press.

King, B.G. & Pearce, N.A. (2010). The Contentiousness of Markets: Politics, Social Movements, and Institutional Change in Markets. *Annual Review of Sociology*, 36, 249-267.

Klein, N. (1999). *No Logo: Taking Aim in the Brand Bullies*. New York, NY, USA: Picador.

Knize, A. & Romaniuk, B. (2010). O Dvukh Tochkakh Zreniya Na Rossiĭskiĭ Les i Lesnoe Khozyaĭstvo. In K. Voronkov (ed.), *Pskovskiĭ Model'nyĭ Les*. Sankt-Peterburg, Russia: Stampo.

Kortelainen, J. & Kotilainen, J. (2006). *Contested Environments and Investments in Russian Woodland Communities*. Helsinki, Finland: Kikimora Publications.

Kortelainen, J., Tysiachniouk, M. & Kotilainen, J. (2010). Transnatsional'nye Seti Upravleniya v Lesnom Sektore: Yadra Global'nogo Dizaĭna, Forumy Peregovorov, Mesta Ryealizatsii. *Zhurnal Sotsiologii i Sotsial'noĭ Antropologii,* 13(5), 10-37.

Kotilainen, J., Kuliasova, A. & Kuliasov, I. (2006). Reshaping of the Environmental Administration. In: J. Kotilainen & J. Kortelainen (eds.), *Contested Environments and Investments in Russian Woodland Communities* (pp. 63-76). Helsinki, Finland: Kikimora Publications.

Kotilainen, J., Tysiachniouk, M., Kuliasova, A., Kuliasov, I. & Pchelkina, S. (2008). The Potential for Ecological Modernisation in Russia: Scenarios from the Forest Industry. *Environmental Politics,* 17(1), 58-77.

Kozyreva, G. (2006). *Problemyi Formirovanya Sotsialnykh Institutov Ustoichivogo Lesoupravlenya.* Petrozavodsk, Russia: Karelsky nauchnyi tsentr, RAN.

Kulyasova, A. (2010). Stolknovenie Biznes-kul'tur: Transformatsiya Post-Sovetskogo Predpriyatiya pri Vkhozhdenii v Mezhdunarodnyĭ Kontsern. *Zhurnal Sotsiologii i Sotsial'noĭ Antropologii,* 13(5), 253-282.

Kuz'minov, I. (2010). Rezul'taty Gosudarstvennykh Initsiativ po Reformirovaniyu Rossiĭskogo Lesnogo Zakonodatel'stva v 2006-2009gg. *Ustoĭchivoe lesopol'zovanie,* 3(25), 28-33.

Kvale, S. (1996). *Interviews: An Introduction to Qualitative Research Interviewing.* London, UK: Sage Publications.

Kvale, S. (2008). *Interviews: Learning the Craft of Qualitative Research Interviewing.* London, UK: Sage Publications.

Laine, A. (2002). *Continuity and Change in 20th Century Russia' in Rise and Fall of Soviet Karelia.* Helsinki, Finland: Kikimora Publications.

Lehmbruch, B. (1999). Managing Uncertainty: Hierarchies, Markets and 'Networks' in the Russian Timber Industry. 1991-1998 *BOFIT Discussion Papers* (Vol. 4). Helsinki, Finland: Institute for Economics in Transition, Bank of Finland.

Lehmbruch, B. (2012). Interests, Ideas and Failed Reform in Russian Forest Governance: Out of the Frying Pan, into the Forest Fire? *Regulation and Governance* (in press).

Lehtinen, A. (2006). *Postcolonialism, Multitude, and the Politics of Nature.* Lanham, MD, USA: University Press of America.

Lemos, M.C. & Agrawal, A. (2009). Environmental Governance and Political Science. In: M.A. Delmas & O.R. Young (eds.), *Governance for the Environment: New Perspectives* (pp. 69-98). Cambridge, UK: Cambridge University Press.

Levin, K., McDermott, C. & Cashore, B. (2008). The Climate Regime as Global Forest Governance: Can Reduced Emissions from Deforestation and Degradation (REDD) Initiatives Pass a 'Dual Effectiveness' Test. *International Forestry Review,* 10(3), 538-548.

Levy, D. L. & Egan, D. (1998). Capital contests: National and transnational channels of corporate influence on the climate change negotiations. *Politics & Society,* 26(3), 337-361.

Lindlof, T.R. & Taylor, B.C. (2002). *Qualitative Communication Research Methods.* Thousand Oaks, CA, USA: Sage Publications.

Linton, A. (2008). A Niche for Sustainability? Fair Labor and Environmentally Sound Practices in the Specialty Coffee Industry. *Globalizations,* 5, 231-245.

Lloyd, S. (1999). *The Last of the Last. The old-growth Forests of Boreal Europe.* Jokkmokk, Sweden: Taiga Rescue Network.

Locke, R., Amengual, M. & Mangla, A. (2009). Virtue out of Necessity?: Compliance, Commitment and the Improvement of Labor Conditions in Global Supply Chains. *Politics & Society,* 37(3), 319-351.

Mahoney, J. (2000). Path Dependence in Historical Sociology. *Theory and Society, 29,* 507-548.

Malets, O. (2010). Mesto i Rol' Sertifikatsionnykh Organov v Sisteme Lesnoĭ Sertifikatsii Lesnogo Popechitel'skogo Soveta. *Zhurnal Sotsiologii i Sotsial'noĭ Antropologii*, 13(5), 107-125.

Malets, O. (2011). From Transnational Voluntary Standards to Local Practices: A Case Study of Forest Stewardship Certification in Russia *MPIFG Discussion Paper 11(7)*. Colon: Max Plank Institute for the Study of Societies.

Malets, O. (2012). *Certifiers and the Impact of Environmental Certification on Production Practices*. Paper presented at the SASE 24th Annual Conference 'Implications for Business, Government and Labor', MIT, Cambridge, USA.

Malets, O. & Tysiachniouk, M. (2009). The Effect of Expertise on the Quality of Forest Standards Implementation: The Case of FSC Forest Certification in Russia. *Forest Policy and Economics*, 11, 422-428.

Marbry, L. (2008). Case Study in Social Research. In P. Alasuutari, L. Bickman & J. Brannen (eds.), *The SAGE Handbook of Social Research Methods* (pp. 214-227). London, UK: Sage Publications.

Marx, A. & Cuypers, D. (2010). Forest Certification as a Global Environmental Governance Tool: What is the Macro-effectiveness of the Forest Stewardship Council? *Regulation & Governance*, 4(4), 408-434.

Mather, A. (2004). Forest Transition Theory and the Reforesting of Scoltand. *Scottish Geographical Journal*, 120, 83-98.

McDermott, C., Cashore, B. & Kanowski, P. (2009). Setting the Bar: an International Comparison of Public and Private Forest Policy Specifications and Implications for Explaining Policy Trends. *Journal of Integrative Environmental Sciences*, 6(3), 217-237.

McDermott, C., Cashore, B. & Kanowski, P. (2010). *Global Environmental Forest Policies: An International Comparison*. London, UK: Earthscan.

McDermott, C., Cashore, B. & Levin, K. (2011a). Building the Forest-Climate Bandwagon: REDD and the Logic of Problem Amelioration. *Global Environmental Politics*, 11(3), 85-103.

McDermott, C., Humphreys, D., Wildburger, C., Wood, P., Marfo, E., Pacheco, P. & Yasmi, Y. (2011b). Mapping of the core actors and issues defining international forest governance. In: J. Rayner, A. Buck & P. Katila (eds.), *Embracing Complexity: Meeting the challenges of international forest governance. A global assessment report. Prepared by the Global Forest Expert Panel on the International Forest Regime. IUFRO World Series no 28* (pp. 19-36). Vienna, Austria: IUFRO.

McNichol, J. (2006). Transnational NGO Certification Programs as New Regulatory Forms: Lessons from Forestry Sector. In: M. Djelic & K. Sahlin-Andersson (eds.), *Transnational Governance: Institutional Dynamics of Regulation* (pp. 349-374). Cambridge, UK: Cambridge University Press.

McNiff, J. & Whitehead, D. (2009). *Doing and Writing Action Research*. London, UK: Sage Publications.

Meidinger, E. (2003). Forest Certification as Environmental Law Making. In: E. Meidinger, C. Elliott & G. Oesten (eds.), *Social and Political Dimensions of Forest Certification* (pp. 293-329). Remagen-Oberwinter, Germany: Forstbuch Verlag.

Meidinger, E. (2007). Beyond Westphalia: Competitive Legalization in Emerging Transnational Regulatory Systems. In: C. Brütsch & D. Lehmkuhl (eds.), *Law and Legalization in Transnational Relations* (pp. 121-143). New York, NY, USA: Routledge.

Meidinger, E. (2008). Competitive Supra-Governmental Regulation: How Could It Be Democratic? *Chicago Journal of International Law*, 8, 513-534.

Meijer, C.P., Verloop, N. & Beijaard, D. (2002). Multi-Method Triangulation in Qualitative Study on Teachers' Practical Knowledge: An Attempt to Increase Internal Validity. *Quality and Quantity*, 36(2), 145-167.

Merger, E., Dutschke, M. & Verchot, L. (2011). Options for REDD+ Voluntary Certification to Guarantee Net GHG Benefits, Poverty Alleviation, Sustainable Forest Management and Biodiversity Conservation. *Forests*, 2, 550-577.

Meuser, M. & Nagel, U. (2009). The Expert Interview and Changes in Knowledge Production. In: A. Bogner, B. Littig & W. Menz (eds.), *Interviewing Experts* (pp. 17-42). Basingstoke, UK: Palgrave Macmillan.

Meyer, N. (2010). *Political Contestation and Information Problems of Public Intervention in Private Standardization: Hard vs. Soft Power.* Paper presented at the Governing through Standards – An International Symposium held at the Danish Institute for International Studies (DIIS), Copenhagen, Denmark.

Micheletti, M. (2003). *Political Virtue and Shopping.* New York, NY, USA: Palgrave Macmillan.

Micheletti, M., Follesdal, A. & Stolle, D. (2006). *Politics, Products, and Markets.* New Brunswick, NJ, USA: Transaction Publishers.

Mol, A.P.J. (2008). *Environmental Reform in the Information Age. The Contours of Informational Governance.* Cambridge, UK: Cambridge University Press.

Mol, A.P.J. (2009). Environmental Deinstitutionalization in Russia. *Journal of Environmental Policy and Planning,* 11(3), 223-241.

Mol, A.P.J. (2012). Carbon Flows, Financial Markets and the Challenge of Global Environmental Governance. *Environmental Development,* 1(1), 10-24.

Mol, A.P.J. & Spaargaren, G. (2006). Towards a Sociology of Environmental Flows. A New Agenda for Twenty-First-Century Environmental Sociology. In: G. Spaargaren, A. P. J. Mol & F. H. Buttel (eds.), *Governing Environmental Flows. Global Challenges for Social Theory* (pp. 39-82). Cambridge, MA, USA: MIT Press.

Newsom, D. & Hewitt, D. (2005). *The Global Impacts of Smart Wood Certification.* TREES Program, Rainforest Alliance.

Nicholls, A. & Opal, C. (2005). *Fair Trade: Market-driven Ethical Consumption.* London, UK: Sage Publications.

North, D. (1990). *Institutions, Institutional Change and Economic Performance.* Cambridge, MA, USA: Harvard University Press.

Olsson, M. (2006). *Is the Russian Virtual Economy Coming to an End? Institutional Change in the Arkhangelsk Forest Sector.* Laxenburg: International Institute for Applied Systems Analysis (IIASA).

Olsson, M. (2008). *Continuity and Change: Institutions and Transition in the Russian forest Sector.* Lulea, Sweden: Lulea University of Technology.

Oosterveer, P. (2006). Environmental Governance of global Flows: The Case of Labeling Strategies. In: G. Spaargaren, A.P.J. Mol & F.H. Buttel (eds.), *Governing Environmental Flows. Global Challenges to Social Theory* (pp. 267-301). Cambridge, MA, USA: The MIT Press.

Oosterveer, P. (2007). *Global Governance of Food Production and Consumption: Issues and Challenges.* Cheltenham, UK: Edward Elgar.

Oosterveer, P. & Spaargaren, G. (2011). Organising Consumer Involvement in the Greening of Global Food Flows: the Role of Environmental NGOs in the Case of Marine Fish. *Environmental Politics,* 20(1), 97-114.

O'Rourke, D. (2003). Outsourcing Regulation: Analyzing Nongovernmental Systems of Labor Standards and Monitoring. *Policy Studies Journal,* 31(1), 1-29.

O'Rourke, D. (2005). Market Movements – Nongovernmental Organization Strategies to Influence Global Production and Consumption. *Journal of Industrial Ecology,* 9(1-2), 115-128.

Overdevest, C. (2010). Comparing Forest Certification Schemes: the Case of Ratcheting Standards in the Forest Sector. *Socio-Economic Review,* 8(1), 47-76.

Overdevest, C. & Zeitlin, J. (2012). Assembling the Experimentalist Regime: Transnational Governance Interactions in the Forest Sector. *Regulation and Governance* (in press).

Ozinga, S. (2005). Footprints in the Forests: Current Practice and Future Challenges in Forest Certification. Available at: http://www.fern.org./sites/fern.org/files/media/documents/document_1890-1900.pdf.

PEFC (2012). *Statistical Figures on PEFC Certification.* Available at: http://pefcregs.info/STATISTICS2.ASP?COUNTRY=Chile&COUNTRY_CODE=24

Pappila, M. (2009). Russian Forest Regulation and the Integration of Sustainable Forest Management. In: S. Nysten-Haarala (ed.), *The Changing Governance of Renewable Natural Resources in Northwest Russia* (pp. 55-76). Cornwall, UK: Ashgate.

Pipponen, M. (1999). Household and Living Conditions in the Forest Village of Matorsy, Russian Karelia. *Working Paper, 99.*

Pisarenko, A.I. & Strakhov, V.V. (2004). *Lesnoe khozyaĭstvo Rossii: ot pol'zovaniya k upravleniyu*. Moskva, Russia: Yurisprudentsiya.

Pisarenko, A. I. & Strakhov, V. V. (2006). Kakaya Lesnaya Politika Nuzhna Rossii? *Lesnoe khozyaĭstvo,* 2, 2-5.

Potoski, M. & Aseem, P. (2009). A Club Theory Approach to Voluntary Programs. In: M. Potoski & P. Aseem (eds.), *Voluntary Programs: A Club Theory Perspective* (pp. 17-40). Cambridge, MA, USA: MIT Press.

Powell, W. & DiMaggio, P. (1991). *The New Institutionalism in Organizational Analysis*. Chicago, IL, USA: Uiversity of Chicago Press.

Powell, W.W., White, D.R., Koput, K.W. & Owen-Smith, J. (2005). Network Dynamics and Field Evolution: The Growth of Interorganizational Collaboration in the Life Sciences. *American Journal of Sociology,* 110(4), 1132-1205.

Ptichnikov, A. (2009). Lesnaya Sertifikatsiya po Skheme Lesnogo Popechitel'skogo Soveta: Obshchie Svedeniya. *Ustoĭchivoe Lesopol'zovanie,* 1(20), 2-6.

Raynolds, L. & Wilkinson, J. (2007). Fair Trade in the Agriculture and Food Sector: Analytical Dimensions. In: L. Raynolds, D. Murray & J. Wilkinson (eds.), *Fair Trade. The Challenges of Transforming Globalisation* (pp. 33-47). New York, NY, USA: Routledge.

Raynolds, L.T. & Murray, D. (2007). Fair Trade: Contemporary Challenges and future Prospects. In: L.T. Raynolds, D. Murray & J. Wilkinson (eds.), *Fair Trade* (pp. 223-234). London, UK: Routledge.

Reed, D., Utting, P. & Mukherjee-Reed, A. (2012). *Business Regulation and Non-State Actors: Whose Standards? Whose Development?* London, UK: Routledge.

Ritzer, G. (2004). *The Globalization of Nothing*. London, UK: Sage.

Rosendal, K. (2001). Overlapping International Regimes. The Case of the Intergovernmental Forum on Forests (IFF) Between Climate Change and Biodiversity. *International Environmental Agreements: Politics, Law and Economics,* 1, 447-468.

Ryzhkov, A.Ye. & Prokazin, N.Ye. (2011). *Sistema dobrovol'noĭ lesnoĭ sertifikatsii PEFC-FCR*. Moskva, Russia: Uchebno-metodicheskiĭ tsentr PEFC-FCR.

Sabel, C. & Zeitlin, J. (2012). Experimentalist Governance. In: D. Levi-Faur (ed.), *The Oxford Handbook of Governance* (pp. 169-183). Oxford, UK: Oxford University Press.

Sassen, S. (1996). *Losing control? Sovereignty in An Age of Globalization*. New York, NY, USA: Columbia University Press.

Sassen, S. (2003). Globalization or Denationalization? *Review of International Political Economy,* 10(1), 1-22.

Sassen, S. (2004). Local Actors in Global Politics. *Current Sociology,* 52(4), 649-670.

Sassen, S. (2005). When National Territory is Home to the Global: Old Borders to Novel Borderings. *New Political Economy,* 10(4), 523-541.

Sassen, S. (2006). *Territory, Authority, Rights: from Medieval to Global Assemblages*. Princeton, NJ, USA: Princeton University Press.

Sassen, S. (2008). Neither Global nor National: Novel Assemblages of Territory, Authority and Rights. *Ethics & Global Politics,* 1(1-2), 61-79.

Sassen, S. (2011). *Cities in a World Economy*. Thousand Oaks, CA, USA: Pine Forge Press.

Sayer, A. (1992). *Method in social science. A realist approach*. London, UK: Routledge.

Scott, W.R. (1995). *Institutions and Organizations*. Thousand Oaks, CA, USA: Sage.

Sears, R., Davalos, L. & Ferras, G. (2001). Missing the Forest for the Profits: The Role of Multinational Corporations in the International Forest Regime. *Journal of Environment and Development*, 10(4), 345-364.

Shmatkov, N. (2012). Otnoshenie Kompaniĭ Lesnogo Sektora i Nepravitel'stvennykh Organizatsiĭ k Lesnomu Kodeksu i k Perspektivam Prinyatiya Natsional'noĭ Lesnoĭ Politiki Rossiĭskoĭ Federatsii. *Ustoĭchivoe lesopol'zovanie*, 1(30), 16-25.

Shmatkov, N. & Kulikova, Ye. (2012). Natsional'naya Lesnaya Politika: Obshchie Podkhody k Razrabotke i Mezhdunarodnyĭ Opyt. *Ustoĭchivoe Lesopol'zovanie*, 1(30), 2-9.

Shutov, I.V. (2003). *Razrushenie i vosstanovlenie lesnogo khozyaĭstva Rossii*. St. Petersburg, Russia: SPbNIILKH.

Sodor, U. & Jarvela, M. (2007) Resilience and Wellbeing of Forestry Settlements in Northwestern Russia. *Final Fieldwork Report*. Available at: http://www.jyu.fi/yhtfil/susfoli/resilience.

Sorensen, E. & Torfing, J. (2005). Network Governance and Post-liberal Democracy. *Administrative Theory & Praxis*, 27(2), 197-237.

Sorenson, E. & Torfing, J. (2007a). Introduction: Governance Network Research: Towards a Second Generation. In: E. Sorenson & J. Torfing (eds.), *Theories of Democratic Network Governance* (pp. 1-21). New York, NY, USA: Palgrave Macmillan.

Sorenson, E. & Torfing, J. (eds.). (2007b). *Theories of Democratic Network Governance*. New York, NY, USA: Palgrave Macmillan.

Spaargaren, G., Mol, A.P.J. & Buttel, F.H. (2000). *Environment and Global Modernity*. London, Uk: Sage.

Spaargaren, G. & Mol, A.P.G. (2008). Greening Global Consumption: Redefining Politics and Authority. *Global Environmental Change*, 18(3), 350-359.

Strauss, A. (1987). *Qualitative Analysis for social Scientists*. Cambridge, UK: Cambridge University Press.

Sun, Y. & Tysiachniouk, M. (2008). 'Caged by Boundaries?' NGO Cooperation at the Sino-Russian Border. In: P. Ho & R. Edmonds (eds.), *China's Embedded Activism* (pp. 171-194). London, UK: Routledge.

Tarasofsky, R. (1999). Introduction. In: *Assessing the International Forest Regime: Gaps, Overlaps, Uncertainties and Opportunities* (pp. 3-14). Environmental Policy and Law Paper N. 37.

Tarasofsky, R. & Downes, D. (1999). Global Cooperation on Forests Through International Institutions. In: *Assessing the International Forest Regime: Gaps, Overlaps, Uncertainties and Opportunities I* (pp. 97-110). Environmental Policy and Law Paper N. 37.

Taylor, M. (2007). Community Participation in the Real World: Opportunities and Pitfalls in New Governance Spaces. *Urban Studies*, 44(2), 297-317.

Taylor, P.L. (2005). In the Market But Not of It: Fair Trade Coffee and Forest Stewardship Council Certification as Market-Based Social Change. *World Development*, 33, 129-147.

Tollefson, C., Gale, F. & Haley, D. (2008). *Setting the Standard: Certification, Governance, and the Forest Stewardship Council*. Vancouver, Canada: University of British Columbia Press.

Tulaeva, S. (2010). Sotsial'nye Posledstviya Konstruirovaniya Senzitivnykh Rynkov dlya Lokal'nykh Soobshchestv (na Primere Sertifikatsii po Skheme FSC Priluzskogo Leskhoza Respubliki Komi). *Zhurnal Sotsiologii i Sotsial'noĭ Antropologii*, 13(5), 186-202.

Tulaeva, S. & Tysiachniouk, M. (2008). Transformatsiya Korporativnoĭ Sotsial'noĭ Otvet-Stvennosti Krupnogo Biznesa Pod Vliyaniem Global'nykh Praktik. *Sotsial'naya Politika i Sotsiologiya*, 4(40), 104-116.

Tysiachniouk, M. (2003). Transnational Environmental Non-Governmental Organizations as Actors of Ecological Modernization in Russian Forest Sector. In M. Tysiachniouk (ed.), *Ecological Modernization of Forest Sector in Russia and the United States* (pp. 8-25). St. Petersburg, Russia: Publishing Group of St. Petersburg University.

Tysiachniouk, M. (2006a). Forest Certification in Russia In B. Cashore, F. Gale, E. Meidinger & D. Newsom (eds.), *Confronting Sustainability: Forest Certification in Developing and Transitioning Countries* (pp. 261-295). New Haven, CT, USA: Yale School of Forestry and Environmental Studies.

Tysiachniouk, M. (2006b). NGO Mezhdu Globalizatsiyeĭ i Lokalizatsiyeĭ: Rol' Global'nykh Protsessov v Mobilizatsii Obshchestvennogo Uchastiya v Lesnykh Poselkakh. *Zhurnal Sotsiologii i Sotsial'noĭ Antropologii,* 9(5), 113-158.

Tysiachniouk, M. (2008a). Adaptatsiya Biznesa Transnatsional'nykh Korporatsiĭ k Rossiĭskomu Kontekstu: na Primere Kompanii 'Stora Enso'. *Ekonomicheskaya Sotsiologiya,* 9(4), 56-72.

Tysiachniouk, M. (2008b). Negosudarstvennye Mekhanizmy Regulirovaniya Transnatsional'nykh Korporatsiĭ: Analiz Rynochnykh Kampaniĭ. *Zhurnal Sotsiologii i Sotsial'noĭ Antropologii,* 3, 111-128.

Tysiachniouk, M. (2008c). Razvitie Demokraticheskikh form Prinyatiya Reshenii v Protsessii Postroeniya Modeli Ustoĭchivogo Lesopol'zovaniya v Priluz'e *Nauchnye doklady MONF (Rol' Grazhdanskogo Obshchestva v Stimulirovanii Korporativnoĭ Sotsial'noĭ Otvetstvennosti)* (Vol. 202, pp. 62-103). Moskva, Russia.

Tysiachniouk, M. (2009). Conflict as a Form of Governance: the Market Campaign to Save the Karelian Forests. In: S. Nysten-Haarala (ed.), *The Changing Governance of Renewable Natural Resources in Northwest Russia* (pp. 169-196). London, UK: Ashgate.

Tysiachniouk, M. (2010a). Institutsional'nye Izmeneniya v Rossiĭskikh Lokal'nostyakh pod Vliyaniem Transnatsional'nykh Aktorov v Protsesse Lesnoĭ Sertifikatsii. *Zhurnal Sotsiologii i Sotsial'noĭ Antropologii,* 13(5), 203-295.

Tysiachniouk, M. (2010b). Markery Doveriya v Tsepochke Postavok: ot Drevesiny na Kornyu v Rossii do Konechnogo Potrebitelya v Yevrope i SSHA. *Ekonomicheskaya Sotsiologiya,* 11(3), 58-83.

Tysiachniouk, M. (2010c). NGO Strategic Partnership in Promoting Sustainable Forest Management in Russia. In S. Nysten-Haarala & K. Pynnoniemi (eds.), *Russia and Europe: from Mental Images to Business Practices. Papers from the VII International Conference of Finnish, Russian and East European Studies* (pp. 87-102). Kotka, Finland: Kymenlaakso University of Applied Sciences.

Tysiachniouk, M. (2010d). Preventing and Overcoming Lack of Trust in the Process of FSC Certification: Designation of Social Value Forests Around the Village Tunguda. *Sustainable Forest Management,* 1(23), 44-50.

Tysiachniouk, M. (2010e). Rol' NPO kak Strategicheskikh Partnerov Biznesa v Postroenii Modelyeĭ Ustoĭchivogo Lesopol'zovaniya. *Zhurnal Sotsiologii i Sotsial'noĭ Antropologii,* 13(5), 154-185.

Tysiachniouk, M. (2010f). Social Movements for the Preservation of Forests in North-West Russia: From Consumer Boycotts to Fostering Forest Certifications. *Russian Analytical Digest,* 79, 20-23.

Tysiachniouk, M. (2010g). VvedenieI. *Zhurnal Sotsiologii i Sotsial'noĭ Antropologii,* 13(5), 5-9.

Tysyachniouk, M. (2010h). Novye Podkhody k Analizu Transgranichnykh Obshchestvennykh Dvizheniĭ v Usloviyakh Globalizatsii. *Zhurnal Sotsiologii i Sotsial'noĭ Antropologii,* 13(5), 38-62.

Tysiachniouk, M. (2012a). Die FSC-Wald-Zertifizierung in Russland: Das Zusammenspiel von Staatlichen und Nicht-staatlichen Akteuren. *Russland Analysen,* 240, 2-7.

Tysiachniouk, M. (2012b). Fostering FSC Forest Certification in Russia: Interplay of State and Non-State Actors. *Russian Analytical Digest,* 114, 8-11.

Tysiachniouk, M. (2012c). Fostering Transparency in the Transnational Supply Chain: From Russian Forest Producers to Consumers in Europe and the USA. *Forest Policy and Economics* (in press).

Tysiachniouk, M. & Meidinger, E. (2007). Using Forest Certifications to Strengthen Rural Communities: Cases from Northwest Russia. In: A. Usha (ed.), *Endangered Species and Forests* (pp. 159-182). Hyderabad, India: The Icfai University Press.

Tysiachniouk, M., Mironova, N. & Reisman, J. (2004). A Historical Perspective on the Movement for Nuclear Safety in Cheliabinsk, Russia. *International Journal of Contemporary Sociology,* 41(1), 41-58.

Tysiachniouk, M. & Reisman, J. (2002). Civil Society and Global Security: Russia's Decision to Import Spent Nuclear Fuel. *Journal of Eurasian Research*, 2, 10-18.

Tysiachniouk, M. & Reisman, J. (2004). Co-managing the Taiga: Russian Forests and the Challenge of International Environmentalism. In: A. Lehtinen & S. Donner-Amnell (eds.), *Politics of Forests* (pp. 157-175). Aldershot, UK: Ashgate.

Tysiachniouk, M. & Reisman, J. (2005). Market of Values Across the Border: Forest Practices on Certified Territories in Northwestern Russia. *Aleksanteri Papers (Understanding Russian Nature: Representations, Values and Concepts)*, 4, 147-176.

Tysiachniouk, M. & Reisman, J. (2006). The Changing Operational Context of Environmental Organisations. In J. Kortelainen & J. Kotilainen (eds.), *Contested Environments and Investments in Russian Woodland Communities* (pp. 101-119). Helsinki, Finland: Kikimora Publications.

Tysiachniouk, M., Reisman, J. & Kotilainen, J. (2006). Environmental Movement in Transformation. In: J. Kortelainen & J. Kotilainen (eds.), *Contested Environments and Investments in Russian Woodland Communities* (pp. 77-84). Helsinki, Finland: Kikimora Publications.

Urry, J. (2000). *Sociology beyond Societies.* London, Uk: Routledge.

Urry, J. (2003). *Global Complexity.* London, UK: Polity Press.

Urry, J. (2007). *Mobilities.* Cambridge, UK: Polity Press.

Vagneron, I. & Roquigny, S. (2011). Value Distribution in Conventional, Organic and Fair Trade Banana Chains in the Dominican Republic. *Canadian Review of Development Studies,* 32(3), 324-338.

Vogel, D. (2005). *The Market for Virtue: The Potential and Limits of Corporate Social Responsibility.* New York, NY, USA: Brookings Institution Press.

Vogel, D. (2008). Private Global Business Regulation. *Annual Review of Political Science,* 11, 261-282.

Vorobiov, D. (1999). Transboundary Movement Toward the Saving of the Karelian Forests. In: M. Tysiachniouk & G.W. McCarthy (eds.), *Towards a Sustainable Future: Environmental Activism in Russia and the United States* (pp. 229-242). St. Petersburg, Russia: Institute of Chemistry of St. Petersburg State University.

Waldrop, M. (1994). *Complexity.* London, UK: Penguin.

Wellman, B. & Hampton, K. (1999). Living Networked on and Offline. *Contemporary Sociology,* 28(6), 648-654.

Wirth, C., Gleixner, G. & Heimann, M. (2009). *Old-growth Forests: Function, Fate and Value.* Berlin, Germany: Springer.

Yanitskiiy, O.N. (1996). *Ekologicheskoe Dvizhenie v Rossii. Kriticheskiĭ Analiz.* Moscow, Russia: Institut sotsiologii RAN.

Yanitsky, O.N. (2000). *Russian Greens in a Risk Society.* Helsinki, Finland: Kikimora Publications.

Yaroshenko, A., Potapov, P. & Turubanova, S. (2001). *Last Intact Forest Landscapes of Northern European Russia.* Moscow, Russia: Greenpeace Russia and Global Forest Watch.

Yin, K.R. (1994). *Case Study Research: Design and Methods.* Newbury Park, CA, USA: Sage Publications.

Yin, K.R. (2003). *Case Study Research: Design and Methods.* Newbury Park, CA, USA: Sage Publications.

Yin, K.R. (2011). *Applications of Case Study Research.* Newbury Park, CA, USA: Sage Publications.

Zagidullina, A. & Romaniuk, B. (2011). Landscape Ecological Planning for Sustainable Forestry in North-West Russia. In: O. Isokaanta & J. Heikkila (eds.), *From Wild Forest Reindeer to Biodiversity Studies and Envrionmental Education.* Helsinki, Finland: Finnish Environmental Institute.

Zaporodsky, O.N. & Murashko, O.A. (2000). How it is Possible to Realize the Constitutional Right for Protection of the Environment and a Traditional Way of Life: Experience in Organizing Specially Protected Territories in Kamchatka. In: *Northern Populations of Russia on the Way into the New Millennium* (pp. 158-166): Association of Low-Population Indigenous Peoples of the North, Siberia, and Far East.

Appendices

Appendix A. List of events, research and expert trips, were participant observation and interviews have been conducted

1. Transnational events

22-25 November, 2005 ENA-FLEG Ministerial Conference, St. Petersburg, Russia.
30-31 July, 2008 1st meeting of technical panel of the High Conservation Value Network, Oxford, UK.
21-24 October, 2008 International 'Dialogue on Forests and Rural Livelihoods', Syktyfkar, Komi Republic, Russia.
2-9 November, 2008 the FSC General Assembly 2008, Cape Town, SAR.
10-15 November, 2008 the study tour of Mondi-Business Paper plantations in South Africa, Durban and surroundings.
7-8 December, 2008 the Landscape Level HCV Mapping & Planning Workshop, Oxford, UK.
1-3 April, 2009 the HCV regional meeting at Siktifkar, Komi Republic, Russia.
15-16 June, 2009 the HCV Network Technical Panel 2nd Meeting, Oxford, UK.
17 June, 2009 Protecting 'High Conservation Values' in production landscapes around the world: an update from the field by the HCV Network's Technical Panel, meeting of the HCV-network technical panel, Oxford, UK.
27-28 April, 2010 the meeting of the Social Chamber, FSC, Bonn, Germany.
15-16 November, 2010 the meeting at the FSC office, Principle and Criteria review workshop, Bonn, Germany.
17-18 November, 2010 the meeting of National Initiatives, at the FSC office, Bonn, Germany.
24-26 January, 2011 the FSC-HCV-network meeting, Bonn, Germany.
25 June-1 July, 2011 the FSC General Assembly 2011, Kota Kinabaly, Malaysia.
25-26 June, 2011 the Principle and Criteria Review Workshop, Kota Kinabalu, Malaysia.
27 June, 2011 the HCV-network side event at the FSC General Assembly 2011, 'Principle 9 Revision and high conservation values in forests and other ecosystems' Kota Kinabalu, Malaysia.

2. Participant observation at the study tours to the US (in the framework of the projects sponsored by the Trust for Mutual Understanding)

Promoting Forest Certification in Russia and North America (2004-2005).
Fostering Effective Partnerships among Russian and North American FSC Experts, Activists and Social Scientists (2007-2008).

3. Participant observation at the study tours to Sweden (in the framework of the projects sponsored by the Swedish institute)

An Interdisciplinary Network for the Study of Multi-Level and Multi-Stakeholder Forest Governance in Russia and Sweden (2009-2010).

An Interdisciplinary Network for the Study of Company-Community Relations in Resource Extracting Sectors in Russia and Sweden (2011-2012, the ongoing project until 2013).

4. Research trips Europe and the US

10-24 September, 2005 Washington, DC/New York, NY/San Francisco, CA, USA
11-14 October, 2007 Helsinki, Finland
1-5 March, 2008 Addelstone/London, UK
20-23 March, 2008 Stockholm, Sweden
11-17 April, 2008 Stockholm, Sweden
20-22 June, 2008 Richmond, Vermont, USA
23-24 June, 2008 New York, USA
20-21 December, 2008 Düsseldorf, Germany
15-29 November, 2008 London, UK
22-29 January, 2009 Helsinki, Finland
18-27 June, 2009 Oxford, London-Hull, UK
20-27 July, 2009 Vienna, Austria
8-16 December, 2010 Brussels, Belgium

5. Participant observation at Russian national events

2005-2012 Participant observations of the work of the FSC-Russia national Initiative/National office, including the national standard setting process, conflict resolution, development of different kind of documents and the FSC policy statements (via e-mail, via Skype and/or conference calls).

20-23 May 2005 the conference of the FSC national initiative, Zvenigorod.

17-18 April 2007 meeting of Russian NGO experts, members of the Russian National Initiative and certification bodies, Moscow.

22-27 August 2007 meeting of Russian NGO experts, members of the Russian National Initiative and certification bodies, Zvenigorod.

15 October-3 November 2007 Russian-US field expedition to the Russian Far East.

20-22 August 2008 meeting of the Russian National Initiative with the FSC certification bodies, Zvenigorod.

27-28 April 2009 meeting of the Russian National Initiative with the FSC certification bodies, Moscow.

20-22 August 2009 meeting of the national initiative with the certification bodies, Zvenigorod.

11-13 November, 2009 the seminar 'HCVF and biodiversity', Moscow.

28-30 March, 2010 5th Conference of the National Initiative, Zvenigorod.

18-19 October 2010 seminar-training for auditors, Pepino, St. Petersburg oblast.
20 October 2010 Global Forest and Trade Network meeting, St. Petersburg.
21-22 October 2010 a research and practice seminar 'Commercial thinning in Karelia' Segezha-petrozavodsk.
4-6 December 2010 courses for Certification bodies, St. Petersburg.
4 April, 2011 Public Council at Rosleshoz.
23 March, 2011 the seminar on PEFC forest certification.
21-22 August, 2011 FSC series of public events, Segezha.
29-30 September, 2011 FSC-Friday, International Conference on FSC in Russia, St. Petersburg.
16-17 February, 2012 the seminar: 'Controlled Wood Risk Assessment'. Suggestions of logging companies to the 7th version of the Russian National Standard', Moscow, Russia.
28 February, 2012 training for NGOs, Petrozavodsk, Karelia Republic.

6. Participant observation at the FSC-Russia Board of directors meetings

6-7 May, 2010 the FSC-Russia Board meeting, Zvenigorod, Russia.
28-29 October, 2010 the FSC-Russia Board meeting, Zvenigorod, Russia.
10-12 April, 2011 the FSC-Russia Board meeting, Moscow, Russia.
10 June, 2011 the FSC-Russia Board meeting, Moscow, Russia.
02 September, 2011 the FSC-Russia Board meeting, Moscow, Russia.
28 October, 2011 the FSC-Russia Board meeting, Moscow, Russia.
09 December, 2011 the FSC-Russia Board meeting, Moscow, Russia.
24 February, 2012 the FSC-Russia Board meeting, Moscow, Russia.
17 April, 2012 the FSC-Russia Board meeting, Moscow, Russia.
29 June, 2012 the FSC-Russia Board meeting, Moscow, Russia.
10 August, 2012 the FSC-Russia Board meeting, Moscow, Russia.

7. Participant observation during field research expeditions to the case study areas in Russia[488]

Case study 1

2-19 October, 2002 Petrozavodsk and Kostomuksha rayon, Karelia Republic.
9-30 November, 2006 Petrozavodsk and Kostomuksha rayon, Karelia Republic.
5-29 June, 2006 Kostomuksha rayon, Karelia Republic.
4-19 August, 2008 Kostomuksha rayon, Karelia Republic.

Case study 2

7-24 July, 2002 Pskov-Strugi Krasnie rayon.
22-30 October, 2002 Pskov-Strugi Krasnie rayon.

[488] Additional data were collected during FSC audits and consulting work.

15 August-16 September, 2002 Pskov-Strugi Krasnie rayon.
18-30 July, 2006 Pskov-Strugi Krasnie rayon.
1-5 November, 2007 Pskov-Strugi Krasnie rayon.
8-15 March, 2008 Pskov-Strugi Krasnie rayon.

Case study 3

8 November-16 December, 2002 Siktifkar, Obiachevo rayon, villages, Komi Republic.
20 March-15 April, 2005 Siktifkar, Obiachevo rayon, villages, Komi Republic.
10-25 October, 2005 Siktifkar, Obiachevo rayon, villages, Komi Republic.
10 February-20 March, 2006 Siktifkar, Obiachevo rayon, villages, Komi Republic.
10-22 April, 2006 Siktifkar, Obiachevo rayon, villages, Komi Republic.
25-28 October, 2008 Siktifkar, Obiachevo rayon, villages, Komi Republic.

Case study 4[489]

17 September-20 October, 2006 Petrozavodsk-Segezha rayon, Medvezhegorsk rayon, Padani rayon, villages.
7-20 September, 2007 Petrozavodsk-Segezha rayon, Medvezhegorsk rayon, Padani rayon, villages.
25-29 March, 2008 Petrozavodsk-Segezha rayon, Medvezhegorsk rayon, Padani rayon, villages.
21-28 February, 2011 Kostomuksha rayon, Medveziegorsk rayon, Pudozh rayon, Karelia Republic.
26-30 July, 2011 Pudozh district, Karelia Republic.
1-5 March, 2011 Zaonezie, Medvezhegorsk rayon, Karelia Republic, Zaonezie.
13-22 May, 2011 Medviezhegorsk rayon, Karelia Republic.

8. Participant observation and interviews during expert work for companies

11-20 August, 2008 consultations with stakeholders in Zaonezhie, Karelia Republic.
13-19 February, 2009 consultations with stakeholders in Karelia Republic (Segezha, Padani, Maslozero, Shugozero, Belomorsk).
20-28 July, 2009 consultations with stakeholders, in Muezerka, Karelia Republic.
03-08 August, 2009 consultations with stakeholders in Kostomuksha, Karelia Republic.
20 February, 2010 training for Segezha PPM company, Segezha, Karelia Republic.
10-25 August, 2010 consultations with stakeholders and public hearings in Maslozero and Tunguda, Karelia Republic, Russia.

9. Participant observation during FSC audits

6-16 July, 2006 SGS-Qualifor main assessment of forest management of the Segezha Pulp and Paper Mill lease (the author was an expert-auditor).

[489] Additional data was collected during FSC audits of the Segezha pulp and Paper Mill lease and consulting work with companies belonging to the Investlesprom holding company, see points 8-9.

17-23 July, 2007 SGS-Qualifor surveillance audit of the Segezha Pulp and Paper Mill lease (the author was an observer).

5-8 December, 2006 SGS-Qualifor audit of Tikhvinski Kompleksniy Lespromchoz (the author served as an expert-auditor).

24-26 October, 2007 surveillance SGS-Qualifor audit of Tikhvinski Kompleksniy Lespromchoz (the author served as an expert-auditor).

19-22 May, 2008 surveillance audit conducted by SGS-Qualifor and ASI audit at the Segezha Pulp and Paper Mill leased territory (the author was an observer).

24-27 February, 2009 surveillance SGS-Qualifor audit of Tikhvinski Kompleksniy Lespromchoz (the author served as an expert-auditor).

15-24 November, 2009 SGS-Qualifor audit of Terney Les in the Russian Far East.

25-29 January, 2010 NEPCon audit of the Segezha Pulp and Paper Mill (the author was an observer).

22 September-7 October, 2010 SGS-Qualifor audit of Gantsewiche Forest enterprise and other forest enterprises belonging to the Ministry of the Forest Management, Minsk, Byelorussia (the author served as an expert-auditor).

25-31 March, 2011 the NEPCon forest management main assessment of Kostomukshskii lespromkhoz (the author was an observer).

13-18 June, 2011 NEPCon forest management main assessment of Muezerskii lespromkhoz (the author was an observer).

Appendix B. Projects, locations and interviews

Project	Location	Interviews (number)
Public-Private Partnerships and Making Environmental Policy in Russia.(2002-2003) sponsored by the John D. and Katherine T. MacArthur Foundation	Moscow	*Federal*: international NGOs (WWF, Greenpeace, IUCN) (#9), representatives of granting foundations (#3), FSC office[1] (#6), other FSC experts[2] (#19), state agencies representatives (#43), Russian NGOs (#22)
	Republic of Karelia: Petrozavodsk	*Case study 1*: state representatives (#14), NGO representatives (#11)
	Pskov region: Pskov, Strugi Krasnie	*Case study 2*: state representatives (#11), NGO representatives (#9), company representatives (#3), local forest management unit representatives (#4), representatives of local administrations (#2), community activists and ordinary citizens (#32), representatives of local museums (#8), librarians (#3), representatives of houses of culture (#3)
	Komi Republic: Siktifkar, Obyachevo, Spassporub, Chernish, Bolshaya Lopia, Malaya Lopia, Bolshaya Pissa, Malaya Pissa	*Case study 3*: state representatives (#24), NGO representatives (#28), company representatives (#29), local forest management unit representatives (#6), representatives of local administrations (#7), community activists and ordinary citizens (#47), representatives of local museums (#14), librarians (#5), representatives of houses of culture (#14)
Governance of Renewable Natural Resources in Northwestern Russia (2004-2007) sponsored by the Finnish Academy	Moscow	*Federal*: TNSs (#6), international NGOs (WWF, Greenpeace) (#5), state agencies representatives (#7), Russian NGO representatives (#9)
	Republic of Karelia: Petrozavodsk (multiple visits), Segezha (multiple visits), Kostomuksha, Muezerka, Kalevala, Vocknavolock, Sudnozero, Padani (multiple visits), Maslozero, Shugozero, Popov Porog, Chernii Porog, Sukkozero, Sudnozero, Shalgovari, other small villages	*Case study 1*: state agencies representatives (#8), NGO representatives (#14), local forest management unit representatives (#2), representatives of local administrations (#5), community activists and ordinary citizens (#14), librarians (#2)
		Case study 4: representatives of certification bodies (#4),[3] TNSs (#6), NGO representatives (#12), company representatives (#11), local forest management unit representatives (#6), representatives of local administrations (#6), community activists and ordinary citizens (#75), librarians (#9), teachers (#2)

Project	Location	Interviews (number)
	Pskov Region: Pskov, Strugi Krasnie, villages	*Case study 2*: state agencies representatives (#15), NGO representatives (#5), company representatives (#4), representatives of local administrations (#4), community activists and ordinary citizens (#17), librarians (#3), teachers (#4), representatives of houses of culture (#6)
	Komi Republic: Siktifcar (multiple visits), Obyachevo (2 visits), Spassporub (1 visit), Chernish (2 visits), Bolshaya Lopia (1 visit), Malaya Lopia (1 visit), villages	*Case study 3*: state agencies representatives (#16), NGO representatives (#7), company representatives (#31), representatives of local administrations (#8), community activists and ordinary citizens (#42), librarians (#17), teachers (#18), representatives of houses of culture (#8)
The Role of Civil Society in Fostering Corporate Social Responsibility Within the Russian Forest Sector(2006-2007) sponsored by the Moscow Public Scientific Foundation	Moscow	*Federal*: TNCs (#7), international NGOs (#3), state agencies representatives (#14), NGO representatives (#13)
	Republic of Karelia	*Case study 1*: state agencies representatives (#5), NGO representatives (#4), company representatives (#14)
	Pskov Region, Pskov, Stugi Krasnie, villages	*Case study 2*: state agencies representatives (#3), company representatives (#2), local forest management unit representatives (#3), representatives of local administrations (#6), community activists and ordinary citizens (#42), librarians (#4), teachers (#18), representatives of houses of culture (#6)
	Komi Republic, Siktifkar, Obiachevo, villages	*Case study 3*: state agencies representatives (#14), NGO representatives (#4), company representatives (#17), representatives of local administrations (#11), community activists and ordinary citizens (#59), librarians (#14), teachers (#9), representatives of houses of culture (#11)
Trust in Finnish-Russian Forest Industry Business Relations (2008-2010) sponsored by the Finnish Academy	Finland, Helsinki, Imatra	*Case study 2*: TNCs (#5)
	Karelia Republic, Petrozavodsk (multiple visits), Segezha (multiple visits), Padani (multiple visits), villages	*Case study 1*: representatives of local administrations (#3), ordinary citizens (#25)
		Case study 4: NGO representatives (#7), company representatives (#11), representatives of local administrations (#7), ordinary citizens (#21), representatives of local museums (#9), librarians (#6), teachers (#8), representatives of houses of culture (#3)

Project	Location	Interviews (number)
Transnationalization of Forest Governance (2008-2011) sponsored by the Finnish Academy	UK: London, Hull, Netherlands, Austria: Vienna, Germany: Frankfurt, Belgium: Brussels and South Africa: Cape Town, Durban	*Transnational:* FSC-International staff (5),[4] International NGOs (#11), pulp and paper associations (#2)
		Case study 2: publishers (#5)
		Case study 3: TNCs (#2)
		Case study 4: TNCs (#3), retailers (#4), buyers of Russian wood (#17), representatives of certification bodies (#12)
An Interdisciplinary Network for the Study of Multi-Level and Multi-Stakeholder Forest Governance in Russia and Sweden (2009-2010) sponsored by the Swedish Institute	Sweden, Stockholm	*Case study 2*: TNCs (#2), international NGOs (#3), representatives of granting foundations (#2)
An Interdisciplinary Network for the Study of Company-Community Relations in Resource Extracting Sectors in Russia and Sweden (2011-2013) sponsored by the Swedish Institute	Moscow	Federal: state agencies representatives (#5
	Karelia	Case study 4: company representatives(#3
Fostering Effective Partnerships Among Russian and North American FSC Experts, Activists and Social Scientists (2007-2008) sponsored by the Trust for Mutual Understanding	USA: New York, Washington	Transnational: international NGOs, (WWF, Rainforest Alliance) (#7)
	USA: Richmond, Vermont	*Case study 2*: publisher (#1)

Project	Location	Interviews (number)
Making Democracy Work: Building Capacity of Ordinary Citizens in Russian Rural Settlements[5] (2006-2007) sponsored by the European Union	Karelia Republic	Case study 1: company representatives (#16), representatives of local administrations (#4), ordinary citizens (#95), representatives of local museums (#4), representatives of houses of culture (#6), librarians (#4), teachers (#11)
		Case study 4: NGO representatives (#4), company representatives (#8), local forest management unit representatives (#4), representatives of local administrations (#15), ordinary citizens (#62), representatives of local museums (#12), librarians (#11), teachers (#7)
Promoting Democracy and Citizens' Rights in Resource-Dependent Communities[6] (2010-2012) sponsored by the European Union	Karelia Republic: Petrozavodsk (multiple visits), Segezha (multiple visits), Kostomuksha (2 visits), Muezerka (2 visits), Kalevala (1visit), Vocknavolock (1 visit), Sudnozero (1visit), Padani (multiple visits), Maslozero (2 visits), Shugozero (2 visits), Popov Porog (1 visit), Chernii Porog (1 visit), Sukkozero (2 visits), Shalgovari (2 visits) Kostamuksha Medvezhegorsk rayon, Zaonezhie rayon, villages	*Case study 4*: NGO representatives (#4), representatives of local administrations (#4), ordinary citizens (#49), teachers (#17), representatives of houses of culture (#5)

[1] Other interviews with FSC office representatives were done during different meetings and seminars (#48), see Appendix A (dates of meetings).
[2] Other interviews with FSC experts were done during the expert work in the FSC system in Russia (#54), see Appendix A (dates of meetings).
[3] Additional and follow-on interviews with certification bodies were done during the meetings of the National Initiative, certification bodies and experts in Moscow (#23), see Appendix A (dates of the meetings).
[4] 13 interviews with FSC staff were done during the expert meetings in Bonn (see Appendix A).
[5] Only short summaries of the interviews were done, no transcripts.
[6] There were no transcripts done in the framework of this project, interviews were not cited in the text, but used for background information.

Appendix C. Interviews conducted for the case studies

Case study	Case 1	Case 2	Case 3	Case 4
Years[1]	2002-2009	2002-2009	2002-2009	2006-2012
Total number of interviews[2]	273	248	470	442
Transnational Corporations[3]		7[4]	8	6
Transnational sawn wood buyers, publishers, and retailers		6		21
Granting agencies		5		
International NGOs	7	3	2	2
Russian NGOs[5]	29	14	39	27
Representatives of certification bodies				16[6]
Representatives of different state agencies, federal representatives in the regions and regional offices[7]	27[8]	29	54	
Representatives of local administrations[9]	12	12	26	32
Local state forest management unit representatives	2	7	6	9
Companies representatives	30	9	77[10]	33
Community activists and ordinary citizens	134	91	148	207
Representatives of local museums	4	8	14	21
Teachers	16	22	27	34
Representatives of houses of culture	6	25	33	8
Librarians	6	10	36	26

[1] For cases 1-3, updates were done with key stakeholders via e-mail in 2009-2012.

[2] Including those that were conducted during different events and consulting work, follow-on interviews were included in numbers.

[3] Total number of interviews with Transnational corporations was 32 (11 interviews were used in Chapter 3 and not in the case studies).

[4] Two of these interviews were used for the case study 1.

[5] Additional 47 interviews were conducted with NGOs not directly involved in the case studies.

[6] NEPCon was certifying companies studied in cases 2-3 as well as case 4, so the auditors were interviewed for all cases 2-4. Additional interviews were conducted for general information on the auditing process and its peculiarities in Russia.

[7] Additional 36 interviews were conducted in Moscow (federal level).

[8] The same interviews were used for the case 4 (both cases are in Karelia republic and governed by the same state agencies).

[9] In cases 3-4, more villages were investigated than in cases 1-2, this determines differences in the numbers of interviews.

[10] The forest management unit involved multiple leases, which determines a large number of interviews with companies.

Summary

The scope of the thesis is on forest governance through the Forest Stewardship Council (FSC) certification scheme. The focus is on transnational processes related to FSC's operation, including standards setting and application of these standards in a particular country. Russia is used as an example of application of global rules and norms in concrete territories. Russia has no major deforestation, although poor forest management is a major cause of forest degradation (Zagidullina *et al.*, 2011). The global importance of sustainable forest management of Russian forests is determined by the existence in its territory of intact forest landscapes, the largest in Europe (Aksenov *et al.*, 2002; Bryant *et al.*, 1997), which are under threat due to economic transition and inappropriate national legislation. Therefore, the introductory chapter starts with explaining the threat to these intact forests coming from the logging industry. According to existing Russian legislation, companies are allowed to log in old growth forests in the event they are located on their leased lands.

Subsequently, transformation and internationalization in the Russian forest sector is highlighted. How the transition from a nationally planned economy to a globalized market economy has changed company-community relations and public participation in forest management is explained. Next, private governance through international certification schemes, with special emphasis paid to FSC certification, is introduced. I provide a brief overview of the existing perspectives used to study certification schemes, and identify the need for developing a new perspective for analyzing certification as a private authority in governance. I explain why, in order to understand transnational private authority in governance, and especially governance through the FSC, it is important to develop new theoretical concepts and vocabulary, which capture both global and local processes and the actors as well as their interplay.

The central objective of this study is to understand the operation of the FSC's voluntary forest certification scheme as a transnational non-state governance institution, both globally and within the nation state. Three research questions were formulated as:

1. How to conceptualize the process of institutionalization of transnational private governance through certification schemes?
2. What is the role of global and local non-state governance actors, especially NGOs, in the process of installing new sustainable forest management practices in Russia?
3. How and to what extent do FSC governance arrangements transform forest management practices in Russia?

Chapter 2 describes the master theoretical concept of this study – the Governance Generating Network (GGN), its origins, its major components, its explanatory power and its relationship with existing concepts in the literature on global governance. The concept of the GGN was developed on the basis of two existing bodies of literature, the sociology of transnational processes (Castells, 1996, 2009; Sassen, 2008; Urry, 2003) and the literature on policy and governance networks (Sørensen & Torfing 2005, 2007). The framework of neo-institutional sociology is used as a supplementary tool for understanding institutional change (DiMaggio, 1991; Fligstein, 2002; Powell *et al.*, 2005).

Following Castells (1996), the study distinguishes between transnational spaces, where the global rules are created, and spaces of places, where they are implemented. GGNs link transnational spaces with the spaces of places. Nodes of design, forums of negotiation and sites of implementation are major components of the GGN. Under the node of design, the association of actors that initiate, design and develop global rules and regulations are explained. Sites of implementation are locales where global rules and norms are transferred and 'translated' into concrete practices. Forums of negotiation are important components of the GGN since they involve different stakeholders in design, adaptation and implementation of new rules. Global standards are developed and implemented by GGNs through four processes: abstraction, standardization, operationalization (translation into national standards), and institutionalization in concrete practices on the ground. The processes of abstraction and standardization are transnational and non-territorialized, while the processes of operationalization, implementation and institutionalization are territorialized and mediated by the local conditions, e.g. the social, political, economic and environmental contexts of forests in different regions of the world.

The chapter proceeds with further developing the GGN concept and by looking at GGNs involved in certification of sustainably produced products. The conceptual model for the study involves three ideal types of GGNs related to certification schemes for fostering sustainability around the globe. In some instances, GGNs are initiated in a bottom-up way, by actors operating in the local spaces of places. In other situations, actors operating in transnational space create and use GGNs to foster – in a top-down way – institutional changes in the space of place by imposing sets of new rules, norms and standards to be implemented locally. Third and finally, there are GGNs in which there is a two-way mode of exchange of information and a constant interaction between local and transnational actors. In this third case, it is argued that new modes of global governance and new ways of decision making are created, which respond to and better fit into the new conditions created by the global network society.

The overall research methodology is presented as an intermezzo to empirical findings. Longitudinal multi-sited research for this project (2002-2012) was conducted on Forest Stewardship Council (FSC)-International and on FSC-Russia and on four case studies selected for this work involving participant observations at the events, expeditions to the field and regular interviews with key informants. As explained, three ideal types of transnational networks were involved in the conceptual model for research; therefore multiple cases studies (Yin, 2003) were selected corresponding to the ideal types. Data were collected using qualitative methodology, involving open-ended interviews, semi-structured interviews, direct observation, participant observation, and analysis of documents.

Chapter 3 is semi-theoretical, semi-empirical, focusing on operationalization of the GGN concept for the analysis of forest governance and the FSC certification scheme. It starts by describing the history of global forest governance and the formation of the major GGNs related to forestry. The FSC is examined from the perspective of a GGN. The main components of the FSC-GGN, namely, the node of design, the forums of negotiations, and the sites of implementation, are analyzed in detail. The process of designing standards by the node is described. The FSC develops several types of standards and delivers several types of certificates. The most common are certificates of forest management (FSC-FM) and chain of custody (FSC-CoC). A FM certificate guarantees that the logging process and other forest operations are done in compliance with the Principles and

Criteria of the FSC Standard, taking into account economic, ecological and social components of sustainable forest utilization. The FSC-CoC guarantees timber legality and shows that the path of wood along the chain of custody has been monitored, from the time of logging through all the stages leading to the customer, including transportation, processing, and manufacturing of goods.

The FSC organizes all kinds of forums that can be either permanent or temporary. The largest of them is the FSC General Assembly (GA), with three chambers where economic, environmental and social actors from 'Southern' and 'Northern' countries are represented. The GA decides how the FSC governance generating network will develop and how the FSC standards will change. The GA forums are the main strategic driver of the FSC. Smaller forums are working groups and conferences, as well as permanent Internet forums for discussing various documents, electing candidates to the Board of Directors and discussing other issues. The FSC national offices organize key forums of adaptation of the FSC standards to local conditions. On the local level, where the adapted standard is transformed into concrete practices, many forums of implementation are organized for this purpose. In these forums, local stakeholders dominate, although actors from transnational spaces are also involved. The chapter proceeds with examining relationships between the FSC-GGN and other GGNs involved in governance.

The following Chapter 4 is on forest governance in Russia, with a major focus on the FSC. It starts with the characteristics of the conditions of forests and forestry in Russia in historical perspective. It characterizes peculiarities of the state forest management authority in the context of economic transition and institutional turbulence. Next, the chapter focuses on the emergence of NGO networks in forest governance. Further focus is on forest certification, including all initiatives existing in Russia, with major focus on the FSC; its emergence, institutional design and infrastructure; and operation of the national FSC office. The chapter develops the analysis of how the FSC global rules coincide with Russian national legislation and on how state and non-state actors interact in the FSC process. Next, the current state of FSC certification and the peculiarities of implementation of new norms as well as roadblocks and challenges are discussed.

The case studies in Chapter 5 start with the preservation of old growth forests in the European part of Russia. The market campaign was initiated by a NGO network in order to save old growth forests in the Republic of Karelia. Market campaigns represent a bottom-up GGN, which involves Western consumer preferences in order to change practices in Russian logging sites. NGO-led market campaigns, despite being conflicts, offer another form of governance that transforms business practices to be more environmentally friendly. By doing so it helps to promote and construct environmentally friendly niche markets.

This campaign was organized by Greenpeace, the Taiga Rescue Network in conjunction with Russian NGOs, in particular the Biodiversity Conservation Center (BCC) and the Socio-Ecological Union (SEU). It also included the Nature Protection Corps, which began as student groups in Russian universities during the Soviet period. Using satellite images, the Forest Club made an inventory and mapped virgin forests in the region. They took this data to the public of Europe, to the Russian government, and to the companies involved in using the forest resources of this area. The Forest Club's message was manifold: they listed companies logging these old-growth forests, as well as those buyers in Europe that accepted wood from these companies. They implored the European consumers to boycott products made with Russia's old-growth wood. They warned timber companies and European buyers, including retail stores, to establish moratoriums on logging these forests. With the Russian government, they tried to initiate a process of creating a specially protected natural area in order to preserve the old-growth timber. The chapter shows that the market campaign stimulated FSC certification all over Russia by providing a threat to the doing of business as usual.

The two subsequent Chapters, 6 and 7, present top-down networks linking transnational spaces with the spaces of places. Both cases represent models of sustainable forest management built with foundational support, in which the FSC took part. Top-down models, with best practices on the ground, served as sites of experimentation. Funding from foundations enabled negotiation, elaboration and testing of different kinds of innovations in specific territories. The cases differed immensely in the FSC implementing agents; in the first case, a NGO partnered with the company and in the second case it partnered with state agencies.

The first case (Chapter 6) is on the TNC-NGO strategic partnership in promoting sustainable forest management in Russia: building the Pskov Model Forest. The Pskov Model Forest was a business strategy of the TNC to adopt its operation to the turbulent environment of a Russian economy in transition. It examines the cooperation between the World Wildlife Fund (WWF) and Stora Enso during the implementation of the Pskov Model Forest project in Northwest Russia. The effectiveness of the partnership and its outcomes for both participants is assessed. For Stora Enso, the project has become one of the tools for adaptation of its business to local contexts. The WWF, as Stora Enso's strategic partner, assisted in resolving problems coming from state authorities, local stakeholders and the local population. It also contributed in 'legitimizing' Stora Enso in the eyes of international stakeholders. Through the project, Stora Enso attempted to create an optimum algorithm for a successful business paradigm in Russia. This algorithm was intended to be applied in all Stora Enso daughter enterprises working in the different regions in Russia. However, instability and turbulence of the organizational fields in the Russian regions has become an obstacle in carrying out this plan in its original form. For its part, the WWF also tried to create a model of sustainable forest management which could be further disseminated all over the country. The project was not sustainable in the long run, when funding ended. However, it produced both innovations in intensive forest management and a new expert community.

The second case explores how the NGO Silver Taiga Foundation (a WWF branch until 2002), with funding coming from the Swiss Agency for Development and Cooperation (SDC), operated top-down and created the Model Forest Priluzie in the Republic of Komi. During the 10 years of the project, Silver Taiga implemented many innovations such as ecological landscape planning for forest operations, preservation of biodiversity, in particular old growth forests and other high conservation value forests, soil erosion prevention, citizen involvement in decisions related to forestry, economic assessment of the forests, and FSC certification of forest management. Preparing the local forest management unit to meet the FSC standards was a difficult task, as this involved not only changing logging practices, but establishing a democratic decision making process in local forest settlements. Silver Taiga tried to achieve this by initiating partnerships between all the forest interests – government, the public and the logging companies. During the project period, Silver Taiga was focusing on dissemination of results through Komi and since 2007 has started to help building new Model Forests in other regions of Russia. The NGO-State partnership in the process of implementation of FSC certification on the ground transformed itself into a NGO-business partnership when grant funding for the project ended.

The last case study (Chapter 8) is on a FSC-GGN applied to a concrete certification case, the Northern Logging Company. It focuses on the 'pure certification' of the holding, the 'two-pronged' GGN constantly linking transnational and local spaces. The certification GGN operated practically without any additional sources of financing. The major channel of translation and institutionalization of the global rules from the node of design to the site of implementation was the company implementing the FSC and NGOs. In this case, the UK buyers of sawn wood were the major drivers for certification of the holding's leased territories. Although the vertical structure of the holding would seem to be an effective transmitter of global rules, it faced two major types of barriers. The first type is related to the turbulent institutional environment in Russia as a whole, and in the Karelia Republic in particular, which was not ready to accept global rules. The second type of barrier was related to the institutional turbulence within the holding itself, as it was acquiring and restructuring former Soviet enterprises and simultaneously modernizing them. The time of certification coincided with the time of formation of the holding company. This process was accompanied by series of bankruptcies of small forest companies, their absorption by the holding and their transformation into subsidiaries of the holding. This led to the loss of ties the enterprises had with local communities. Certification smoothed these tensions only to a certain extent. The chapter gives special attention to the role of auditors and NGOs as agents of institutional change that directed the company toward a sustainable path.

The concluding Chapter 9 describes both the theoretical and empirical contributions of this study to the fields of sociology and political science. It starts by revisiting the major analytical concept of this volume, the GGN, theorizes it further and explains how it works in practice. It also compares the GGN concept and its explanatory power with other conceptualizations of certification schemes. The chapter proceeds by revisiting the conceptual framework for the study involving different types of networks, summarizes empirical findings and their diversion from the previously identified ideal types. Moreover, it explains what aspects of sustainable forest management can be achieved using the FSC certification scheme and suggests why other aspects fail to be institutionalized. The chapter concludes by highlighting perspectives for future research and scholarship in the field.

About the author

Maria Tysiachniouk holds a Master of Science in Environmental Studies from Bard College, NY, USA, a PhD in Biology from the Russian Academy of Sciences and a Certificate in Nonprofit Studies from Johns Hopkins University. She has taught at Herzen Pedagogical University in St. Petersburg, St. Petersburg State University, Johns Hopkins University, Dickinson College, PA, Ramapo College of New Jersey, Towson University, and short courses at several universities in Europe. Since November 2004 she was enrolled the PhD-Sandwich program at Wageningen university Environmental Policy Group to study global governance through FSC certification. Since then she has combined her PhD work with multiple research projects and practitioners work in the FSC system. Maria Tysiachniouk has written more than hundred seventy publications on topics related to transnational environmental governance, edited several books and has had fieldwork experience in several countries and regions. She is currently chair of the Environmental Sociology group at the Center for Independent Social Research, St. Petersburg, Russia and doing intense field research on global governance of natural resources, including forests, mining and oil.

Selected publications

Tysiachniouk, M. (2012) 'Fostering Transparency in the Transnational Supply Chain: From Russian Forest Producers to Consumers in Europe and the USA.' *Forest Policy and Economics,* in press (available on line).

Tysiachniouk, M. (2012) 'Fostering FSC Forest Certification in Russia: Interplay of State and Non-State Actors.' *Russian Analytical Digest,* 114, 8-11.

Tysiachniouk, M. (2012) 'Fostering Sustainability Through the Supply Chain: Soft Laws and Trust Markers as Drivers of Social Change in Russian Communities.' In *Politics of Development in Barents Region*, M. Tennberg (ed.), pp. 222-264. Rovaniemi, Finland: Lapland University Press.

Tysiachniouk, M. (2010) 'NGO Strategic Partnership in Promoting Sustainable Forest Management in Russia.' In *Russia and Europe: from Mental Images to Business Practices. Papers from the VII International Conference of Finnish, Russian and East European Studies*, S. Nysten-Haarala & K. Pynnoniemi (eds.), pp. 87-102. Kotka, Finland: Kymenlaakso University of Applied Sciences.

Tysiachniouk, M. (2010) 'Social Movements for the Preservation of Forests in North-West Russia: From Consumer Boycotts to Fostering Forest Certifications.' *Russian Analytical Digest,* 79, 20-23.

Tysiachniouk, M. (2010) 'Markers of Trust in the Supply Chain: From Forest Stands in Russia to End Consumers in Europe and the USA.' *Economic Sociology,* 3, 58-83.

Tysiachniouk, M. (2009) 'Conflict as a Form of Governance: the Market Campaign to Save the Karelian Forests.' In *The Changing Governance of Renewable Natural Resources in Northwest Russia*, S. Nysten-Haarala (ed.), pp. 169-196. London, UK: Ashgate.

Malets, O. & Tysiachniouk, M. (2009) 'The Effect of Expertise on the Quality of Forest Standards Implementation: The Case of FSC Forest Certification in Russia.' *Forest Policy and Economics,* 11, 422-428.

Keskitalo, E., Sandstrom, C., Tysiachniouk, M. & Johansson, J. (2009) 'Local Consequences of Applying International Norms: Differences in the Application of Forest Certification in Northern Sweden, Northern Finland, and Northwest Russia.' *Ecology and Society,* 14(2). Available at: http://www.ecologyandsociety.org/vol14/iss2/art1/.

Kotilainen J., Tysiachniouk, M., Kuliasova, A., Kuliasov, I. & Pchelkina, S. (2008) 'The Potential for Ecological Modernisation in Russia: Scenarios from the Forest Industry'. *Environmental Politics*, 17(1), 58-77.

Sun, Y., & Tysiachniouk, M. (2008) 'Caged by Boundaries?' NGO Cooperation at the Sino-Russian Border'. In *China's Embedded Activism*, P. Ho & R. Edmonds (eds.), pp. 171-194. London, UK: Routledge.

Edelstein, M., Tysiachniouk, M. & Smirnova, L. (2007) 'Cultures of contamination: Legacies of Pollution in Russia and the US. *Research in Social Problems and Public Policy*, volume 14. Amsterdam, the Netherlands: Elsevier Ltd.

Tysiachniouk, M. & Meidinger, E. (2007) 'Using Forest Certifications to Strengthen Rural Communities: Cases from Northwest Russia'. In *Endangered Species and Forests*, A. Usha (ed.), pp. 159-182. Hyderabad, India: The Icfai University Press.

Tysiachniouk, M. (2006) 'Forest Certification in Russia'. In *Confronting Sustainability: Forest Certification in Developing and Transitioning Countries*, B. Cashore, F. Gale, E. Meidinger & D. Newsom (eds.), pp. 261-295. New Haven, CT, USA: Yale School of Forestry and Environmental Studies.

Tysiachniouk, M. & Reisman, J. (2006) 'The Changing Operational Context of Environmental Organisations'. In *Contested Environments and Investments in Russian Woodland Communities*, J. Kortelainen & J. Kotilainen (eds.), pp. 101-119. Helsinki, Finland: Kikimora Publications.

Tysiachniouk, M., Reisman, J. & Kotilainen, J. (2006) 'Environmental Movement in Transformation'. In *Contested Environments and Investments in Russian Woodland Communities*, J. Kortelainen & J. Kotilainen (eds.), pp. 77-84. Helsinki, Finland: Kikimora Publications.

Tysiachniouk, M., Mironova, N. & Reisman, J. (2004) 'A Historical Perspective on the Movement for Nuclear Safety in Cheliabinsk, Russia'. *International Journal of Contemporary Sociology*, 41(1), 41-58.

Tysiachniouk, M. & Reisman, J. (2004) 'Co-managing the Taiga: Russian Forests and the Challenge of International Environmentalism'. In *Politics of Forests*, A. Lehtinen and S. Donner-Amnell (eds.), pp. 157-175. Aldershot, UK: Ashgate.

The research described in the thesis was supported by Wageningen University, by the Finnish Academy and by the European Union. These funding agencies are gratefully acknowledged.

Printed in the United States
by Baker & Taylor Publisher Services